Fractal Horizons

Fractal Horizons

The Future Use of Fractals

Edited by Clifford A. Pickover

St. Martin's Press
New York

ISBN 0-312-12599-2

Library of Congress Cataloging-in-Publication Data

Fractal horizons / edited by Clifford A. Pickover. — 1st ed.
        p.   cm.
    Includes bibliographical references (p.   -) and index.
    ISBN 0-312-12599-2 (cloth)
    1. Fractals.   I. Pickover, Clifford A.
QA614.86.F6845 1996b
514'.74—dc20   96-7352
                                                                          CIP

Design by Acme Art, Inc.

First Edition: August 1996
10 9 8 7 6 5 4 3 2 1

This book is dedicated to my son, Alan.

CONTENTS

Part I. Fractals in Education

Part II. Fractals in Art

Part III. Fractal Models and Metaphors

GLOSSARY

714—The number of home-runs hit by baseball player Babe Ruth during his major league career.

Abscissa—The horizontal coordinate of a point in a plane rectangular coordinate system (the x-axis).

Acoustics—The study of sound.

AD—See *audioactive development*.

Address—For the fractal A of an *IFS* $T_1 \ldots , T_N$, the address of a region of A is the sequence of transformations determining the region. For example, $T_1(A)$ has address 1, and $T_2(T_3(A))$ has address 23.

Affine contraction—An *affine transformation* T that reduces the distance between points. That is, the distance between any pair of points (x, y) and (x', y) is greater than the distance between $T(x, y)$ and $T(x', y)$.

Affine transformation—The sum of a linear transformation and a translation. In Euclidean spaces, this is a matrix multiplication plus a vector addition. Affine transformations are convenient to use because they are relatively simple. Generally speaking, affine transforms consist of a linear combination of rotations, translations, magnifications (or minifications), or skew changes. An affine transformation can be used to convert one geometric shape into another by the action of translation, rotation, or uniform stretching or shrinking. Affine transformations map straight lines into straight lines and parallel lines into parallel lines but may alter the distance between points and the angles between lines. There will still be an "affinity" between the original shape and the transformed shape. Affine functions may be defined by a rule of the form $T(x, y) = (ax + by + e, cx + dy + f)$, where $a, b, \ldots,$ and f are constants.

Algebraic function—A function containing or using only algebraic symbols and operations such as $2x + x^2 + \sqrt{2}$.

Algebraic operations—Operations of addition, subtraction, multiplication, division, extraction of roots, and raising to integral or fractional powers.

Algorithmic complexity—A measure of the complexity of a data string, defined as the length in bits of the shortest program required for a computer to produce the string.

Amino acid—Basic building block of proteins.

Amplitude (of a wave)—The absolute value of the maximum displacement from zero value during one period of an oscillation. The "height" of the wave.

Analog-to-digital converter—Electronic device that transforms continuous signals into signals with discrete values.

Asymptotic fractal—A geometric object in which the effect of resolution on length provides an Euclidean asymptote (finite length) at high resolution and a fractal asymptote at low resolutions.

Attractor—An object in state space to which trajectories are eventually attracted. A geometrical form in phase space showing the characteristics of the dynamic system. The long-term behavior of a system. An attractor may be thought of as an underlying orbit into which a moving object is drawn or an underlying point set to which a set of points converges. If two objects or points are initially very close together, but after being drawn to the attractor they are at an unpredictable distance from each other, then the attracting system may be called chaotic, or "strange." Dissipative dynamical systems evolve to a set of points known as the attractor of the system. *Predictable attractors* correspond to the behavior to which a system settles down or is "attracted" (for example, a point or a looping closed cycle). The structure of these attractors is simple and well understood. A *strange attractor* is represented by an unpredictable trajectory where a minute difference in starting positions of two initially adjacent points leads to totally uncorrelated positions later in time or in the mathematical iteration. The structure of these attractors is very complicated and often not well understood.

Audioactive decay (AD)—See *audioactive development*.

Audioactive development (AD)—A succession of strings, where each string is the verbal description of the previous string. For example: "b," "a b," "an a, a space, and a b," and so on.

Autocorrelation—Describes the general dependence of the values of the data at one time on the values at another time.

Autonomous—The behavior of an autonomous dynamical system is expressed by an equation that is independent of time. If a time-dependent term is added, this represents an "external influence" that drives the system away from this equilibrium, for example, by adding or subtracting energy. Systems with a time-dependent term are nonautonomous. (An unsteady fluid flow is such a system.)

Basin of attraction—For *stable periodic points* $x_1, x_2, \ldots, x_N$, the basin of attraction is the set of all points x_0 converging to this orbit under iteration of $f(x)$. The basin of attraction can be thought of as the set of initial conditions for which a dynamical system will evolve to a particular attractor.

BCSK chain condition—The probability of going from x_0 to x in a time interval t can be determined from the product of the probability of going from x_0 to

z in a time τ and from z to x in the remaining time $(t-\tau)$ where $x_0 \geq z \geq x$
This condition is expressed as

$$P(x, t \mid x_0) = \int P(x, t \mid z, \tau)\, P(z, \tau \mid x_0)\, dz$$

where we integrate (sum over) all the states z between x_0 and x. The solutions to this chain condition are said to be Markovian.

Bifurcation—The radical change in a system's behavior corresponding to a change in form of its attractor. It occurs when the system's parameters are changed. Any value of a parameter at which the number and/or stability of steady states and cycles changes is called a bifurcation point, and the system is said to undergo a bifurcation.

Bifurcation diagram—A graph consisting of points (r, x_{N+1}), ..., (r, x_{N+M}), for r in some range of values. Here r is a parameter in a family of functions (such as the *Logistic Map* or the *Tent Map*) and the points x_i are from the orbit of a point x_0 under iteration of this function.

Binomial coefficients—The coefficients in the expansion of $(x+y)^n$. For example, $(x+y)^2 = x^2 + 2xy + y^2$ so that the binomial coefficients of order 2 are 1, 2, and 1.

Box-counting dimension—See *Dimension, box-counting.*

Cancer—An uncontrolled cell proliferation that usually invades and destroys adjacent tissues with fatal course if untreated and that persists irrespective of the continued presence of the etiological agent.

Cantor set—A fractal that can be generated by starting with a unit line segment and successively eliminating the middle third of all remaining segments. Other Cantor sets can be constructed.

Carcinogenesis—The sequence of events occurring in the development of cancer.

Cellular automata—Dynamic systems in which space and time are discrete. A typical cellular automata system uses a rectangular grid, where the action on any square of the grid depends on the action on the immediately adjacent squares. Cellular automata can be thought of as mathematical models composed of a large number of simple identical elements that interrelate their status in parallel, by means of local interactions determined by a fixed set of rules. Cellular automata are a class of simple mathematical systems that are becoming important as models for a variety of physical processes. Though the rules governing the creation of cellular automata are simple, the patterns they produce are complicated and sometimes seem almost random, like a turbulent fluid flow or the output of a cryptographic system. Cellular automata act on a discrete space or grid as opposed to a continuous medium.

Center—See *Limit* and *Fixed point.*

Central limit theorem (CLT)—Pertains to the laws that are characteristic of sums of a large number of independent random variables each of which has only a small effect on the sum. Each of the random variables in the sum has a finite mean and a finite variance. If $x_1, x_2, \ldots x_N$ are the random variables, $<x_k>$ is the mean of the kth variable and $\sigma_k{}^2$ its variance, then the CLT proves that the quantity

$$\frac{1}{\sigma_N} \sum_{k=1}^{N} (x_k - <\bar{x}_k>)^2 \quad \text{with} \quad \sigma_n^2 = \sum_{k=1}^{N} \sigma_k^2$$

converges to a Gaussian distribution in the limit $N \to \infty$. The condition imposed in the theorem is that the second moment of the distribution is finite regardless of the statistics of the individual x_k's.

Chaos—Irregular, unpredictable behavior of a deterministic nonlinear dynamical system caused by a sensitivity to initial conditions. For example, very slight differences in a natural habitat may lead to completely different animal population ratios after several generations. Some physicists have referred to chaos as the seemingly paradoxical combination of randomness and structure in certain nonperiodic solutions of dynamical systems. Chaotic behavior sometimes can be defined by a simple formula. Some researchers believe that chaos theory offers a mathematical framework for understanding much of the noise and turbulence that is seen in experimental science.

Chaos game—For fixed points $(a_1, b_1), \ldots, (a_n, b_n)$ in the plane, the process of generating a sequence of points $(x_1, y_1),(x_2, y_2), \ldots$, starting from a point (x_0, y_0), by the rule $(x_{i+1}, y_{i+1}) = ((x_i + a_j)/2, (y_i + b_j)/2))$, where j is chosen uniformly and randomly from $\{1, 2, \ldots, n\}$.

Chaotic trajectory—Exhibits three features: (1) The trajectory, or motion, stays within a bounded region—it does not get larger and larger without limit. (2) The trajectory never settles into a periodic pattern. (3) The motion exhibits a sensitivity to initial conditions. See also *Chaos*.

Chimera—An organism composed of two or more genetically differentiable cell types.

Close-return plot—For a time series $x_1, x_2, \ldots, x_n$, and a level of error, ε, a plot of points at location (i, j) if $|x_i - x_{i+j}| < \varepsilon$, where $1 \le i \le n$ and $1 \le j \le n - i$. A horizontal band of points plotted at height j suggests the time series has entrained on an approximate j-cycle. Close-return plots assay chaos by detecting the presence of unstable cycles.

Coherence—The constancy over time of the relative phase between two sources of light. Coherent beams are required to generate a stable interference pattern.

Collage theorem—A method for finding *affine contractions* to generate a fractal, A, by *iterated function system*. First, decompose the fractal A into its component pieces $a_1, \ldots, a_n$. Then find affine contractions $T_1, \ldots, T_N$ satisfying $T_1(A) = A_1, \ldots, T_N(A) = A_N$. The collage theorem guarantees that even if the A_i give only an approximate decomposition of A, the IFS $T_1, \ldots, T_N$ will produce a picture close to A.

Complex number—A number containing a real and imaginary part, and of the form $a + bi$, where $i = \sqrt{-1}$.

Compression ratio—A measure of the quantity of information reduction produced, for example, by the analysis part of a sound model.

Conceptual slippage—Extension or distortion of a concept to fit some new situation analogy, metaphor. (See Hofstadter and Mitchell references in chapter 7.)

Connective tissue—A composite mass of intercellular matrix, fibers, and cells that provide structural support to the body.

Conservative dynamical systems—In mechanics, conservative dynamical systems, also known as Hamiltonian dynamical systems, are frictionless. These systems do not entail a continual decrease of energy. See also *Dissipative dynamical systems.*

Contraction mapping—A function whose domain and range are in the same space, and that satisfies the property that for all pairs of points in the domain, the distance between the images of the points under the mapping is less than the distance between the original points. In other words, a contraction mapping shrinks space, so that every point finds itself closer to every other point.

Converge—To draw near to. A variable is sometimes said to converge to its limit.

Cornea—The anterior convex and transparent tunic of the eyeball.

Cycle—Describes predictable periodic motions, such as circular orbits. In phase plane portraits, the behavior often appears as smooth closed curves.

Damp—To diminish progressively in amplitude of oscillation.

Dendritic branching (of neurons)—Neurons have ramified tendrils whose branches splay out filling the available space in a fractal manner.

Depth first search—A method of traversing a tree data structure by always going as far down a limb as possible—all the way to a leaf —before backing up. Usually it is implemented with a stack.

Deterministic IFS—The process of generating a sequence of sets $S_1, S_2, \ldots$, where $S_{i+1} = (T_1(S_i) \cup T_2(S_i) \cup \ldots \cup T_n(S_i))$, T_1, where $T_2, \ldots, T_n$ are affine contractions. The sequence of S_i converges to the unique fractal determined by the IFS $T_1, \ldots, T_n$.

Differential equations—Equations often of the form $dx_i / dt = f_i(x)$ where $x_i(t)$ represents the ith variable and the function $f_i(x)$ gives the time, or spatial, evolution of $x_i(t)$. Mathematical models in the physical and biological sciences often are formulated as differential equations.

Diffraction grating—A flat reflective surface etched with dense parallel lines that splits white light into its constituent colors.

Diffusion—The process by which fluids and solids mix intimately with one another due to the kinetic motions of thermally agitated particles such as atoms, molecules, or groups of molecules.

Diffusion, anomalous—In a fluid, the mean square separation of any two particles initially close together and undergoing diffusion increases with time t as t^α with $\alpha = 1$. The diffusion becomes anomalous when $\alpha \neq 1$ and can be caused by a number of mechanisms.

Diffusion coefficient—The mass of substance transported in unit time across unit area of fluid, divided by the concentration gradient across the area. Sometimes symbolized by D, it is measured in meter^2/second. This term is synonymous with diffusivity.

Diffusion limited aggregation (DLA)—A model in which particles are released far from a nucleation center, random walk in the medium, and eventually attach themselves to the structure that blossoms from the initial nucleation site. A DLA can be thought of as a collection of particles, usually modeled on a computer using a two-or three- dimensional array, produced by allowing simulated sticky particles to diffuse, encounter one another, and join together.

Digital audio—A sequence of numbers that represent a sound on a computer.

Dimension, box-counting—Consider a geometrical object A in n-dimensional space. Denote by $N(\varepsilon)$ the minimum number of n-dimensional cubes of side length ε needed to contain A. The box-counting dimension of A is $\lim_{\varepsilon \to 0} \log(N(\varepsilon))/\log(1/\varepsilon)$.

Dimension, capacity—See *Dimension, fractal*

Dimension, correlation—A measure of the probability that two points in a fractal set are near one another.

Dimension, fractal—A measure of a fractal's space-filling properties.

Dimension, mass—Covering an object with concentric circles or spheres, denote by $M(r)$ the mass of the object contained within the circle or sphere of radius r. When the scaling relation $M(r) = K \times r^d$ holds, the exponent d is the mass dimension of the object. Mass dimension measures the distribution of mass and consequently of gap sizes.

Dissipation—Loss of mechanical energy in a dynamical system, causing the orbit to approach an attractor.

Dissipative dynamical systems—Systems typical of the macroscopic engineering world in which some resisting source causes energy loss. In dissipative dynamical systems, the volume of phase space occupied by an ensemble of starting points decreases with time. See also *conservative dynamical systems*.

DNA—Deoxyribonucleic acid. This macromolecule is the genetic material of the cell.

Dragon—A fractal dragon can be produced by repeatedly folding a long strip of paper in half, taking care always to fold the paper in the same direction. The first few folds produce the general humped appearance of a dragon. Certain Julia sets are also called dragons because of their fantastic twisted shapes.

Driven IFS—A variation of the *random iterated function system* where the transformations are applied in an order determined by some dynamical process or time series of data (placing the data into bins, for example, to select the transformation from the time series).

Dynamical system—An interrelated set of quantities that vary with time. In other words, a dynamical system evolves in time according to a prescribed mathematical form. It also can be a model containing the rules describing the way a given quantity undergoes a change through time or iteration steps. For example, the motion of planets about the sun can be modeled as a dynamical system in which the planets move according to Newton's laws. A discrete dynamical system can be represented mathematically as $x_{t+1} = f(x_t)$. A continuous dynamical system can be expressed as $dx/dt = f(x,t)$.

Embedding—A method of accessing the attractor of a dynamical system via a time series generated by that system.

Embedding space—The space in which an attractor resides when reconstructed from time-delay variables in which each point in the time series is plotted versus a number of its preceding points.

Epithelial dysplasia—Cellular and architectural atypia of the epithelial tissue with a potential for malignant transformation.

Epithelial tissue—The cellular covering of the skin and mucous membranes.

FBm—See *Fractional Brownian motion.*

Feedback—The return to the input of a part of the output of a system.

Fibonacci sequence—The sequence 1, 1, 2, 3, 5, 8, 13 . . . , $(F_n = F_{n-2} + F_{n-1})$, which governs many patterns in the plant world. Each term is the sum of the last two.

FIF—See *Fractal Interpolation Functions.*

Finite difference equations—Equations often of the form $x_i(t+1) = f_i(x(t))$ where $x_i(t)$ represents the value of the *ith* component at a time, or other coordinate, t.

Finite element method—A modeling technique in which an object is represented by many interlinked elements, such as triangles. The object's physical properties, such as shape, strength, and flexibility, are approximated on these elements. Bending, twisting, and other deformations in response to external forces can then be measured and visualized by performing appropriate calculations to determine how each element stretches, pushes against its neighbors, and so on.

First passage time distribution—The probability that an event η occurs for the first time in the interval $(t, t+dt)$ is given by $F(\eta, t)$. The probability density is the negative time derivative of $F(\eta,t)$ and is called the first passage time distribution.

Fixed point—A point that is left unchanged by the evolution of a dynamical system. A point that is invariant under a mapping (that is, $x_t = x_{t+1}$ for discrete systems, or $x = f(x)$ for continuous systems). A center is a particular kind of fixed point. For a center, nearby trajectories neither approach nor diverge from the fixed point. In contrast to the center, for a hyperbolic fixed point, some nearby trajectories approach and some diverge from the fixed point. A saddle point is an example of a hyperbolic fixed point. An unstable fixed point (or repulsive fixed point or repelling fixed point) x of a function occurs when $f'(x) > 0$. A stable fixed point (or attractive fixed point) x of a function occurs when $f'(x) < 0$. For cases where $f'(x) = 0$, higher derivatives need to be considered.

Fourier analysis—The separation of a complex wave into its sinusoidal components.

Fractal, escape-time—A fractal produced by plotting contours representing the number of iterations required for the orbit of a dynamical system to escape beyond some boundary.

Fractal, orbital—A fractal produced by the evolution of a chaotic dynamical system.

Fractal dimension—See *Dimension, fractal.*

Fractal Interpolation Functions (FIFs)—A form of *iterated function system* that is a single-valued function of one variable.

Fractals—Objects (or sets of points, or curves, or patterns) that exhibit increasing detail ("bumpiness") with increasing magnification. Many interesting fractals are self-similar. B. Mandelbrot informally defines fractals as "shapes that are equally complex in their details as in their overall form. That is, if a piece of a fractal is suitably magnified to become of the same size as the whole, it should look like the whole, either exactly, or perhaps only after slight limited deformation." Otto Rössler once said, "Nobody knows what chaos really is." Fractals and chaos are connected because chaotic behavior often yields fractal patterns. Any attempts at defining fractals are bound to be incomplete. Fractals are structures that have some or all of the following characteristics: self-similarity (parts resemble the whole), fractionalization ("brokenness" or discreteness), fractal (noninteger) dimension (for example, a wiggly broken line in the plane has a dimension between 1 and 2), and power-law behavior (relations of elements of fractal systems are often in the form $y = b \times x^a$. The power a may not be an integer.) Due to "self-similarity," if one examines a fractal at increasing levels of magnification, the structure continues to have the same appearance because each component is geometrically or statistically similar to the whole object.

Fractional Brownian motion (fBm)—Brownian motion is a continual, random, irregular motion of minute particles suspended in a liquid or gas, caused by the impact of the surrounding molecules. Particles move by a process called diffusion. When forces act to increase the time dependence of the variance beyond linear, but do not change the statistical distribution, the process is called fBm.

Full-immersion environment—A computer simulation system that generates a sufficiently rich set of multisensory signals to convey the effect of a physical setting. A person experiences the signals through a headset with a visual screen and gloves that provide kinesthetic feedback.

Gasket—A piece of material from which sections have been removed. Mathematical gaskets, such as Sierpinski gaskets, can be generated by removing sections of a region according to some rule. Usually the process of removal leaves pieces that are similar to the initial region; thus the gasket may be defined recursively.

Gaussian distribution—A bell-shaped distribution of numbers about a mean. Numbers closer to the mean occur more frequently than numbers farther away. Many natural measures (people's height, for example) have an approximately Gaussian distribution. Gaussian random numbers can be obtained by summing several uniformly distributed random numbers.

Gaussian white noise—White noise that is subsequently altered so that it has a bell-shaped distribution of values. In this book, Gaussian noise is often approximated by summing random numbers. The following formula generates a Gaussian distribution: $r = \sqrt{-\log(r_1)} \times \cos(2\pi r_2)\,\sigma + \mu$, where r_1 and r_2 are two random numbers from a (0,1] uniform distribution, and σ and μ are the desired standard deviation and mean of the Gaussian distribution.

Gestalt—A pattern of phenomena so integrated as to constitute a functional unit with properties not derivable by summation of its parts.

Gleichniszahlenreihe (GZR)—German term. "Gleichnis" means "parable" in the sense of "He spoke in parables." "Zahl" means "number," and "Reihe" means "sequence." The GZR is an audioactive decay that starts with 1. This sequence has been studied extensively by mathematician John Conway.

Glossitis—Inflammation of the tongue.

GM model—A simple random walk process initiated between two barriers, one reflecting and the other absorbing—used to determine the first passage time distribution to the absorbing boundary. This distribution was then used by Gerstein and Mandelbrot to model the interspike interval distribution in the discharge of neurons.

Graphical iteration—See *Iteration, graphical.*

GZR—See *Gleichniszahlenreihe.*

Handle—A "T"-shape that FractalVision software uses to show the current part of a template.

Helix—A space curve lying on a cylinder (or sphere, or cone) that maintains a constant distance from a central line (that is, a "spiral extended in space").

Henon map—Defines the point (x_{n+1}, y_{n+1}) by the equations $x_{n+1} = 1.4 + 0.3y_n -x_n^2, y_{n+1} = x_n$. Note that there are various expressions for the Henon map, including $x_{n+1} = 1 + y_n - \alpha x_n^2, y_{n+1} = \beta x_n$.

Herpes simplex—A disease caused by the Herpes simplex virus, characterized by small vesicles developing on the lips, skin, oral mucosa, genital mucosa, eye, brain, or meninges. Rupture of the vesicles leaves an ulcer.

Hologram—A record of the interference pattern between two coherent beams of light.

Homeomorphism—This is best explained by an example. Consider a circle inside a square. Draw a line from the circle's center out through the circle and square. The line intersects the circle at point P and the square at P'. The mapping $g:P \to P'$ assigns to each point of the circle a point of the square and vice versa. In addition, two adjacent points on one shape are mapped to two adjacent points of the other. The mapping g is a homeomorphism. The circle and triangle are homeomorphic.

Histopathology—The study of the structural alterations of cells and tissues caused by disease.

IFS—See *Iterated function system.*

Internet—The global collection of computer networks generated by universities, corporations, and governments. A set of network "protocols" is sufficiently standardized that millions of computer users can connect to each other around the planet. Two hundred million users of the Internet are expected before the twenty-first century.

Intron (gene)—A DNA sequence containing large segments for which there is no discernible coding function.

Invariant curve—Generalization of a *fixed point* to a line (in this case, a curve is invariant under the map or flow).

IS—See *Iterated system.*

Iterate—A system state or string that is derived from its predecessor by an iteration rule.

Iterated function system (IFS)—A scheme for describing and manipulating complex fractal attractors with simple systems. IFS are also known as a set of rotations and translations in two dimensions (affine transformations) that are useful in generating fractal aesthetic or natural forms, and in data compression, neural network theory, and other areas. Iterated functions systems can also be thought of as a collection of functions whose domain and range is in a given space. Since the domain and range are in the same space, the functions can be applied successively, in any order. If the functions satisfy some minor conditions (in particular, they are all contraction mappings), they define a fractal, the union of all the limit points of all possible compositions of the functions. An IFS can be thought of as a collection of affine contractions, $T_1, \ldots, T_n$. Associated with each IFS is a unique set A of points satisfying $A = T_1(A) \cup \ldots \cup T_N(A)$. Iterative function systems also can be represented by a discrete dynamical system of the form

$$S_{t+1} = \cup \ F_i(S_t)$$

where the union goes from $i = 1$ to n. The S_t are sets of points. Each iteration replaces S_t with the union of n copies of S_t, transformed by functions F_i. The F_i are affine transformations (scalings, rotations, translations and shears, or compositions thereof). If the F_i all reduce the extent of the set, then as t approaches infinity, S_t approaches a limit set called the attractor of the iterative function system. The attractor is independent of the starting set S_0 and is usually a fractal.

Iterated function system, random—The process of generating a sequence of points $(x_1, y_1), (x_2, y_2), \ldots$, starting from a point (x_0, y_0), by the rule $(x_{i+1}, y_{i+1}) = T_j(x_i, y_i)$, where each of $T_1(x,y), \ldots, T_n(x,y)$ is an *affine contraction* and j is selected randomly with assigned probabilities from $\{1, 2, \ldots, n\}$. Also see *Random iteration algorithm.*

Iterated function system pattern—A pattern produced by the successive application of two or more affine transformations.

Iterated map—A rule for repeatedly advancing a dynamical system in discrete steps.

Iterated system (IS)—A model where the next state of a system is determined from previous states. (Also see String-rewriting system.) An iterated system may be a computer or robot, perceiving its environment (input data), processing these data, learning from them, and taking actions that are appropriate to the situation (output data, behavior) that may influence its environment.

Iteration—In mathematics, this refers to the repeated application of a function f, beginning at an initial point x. The result is a series of values: x, $f(x)$, $f(f(x))$, $f(f(f(x)))$, ... This series may display various behaviors. For example, it may grow large without bound, converge to an attracting point, or oscillate between two or more values. The theory of iteration of functions has flourished during the last three decades as computer power has become available to mathematicians. An iteration can also be thought of as a repetition of an operation or set of operations. In mathematics, composing a function with itself, such as in $f(f(x))$, can represent an iteration. The computational process of determining x_{i+1} given x_i is called an iteration.

Iteration, graphical—A visual process for generating the orbit of a number x_0 under repeated application of a function $f(x)$. Graphical iteration begins by drawing the vertical line connecting (x_0, x_0) to (x_0, x_1) and then drawing the horizontal line connecting (x_0, x_1) to (x_1, x_1). The general step of the process connects (x_i, x_i) to (x_i, x_{i+1}) and (x_i, x_{i+1}) to (x_{i+1}, x_{i+1}).

Jacobian—A measure of the expansion or contraction of the volume of a cluster of points as they move under the influence of a dynamical process.

Julia set—For a given function in the plane of complex numbers, there is a set of points that is bounded under iteration. This set is called the filled Julia set for that function. The boundary of the filled Julia set is called simply the Julia set. Julia sets are named for the French mathematician Gaston Julia, who studied their properties during the 1920s. A Julia set can be thought of as the smallest closed set in the complex plane that contains all the repelling periodic points of a complex mapping. It also can also be thought of as the set of all points that do not converge to a fixed point or finite attracting orbit under repeated applications of the map. Most Julia sets are fractals, displaying an endless cascade of repeated detail. An alternate definition: Repeated applications of a function f determine a trajectory of successive locations x, $f(x), f(f(x)), f(f(f(x))), \ldots$ visited by a starting point x in the complex plane. Depending on the starting point, this results in two types of trajectories, those that go to infinity and those that remain bounded by a fixed radius. The Julia set of the function f is the boundary curve that separates these regions.

Knight's tour—the movement of a knight on a chess board so that all 64 squares are jumped on only once.

Lévy stable distribution—The most general solution to the BCSK chain condition is the Lévy stable distribution. It has the integral representation given in equation 14.23 in chapter 14. It describes statistics of fractal random processes.

LIFE—A cellular automata simulation in which a square on a grid is considered "alive" or "dead" according to a simple set of rules: (1) If a cell is alive and has two or three neighbors that are alive, then the cell will be alive in the next generation, (2) if a cell is alive and has fewer than two living neighbors or more than three living neighbors, then the cell will be dead in the next generation, (3) if a cell is dead and has exactly three living neighbors, the cell will be alive during the next generation. When the generations are calculated rapidly for an area of several hundred cells and the result displayed on a computer, a series of animated shapes moves and flickers across the screen. LIFE was devised by the British mathematician John Conway. It is a square grid cellular automaton that exhibits a variety of periodic and growing configurations, and has been shown to be capable of universal computation.

Limit—In general, the ultimate value toward which a variable tends.

Lindenmayer system (L-system)—String-rewriting system, named after Danish theoretical biologist Aristid Lindenmayer who discovered its applicability for modeling plant development using recursive grammars. A Lindenmayer system is a set of parallel production rules that generates a string of characters. ("Production rules" are rules for converting character strings into other character strings. For example, we might have rules converting "A" to "AAB" and "C" to "ACC." Acting in parallel—that is, at the same time—these rules convert "ABCBA" to "AABBACCBAAB," then to "AABAABBBAABACCACCBAABAABB," and so on.) The character string may be interpreted as directional information by letting each character represent a different movement: turn left 60 degrees, move forward one unit, raise red pen, lower green pen.

Linear transformation—A relation where the output is directly proportional to the input. A function satisfying two conditions:

$$(1)\ F(\vec{p}+\vec{q}) = F(\vec{p}) + F(\vec{q})\ \text{ and}$$

$$(2)\ F(r\vec{p}) = rF(\vec{p}).$$

Logistic map: A one-parameter family of functions $L_r(x) = rx(1-x)$. For $0 \le r < 1$, all points x in the interval $[0, 1]$ iterate to 0; for $1 < r < 3$, almost all points x in the interval $[0, 1]$ iterate to the point $1 - (1/r)$; for $3 < r \le 4$, points undergo a variety of periodic and chaotic behaviors, depending on the particular value of r. This nonlinear equation also can be written as $x_{n+1} = kx_n(1 - x_n)$; it is called the logistic equation, and it has been used in ecology as a model for simulating population growth.

Lorenz attractor— $\dot{x} = -10x + 10y$, $\dot{y} = 40x - y - xz$, $\dot{z} = -8z/3 + xy$. Initially the system starts anywhere in the three-dimensional phase space, but as transients die away the system is attracted onto a two-lobed surface. For a more general formulation, the Lorenz equations are sometimes written with variable coefficients.

Lotka-Volterra equations—The Hamiltonian system defined by ($dx/dt = ax - bxy$, $dy/dt = -cy + dxy$), which was one of the first predator-prey equations to predict cyclic variations in population.

L-system—See *Lindenmayer system.*

Lyapunov exponent—A measure of the average rate of exponential separation of nearby trajectories in a dynamical system. The Lyapunov exponent also can be thought of as a quantity, sometimes represented by the Greek letter λ, used to characterize the divergence of trajectories in a chaotic flow. For a one-dimensional formula, such as the logistic equation,

$$\Lambda = \lim_{N \to \infty} 1/N \sum_{n=1}^{N} \ln | dx_{n+1} / dx_n |$$

Mandelbrot set—Named for Benoit Mandelbrot, the mathematician who also coined the term "fractal." The Mandelbrot set is the set of all complex numbers c such that iteration of the function $f(x) = x^2 + c$ starting at $x = 0$ does not go to infinity. Sometimes the Greek letter μ is used to denote the constant. For example, for each complex number μ let $f_\mu(x)$ denote the polynomial $x^2 + \mu$. The Mandelbrot set is defined as the set of values of μ for which successive iterates of 0 under f_μ do not converge to infinity. An alternate definition: The set of complex numbers μ for which the *Julia set* of the iterated mapping $z \to z^2 + \mu$ separates disjoint regions of attraction. When μ lies outside this set, the corresponding Julia set is fragmented. The term "Mandelbrot set" was originally associated with this quadratic formula, although the same construction gives rise to a (generalized) Mandelbrot set for any iterated function with a complex parameter.

Manifold—Curve or surface. The classical *attractors* are manifolds (they're smooth). *Strange attractors* are not manifolds (they're rough and fractal).

Markov process—A sequence of random events that is described by a probability density in such a way that each event can at most depend on the event immediately preceding. It also can be thought of as a stochastic process in which the "future" is determined by the present.

Mean first passage time—The average time for an event η to occur for the first time is the first moment of the first passage time distribution

$$< t;\eta > = \int_0^\infty tF(\eta,t)\, dt$$

Mean survival probability—If the threshold level η has a distribution of possible values characterized by $p(\eta)$, then the mean survival probability is given by

$$F(t) = \int_0^\infty tF(\eta,t)p(\eta)\, dy$$

which averages the survival probability over all possible thresholds.

Metastasis—Spread of a disease (cancer) to a distant location in the body. It may be spread through the blood, lymph system, or cerebrospinal fluid.

Moment—The weighted average of the variate x to a specified power. The n th moment of distribution $P(x)$ is then

$$<x^n> = \int_{-\infty}^\infty x^n P(x)dx.$$

The mean value is given by $n = 1$. The second moment is given by $n = 2$, and so on.

Multi-user domain—A computer linked to the Internet and programmed to allow a group of users to log in and communicate with each other simultaneously. The program also may provide a running description of a simulated environment in which the group's members can converse and can interact with each other.

Newton's method—A numerical method for approximating a solution x^* of the equation $f(x) = 0$, for a differentiable function $f(x)$. From a starting guess x_0, Newton's method generates a sequence of points x_1, x_2, . . . , where $x_{i+1} = x_i - f(x_i)/f'(x_i)$. Newton's method can be thought of as a technique for approximating roots of equations. Suppose the equation is $f(x) = 0$, and a_1 is an approximation to the roots. The next approximation, a_2, is found by $a_2 = a_1 - f(a_1)/f'(a_1)$, where f' is the derivative of f.

Nonlinear equation—An equation where the output is not directly proportional to the input. An equation that describes the behavior of most real-world problems. The response of a nonlinear system can depend crucially on initial conditions.

Nonlinearity—A mathematical relation in which the output is not proportional to the input.

Noise, $1/f$—A signal having (statistically) fractal properties.

Nucleus—A membrane bounded compartment in eukaryotic cells that contains the genetic material (*DNA*).

Orbit—The sequence of points x_0, $x_1 = f(x_0)$, $x_2 = f(x_1)$, ...

Pascal triangle—A triangular array of integers named for the French mathematician and philosopher Blaise Pascal. The first six rows are:

1
1 1
1 2 1
1 3 3 1
1 4 6 4 1
1 5 10 10 5 1.

Many of the interesting properties of this array, are apparent while others require deeper analysis. The Chinese mathematician Chu Shih-chieh studied the properties of this array in 1303, 350 years before Pascal.

Pathology—The study of causes, development, structural and functional changes, and effects of disease.

Percolation—A critical phenomenon of diffusion in disordered media.

Perfect number—An integer that is the sum of all its divisors excluding itself. For example, 6 is a perfect number since $6 = 1 + 2 + 3$.

Period—The time taken for one cycle of vibration of a wave.

Periodic—Recurring at equal intervals of time.

Periodic cycle—An orbit that repeats after some finite number of iterations.

Phase portrait—The overall picture formed by all possible initial conditions in the $(x, \dot{x})$ plane. Consider the motion of a pendulum, which comes to rest due to air resistance. In the abstract two-dimensional *phase space* (with coordinates x representing the displacement, and $\dot{x}$ the velocity) motions appear as noncrossing spirals converging asymptotically toward the resting, fixed state. This focus is called a point attractor that attracts all local transient motions.

Piaget—Jean Piaget was a Swiss psychologist who showed that cognition develops in stages. Many of his experiments explored the concept of conservation. For example, a child will choose a tall glass of fruit juice over a wide cup of juice, even though the child sees the same amount of liquid poured back and forth between the two vessels. At this early stage, the child does not understand that the volume of liquid is conserved. Piaget's theories are important to mathematics teachers who seek to introduce appropriate concepts at the right stages of cognitive development.

Pixel—Picture element, also known as "pel." A digital image is made up of an array of pixels, which are usually square or rectangular. Since all pixels are the same size, their width is frequently used as a distance scale also, as in "two points less than a pixel apart." A pixel is the smallest dot (usually a tiny square) that can be displayed on a computer screen.

Pleomorphism—The characteristic of occurring in a variety of forms, often a variety of shapes and sizes.

Poincaré map—Is established by cutting trajectories in a region of phase space with a surface one dimension less than the dimension of the phase space.

Polynomial—An algebraic expression of the form $a_0 x^n + a_1 x^{n-1} + \ldots a_{n-1}x + a_n$, where n is the degree of the expression and $a_0 \neq 0$.

Probability density—The probability that an event in an interval $(x, x+dx)$ occurs is given by $P(x)dx$, where $P(x)$ is the probability density and the derivative of $F(x)$ with respect to x.

Probability distribution—A probability is the numerical value of the chance of occurrence of one or more possible results of an unpredictable event, such as the rolling a particular number with a die. If the set of possible values is a continuous variable x and $F(x)$ is the function that specifies these values, it is called the probability distribution.

Purine—A colorless, crystalline organic base containing nitrogen and related to uric acid, or any group of compounds naturally derived from it, such as the nucleic acid components adenine and guanine.

Pyrimidine—A liquid or crystalline organic base with a strong odor, whose molecular arrangement is a six-membered ring containing atoms of nitrogen, or any of various compounds naturally derived from it, such as the nucleic acid components cytosine or uracil.

Pythagorean theorem—A theorem in geometry: the square of the length of the hypotenuse of a right triangle equals the sum of the squares of the lengths of the other two sides.

Quadratic mapping—Also known as the *logistic map,* this famous discrete dynamical system is defined by $x_{t+1} = cx_t(1 - x_t)$.

QRS pulse—The classical shape of a single beat of an electrocardiogram heart scan: the large upsweep of the time trace, followed by a dip below the baseline and ending with a small trailing blip.

Quasi-periodicity—Informally defined as a phenomenon with multiple periodicity. One example is the astronomical position of a point on the surface of the earth, since it results from the rotation of the earth about its axis and the rotation of the earth around the sun.

Quaternion—A four-dimensional "hyper" complex number of the form $Q = a_0 + a_1 i + a_2 j + a_3 k$.

Random iterated function system—See *Iterated function system, random.*

Random iteration algorithm—A method of plotting a fractal defined with an *iterated function system.* In the random iteration method, one picks an arbitrary starting point and successively applies functions randomly chosen from those making up the IFS. The series of points thus obtained can be plotted yielding a good approximation of the fractal. A random iteration algorithm can be thought of as a procedure in which one of several *affine transformations* is chosen randomly and applied sequentially. Also see *Iterated function system, random.*

Raster scan—A method of converting an image to or from a sequence of numerical values in line-by-line succession.

Rational function—A quotient of two polynomials. The study of iteration of rational functions forms much of the foundation of fractal mathematics.

Ray tracing—A technique used in computer graphics to produce images of near-photographic realism from a textual description of the objects in a scene. The rays used travel in the opposite direction from real light rays. Starting from the observer, the computer projects a ray through each pixel of the screen into the scene, to see which objects it intersects. The nearest such object is found, and its illumination (from light sources defined as part of the scene) is calculated. The color of the pixel is determined from the illumination plus the surface characteristics of the object. If the surface is reflective, the reflected ray must be traced further to find what object is reflected in the surface. If the object is transparent, it may be necessary to trace both reflected and refracted rays.

Recursive—An object is said to be recursive if it partially consists of or is defined in terms of itself. A recursive operation invokes itself as an intermediate operation.

Renormalization—A process in which some property of a map is rescaled so as to reproduce the same map.

Repeller—A set of points for which nearby orbits move away from the set.

Retina—The back of the eye where light-sensitive cells (rods and cones) receive incoming light (like the film in a camera).

Seed—A polygon in a template used by FractalVision software that serves as the basis for all other parts (copies) of the template. In chapter 6, the seed is a thin rectangle.

Self-affinity—The property of an object produced by the repeated application of *affine transformations.*

Self-similar—A shape made up of copies of itself, each similar to (though smaller than) the original. Self-similarity can be thought of as the property of looking the same under repeated magnifications.

Semantic net—Relations of various words and concepts often depicted in a graph. Used in creativity techniques, linguistics and artificial intelligence.

Senile depth first search—Same as *depth first search,* except that stack overflows are allowed and ignored.

Sensitive dependence on initial conditions—The property of chaotic dynamics by which the distance between nearby points increases (initially) as the dynamics proceeds.

Sequence—An ordered succession of strings where each string is connected by rules (or themes) to previous (or following) strings of the sequence. Examples of sequences or systems that produce sequences: string-rewriting systems, mathematical sequences, movies, literature, and musical compositions.

Shear—Distortion of an object in which portions are translated different distances in the same direction.

Sierpinski gasket—A fractal constructed recursively from a triangle by the following process: (1) Find the midpoint of each edge of the triangle; (2) draw lines connecting the midpoints; and (3) remove the middle triangle from among the four smaller triangles formed. See *Gasket.*

Sierpinski tetrahedron—The three-dimensional analog of the *Sierpinski triangle.*

Sierpinski triangle—Named after the Polish mathematician Waclaw Sierpinski. It also is called the *Sierpinski gasket.*

Signal-to-noise ratio (SNR)—measure of the degradation of a signal when, for example, represented by a model.

SNR—See *Signal-to-noise ratio.*

Sound model—A means of representing a sound so that it may be manipulated.
Squamous cell carcinoma—A type of epithelial malignant tumor.

SRS—See *String-rewriting system.*

Stable periodic points (or stable periodic orbit)—Points $x_1, x_2, \ldots, x_n$ periodic for a function $f(x)$ in the sense that $f(x_1) = x_2, f(x_2) = x_3, \ldots, f(x_n) = x_1$, and stable in the sense that if x is near x_i, then $f(x)$ is nearer to x_{i+1}.

State—The complete description of a system at one moment in time.

State space—The abstract space in which the state of a system is represented by a single point.

Steady state—Also called equilibrium point or *fixed point.* A set of values of the variables of a system for which the system does not change as time proceeds. Strange attractor—A geometric object that embodies the dynamics of a chaotic system and that is a fractal. A strange attractor can be thought of as the attractor of a dissipative chaotic system, usually a fractal. See *Attractor.*

String—A linear arrangement of characters (of an alphabet). Numbers and notes, words, sentences, novels, and, of course, beads are strings. Examples for strings: "1211," "this is a string," and "CDEFGABc."

String-rewriting system—A set of rules by which one can derive a string from another. The strings are understood to consist of characters of a certain alphabet. Examples of string-rewriting systems: *Gleichniszahlenreihe and Lindenmayer systems.*

Survival probability—If $P(x,t)$ is the probability density for a random variable $X(t)$, then the probability that the variable has achieved a value greater than η in the time t is

$$F(\eta,t) = \int_{\eta}^{\infty} P(x,t)dx.$$

The function $F(\eta,t)$ is also the probability that the process has survived for a length of time t; it is the survival probability.

Template—The seed, along with all copies of it, that defines how a fractal is generated using the FractalVision software.

Tent Map—A one-parameter family of functions $T_r(x) = (r/2) - r|x - \frac{1}{2}|$. For $0 < = r < 1$, all points x in the interval $[0, 1]$ iterate to 0; for $1 < r < = 2$, almost all points x in the interval $[0, 1]$ exhibit chaotic behavior under iteration.

Tesselation—A division of a plane into polygons, regular or irregular.

Time series—A function of time such as an audio signal trajectory. A time series also can be thought of as the path of a system's state through state space, or a series of numerical values at equal time intervals representing the evolution of a dynamical system.

Time series graph—For a sequence of data values $x_1, x_2, \ldots, x_n$, the graph consisting of the points $(1, x_1), (2, x_2), \ldots, (n, x_n)$.

Trajectory—A sequence of points in which each point produces its successor according to some mathematical function.

Transcendental functions—Functions that are not algebraic; for example, circular, exponential, and logarithmic functions.

Transfinite number—An infinite cardinal or ordinal number. The smallest transfinite number is called aleph-nought (written as $\aleph_0$), which counts the number of integers.

Transformation—The operation of changing (as by rotation or mapping) one configuration or expression into another in accordance with a mathematical rule.

Unstable periodic points (or unstable periodic orbits)—Points $x_1, x_2, \ldots, x_n$ periodic for a function $f(x)$ in the sense that $f(x_1) = x_2$, $f(x_2) = x_3, \ldots, f(x_n) = x_1$, and unstable in the sense that if x is near x_i, then $f(x)$ is farther from x_{i+1}.

Vascularization—The formation and development of blood vessels in a tissue or organ.

Variance—The variance of a distribution σ^2 is a measure of the width of the distribution in terms of the first two moments:

$$\sigma^2 = <x^2> - <x>^2$$

where the brackets denote the average over the probability density $P(x)$.

Virtual reality—A term used to describe a complex computer simulation of a "real-world" environment.

Visual cortex—The part of the brain where pictures are formed and "seeing" is believed to happen.

PREFACE

These days computer-generated fractal patterns are everywhere. From squiggly designs on computer art posters to illustrations in the most serious of physics journals, interest continues to grow among scientists and, rather surprisingly, artists and designers. The word "fractal" was coined in 1975 by mathematician Benoit Mandelbrot to describe an intricate-looking set of curves, many of which had never been seen before the advent of computers with their ability to perform massive calculations quickly. Fractals often exhibit self-similarity, which means that various copies of an object can be found in the original object at smaller size scales. The detail continues for many magnifications—like an endless nesting of Russian dolls within dolls. Some of these shapes exist only in abstract geometric space, but others can be used as models for complex natural objects such as coastlines and blood vessel branching. Interestingly, fractals provide a useful framework for understanding chaotic processes and for performing image compression. The dazzling computer-generated images can be intoxicating, motivating students' interest in math more than any other mathematical discovery in the last century.

Fractal Horizons takes many of these ideas forward, giving an account of the state of the art and speculating on advances in the twenty-first century. Since the book is filled with beautiful images, a strange array of topics on art and science, and computer/mathematical recipes, it should have broader appeal than most scientific books. I envision that the book will appeal to computer artists and traditional artists, computer hobbyists, mathematicians, humanists, fractal enthusiasts, scientists, and anyone fascinated by unusual ideas and optically provocative art. The book is divided into six broad sections: "Fractals in Education," "Fractals in Art," "Fractal Models and Metaphors," "Fractals in Music and Sound," "Fractals in Medicine," and "Fractals and Mathematics." Some contributors describe the challenges of using fractals in the classroom. Others discuss new ways of generating art and music, the use of fractals in clothing fashions of the future, fractal holograms, fractals in medicine, fractals in boardrooms of the future, fractals in chess, and more. The glossary found at the beginning of the book should help ease new readers into unfamiliar waters. Most of the ideas expressed in this book are practical and are either being implemented currently or will be implementable within the next decade. My goal, therefore, is to provide information that students, laypeople, scientists, programmers, and artists will find of practical value today as they begin to explore an inexhaustible reservoir of magnificent shapes, images, and ideas.

Clifford A. Pickover
Yorktown Heights, New York

SOME RESOURCES

Books

P. Anthony, *Fractal Mode* (New York: Ace, 1992). (Science fiction.)

M. Barnsley, *Fractals Everywhere* (New York: Academic Press, 1988).

M. Batty and P. Longley, *Fractals Cities: A Geometry of Form and Function* (San Diego: Academic Press, 1994). (Topic: the development and use of fractal geometry for understanding and planning the physical form of cities; shows how fractals enable cities to be simulated through computer graphics.)

J. Briggs, *Fractals* (New York: Simon and Schuster, 1992).

K. Falconer, *Fractal Geometry* (New York: Wiley, 1990).

J. Feder, *Fractals* (New York: Plenum, 1988).

P. Fischer and W. Smith, *Chaos, Fractals, and Dynamics* (New York: Marcel Dekker, 1985).

L. Glass and M. Mackey, *From Clocks to Chaos: The Rhythms of Life* (Princeton, NJ: Princeton University Press, 1988).

J. Gleick (1987), *Chaos* (New York: Viking, 1987).

B. Kaye, *A Random Walk through Fractal Dimensions* (New York: VCH Publishers, 1989).

H. Lauwerier, *Fractals* (Princeton, NJ: Princeton University Press, 1990).

B. Mandelbrot, *The Fractal Geometry of Nature* (New York: Freeman, 1984).

F. Moon, *Chaotic Vibrations* (New York: Wiley, 1987).

D. Peak and M. Frame, *Chaos Under Control: The Art and Science of Complexity* (New York: Freeman, 1994).

H. Peitgen and P. Richter, *The Beauty of Fractals* (Berlin: Springer, 1986).

H. Peitgen and D. Saupe, *The Science of Fractal Images* (Berlin: Springer, 1988).

I. Peterson, *Islands of Truth* (New York: Freeman, 1990).

C. Pickover, *Computers, Pattern, Chaos and Beauty* (New York: St. Martin's Press, 1990).

C. Pickover, *Chaos in Wonderland: Visual Adventures in a Fractal World* (New York: St. Martin's Press, 1994).

C. Pickover, *Computers and the Imagination* (New York: St. Martin's Press, 1991).

C. Pickover, *Keys to Infinity* (New York: Wiley, 1995).

C. Pickover, *Mazes for the Mind: Computers and the Unexpected* (New York: St. Martin's Press, 1992).

C. Pickover, *The Pattern Book: Fractals, Art and Nature* (River Edge, New Jersey: World Scientific, 1995).

E. Rietman, *Exploring the Geometry of Nature* (Blue Ridge Summit, PA: Windcrest, 1989).

M. Schroeder, *Number Theory in Science and Communication* (New York: Springer, 1984). (This book is recommended highly. An interesting book, by a fascinating author.)

M. Schroeder, (1991), *Fractals, Chaos, Power Laws* (New York: Freeman, 1991).

C. Sprott, *Strange Attractors: Creating Patterns in Chaos* (New York: M&T Books, 1993). (A book by the guru of strange attractors and their graphics.)

C. Stevens, *Fractal Programming in C* (New York: M&T Books: 1989).

I. Stewart, *Does God Play Dice?* (New York: Basil Blackwell, 1980).

T. Wegner and M. Peterson, *Fractal Creations* (Mill Valley: Waite Group Press, 1991).

T. Wegner, B. Tyler, M. Peterson, and P. Branderhorst, *Fractals for Windows* (Mill Valley: Waite Group Press, 1991).

B. West, *Fractal Physiology and Chaos in Medicine* (River Edge, New Jersey: World Scientific, 1990).

Newsletters, Organizations, and Related

AMYGDALA, a newsletter on fractals. Write to AMYGDALA, Box 219, San Cristobal, NM 87564 for more information.

ART MATRIX, creator of beautiful postcards and videotapes of exciting mathematical shapes. Write to ART MATRIX, PO Box 880, Ithaca, NY 14851 for more information.

Bourbaki Software: FracTools (fractal generation software), *A Touch of Chaos* (fractal screensaver for windows), *FracShow CD* (fractal slide shows on CD-ROM for DOS and Windows). Contact: Bourbaki Inc., PO Box 2867, Boise, ID 83701.

Chaos Demonstrations, a program that contains examples of Julia sets and over 20 other types of chaotic systems. This peer-reviewed program won the first annual *Computers in Physics* contest for innovative educational physics software. It is published by Physics Academic Software and is available from The Academic Software Library, Box 8202, North Carolina State University, Raleigh, NC 27695-8202. Chaos Demonstrations is by J. C. Sprott and G. Rowlands, and it requires an IBM PC or compatible.

"Chaos and Graphics Section" of *Computers and Graphics.* (This section is devoted to fractals and chaos in art and science.) Publishing, subscription, and advertising office for *Computers and Graphics.* Elsevier Science, Inc, 660

White Plains Road, Tarrytown, NY 10591-5153. E-mail: ESUK. USA@ELSEVIER.COM.

Fractal Calendar. Address inquiries to J. Loyless, 5185 Ashford Court, Lilburn, GA 30247.

Fractal Discovery Laboratory. Designed for a science museum or school setting. "Entertaining for a four-year-old, and fascinating for the mathematician." Earl Glynn, Glynn Function Study Center, 10808 West 105th Street, Overland Park, KS 66214-3057.

Fractal Frenzy CDROM, 2,010 beautiful fractals. Also includes software for creating fractals. GIFs, DOS, Windows, Mac, Unix. Contact: Walnut Creek CDROM, 4041 Pike Lane Suite D-388, Concord, CA 94520, Tel: 1-800-731-7177.

Fractal Postcards. The Mathematical Association, 259 L Leichester, LE2 3BE UK.

Fractal Report, a newsletter on fractals. Published by J. de Rivaz, Reeves Telecommunications Lab. West Towan House, Porthtowan, Cornwall TR4 8AX, United Kingdom.

Great Media Company, PO Box 598, Nicasio, CA 94946. This fine company distributes books, videos, prints, and calendars.

HOP—Fractals in Motion. Software that produces a large variety of novel images and animations. HOP features Fractint-like parameter files, GIF read/write, MAP palette editor, a screensaver for DOS, Windows, and OS/2, and more. Math coprocessor (386 and above) and SuperVGA required. The program is written by Michael Peters and Randy Scott. $30 shareware. Download locations: Compuserve GRAPHDEV Forum, Lib 4 (HOPZIP.EXE), The WELL ibmpc/graphics (HOPZIP.EXE), rever.nmsu.edu under /pub/hop, slopoke.mlb.semi.harris.com, ftp.uni-heidelberg. de under /pub/msdos/graphics, spanky.triumf.ca [128.189. 128.27] under [pub.fractals.programs.ibmpc], HOP WWW page: http://rever.nmsu.edu/~ras/hop. Subscriptions and information requests for the HOP mailing list should be sent to: hop-request@ acca.nmsu.edu. To subscribe to the HOP mailing list, simply send a message with the word "subscribe" in the Subject: field. For information, send a message with the word "INFO" in the Subject: field.

Recreational and Educational Computing Newsletter. Dr. Michael Ecker, 909 Violet Terrace, Clarks Summit, PA 18411.

StarMakers Rising (fractal posters). 6801 Lakeworth Road, Suite 201, Lakeworth, FL 33467.

Strange Attractions. A store devoted to chaos and fractals (fractal art work, cards, shirts, puzzles, and books). For more information, contact: Strange Attractions, 204 Kensington Park Road, London W11 1NR England.

YLEM—Artists Using Science and Technology. This newsletter is published by an organization of artists who work with video, ionized gases, computers, lasers, holograms, robotics, and other nontraditional media. It also includes artists who use traditional media but who are inspired by images of electromagnetic phenomena, biological self-replication, and fractals. Contact: YLEM, Box 749, Orinda, CA 94563.

Recent Unconventional Articles

This list is not meant to be comprehensive. It is included for researchers in the field who may not be aware of some of the following unusual applications of fractals.

M. Batty and P. Longley, "Urban Shapes as Fractals," *Area* 19 (1987): 215-221.

M. Batty, P. Longley, and A. Fotheringham, "Urban Growth and Form: Scaling, Fractal Geometry and Diffusion-Limited Aggregation," *Environment and Planning A* 21 (1989): 1447-1472.

A. Clarke, *Ghost from the Grand Banks* (New York: Bantam, 1989). (A female character becomes insane after exploring the Mandelbrot set.)

J. Cutting and J. Garvin, "Fractal Curves and Complexity," *Perception and Psychophysics* 42 (1987): 365-370.

R. Dixon, "The Pentasnow Gasket and Its Fractal Dimension." *Fivefold Symmetry*, ed. I. Hargittai (River Edge, NJ: World Scientific, 1992).

G. Entsminger, "Stochastic Fiction—Fiction from Fractals," *Micro Cornucopia* 49 (September-October 1989): 96.

L. Fogg, "PostScriptals: Ultimate Fractals Via Postscript," *Micro Cornucopia* 49 (September-October 1989): 16-22. (Discusses the "ultimate" fractal at a resolution of 2540 dots per inch.)

K. Hsu and A. Hsu, "Fractal Geometry of Music," *Proceedings of the National Academy of Science* 87 (3) (1990): 938-941.

A. Lakhtakia, "Fractals and The Cat in the Hat," *Journal of Recreational Math* 23 (3) (1990): 161-164.

A. Lakhtakia, R. Messier, V. Vasandara, and V. Varadan, "Incommensurate Numbers, Continued Fractions, and Fractal Immittances," *Naturforsch* 43A (1988): 943-955.

G. Landini, "A Fractal Model for Periodontal Breakdown in Periodontal Disease," *Journal of Periodontal Research* 26 (1991): 176-179.

L. Nottale, "The Fractal Structure of the Quantum Space-Time." In A. Heck and J. Perdang, *Applying Fractals in Astronomy* (New York: Springer, 1991).

C. Pickover, "Fractal Fantasies," *BYTE* (March 1993): 256. (Describes a game played on a fractal playing board.)

C. Pickover, "Is the Fractal Golden Curlicue Cold?" *The Visual Computer* 11 (6) (1995): 309-312. (On the thermodynamics of fractal curlicues.)

R. Pool, "Fractal Fracas," *Science* 249 (July 27, 1990): 363-364. ("The math community is in a flap over the question of whether fractals are just pretty pictures, or more substantial tools.")

C. Taylor, "Condoms and Cosmology: The 'Fractal' Person and Sexual Risk in Rwanda," *Social Science and Medicine* 31 (6) (1990): 1023-1028.

Other Fractal Articles

J. Aqvist and O. Tapia, "Surface Fractality as a Guide for Studying Protein-Protein Interactions," *Journal of Molecular Graphics* 5 (1) (1987): 30-34.

M. Barnsley and A. Sloan, "Chaotic Compression (a New Twist on Fractal Theory Speeds Complex Image Transmission to Video Rates)," *Computer Graphics World* (November 1987): 107-108.

M. Batty, "Fractals—Geometry Between Dimensions," *New Scientist* (April 1985): 31-40.

D. Boyd, "The Residual Set Dimensions of the Apollonian Packing," *Mathematika* 20 (1973): 170-174. (Very technical reading.)

R. Brooks and J. Matelski, "The Dynamics of 2-Generator Subgroups of PSL(2,C)," in *Riemann Surfaces and Related Topics: Proceedings of the 1978 Stony Brook Conference*, ed. I. Kyra and B. Maskit, (Princeton, NJ: Princeton University Press, 1981). (This 1978 paper contains computer graphics and mathematical descriptions of both Julia and Mandelbrot sets.)

S. Casey, "Formulating Fractals," *Computer Language* 4 (4) (1987): 28-38.

J. Collins and C. De Luca, "Open-Loop and Closed-Loop Control of Posture: A Random-Walk Analysis of Center-of-Pressure Trajectories," *Experimental Brain Research* 95 (1993): 308-318.

J. Collins and C. De Luca, "Random Walking During Quiet Standing," *Physical Review Letters* 73 (1994): 764-767. (The Collins papers deal with fractal like measures and the postural control system.)

R. Devaney and M. Krych, "Dynamics of exp(z)," *Ergodic Theory & Dynamical Systems* 4 (1984): 35-52.

A. Dewdney, "Computer Recreations," *Scientic American* 253 (1985): 16-24.

A. Douady and J. Hubbard, "Iteration des polynomes quadratiques complexes," *Comptes Rendus* (Paris) 2941 (1982): 123-126.

F. Family, "Introduction to Droplet Growth Processes: Simulation Theory and Experiments," in *Random Fluctuations and Pattern Growth: Experiments and Models*, ed. H. Stanley and N. Ostrowsky, (Boston: Kluwer, 1988).

P. Fatou, "Sur les equations fonctionelles." *Bulletin Society Mathematics Fr.* 47 (1919/1920): 161-271.

A. Fournier, D. Fussel, and L. Carpenter, "Computer Rendering of Stochastic Models," *Communications of the ACM* 25 (1982): 371-378. (How to create natural irregular objects.)

M. Gardner, "White and Brown Music, Fractal Curves, and 1/f Fluctuations," *Scientic American* (April 1978): 16-31.

M. Gardner, "In Which Monster Curves Force Redefinition of the Word Curve," *Scientific American* 235 (December 1976): 124-133.

J. Gordon, A. Goldman, and J. Maps, "Superconducting-Normal Phase Boundary of a Fractal Network in a Magnetic Field," *Physical Review Letters* 56 (1986): 2280-2283.

C. Grebogi, E. Ott, and J. Yorke, "Chaos, Strange Attractors, and Fractal Basin Boundaries in Nonlinear Dynamics," *Science* 238 (1985): 632-637. (A great overview, with definitions of terms used in the chaos literature.)

C. Grebogi, E. Ott, and J. Yorke, "Attractors on an N-Torus: Quasiperiodicity Versus Chaos," *Physica* 15D (1985): 354-373. (Contains some gorgeous diagrams of dynamical systems.)

M. Hirsch, "Chaos, Rigor, and Hype," *Mathematical Intelligencer* 11 (3) (1989): 6-9. (pp. 8 and 9 include James Gleick's response to the article.)

N. Holter, A. Lakhtakia, V. Varadan, V., Vasundara, and R. Messier, "On a New Class of Planar Fractals: The Pascal-Sierpinski Gaskets," *Journal of Physics A: Mathematics General* 19 (1986): 1753-1759.

G. Julia, "Memoire sur l'iteration des fonctions rationnelles," *Journal of Mathematics, Pure Applications* 4 (1918): 47-245.

L. Kadanoff, "Fractals: Where's the Physics?" *Physics Today* (February 1986): 6-7.

M. La Brecque, "Fractal Symmetry," *Mosaic* 16 (1985): 10-23.

A. Lakhtakia, V. Vasundara, R. Messier, and V. Varadan, "On the Symmetries of the Julia Sets for the Process $z \rightarrow z^p + c$," *Journal of Physics A: Mathematics General* 20 (1987): 3533-3535.

A. Lakhtakia, V. Vasundara, R. Messier, and V. Varadan, "The Generalized Koch Curve," *Journal of Physics A: Mathematics General* 20 (1987): 3537-3541.

A. Lakhtakia, V. Vasundara, R. Messier, and V. Varadan, "Self-Similarity Versus Self-Affinity: The Sierpinski Gasket Revisited," *Journal of Physics A: Mathematics General* 19 (1986): L985-L989.

A. Lakhtakia, V. Vasundara, R. Messier, and V. Varadan, "Fractal Sequences Derived from the Self-Similar Extensions of the Sierpint.ski gasket," *Journal of Physics A: Mathematics General* (1988): 1925-1928.

B. Mandelbrot, "On the Quadratic Mapping $z \rightarrow z^2 - \mu$ for complex μ and z: The Fractal Structure of Its M set, and Scaling," *Physica* 17D (1983): 224-239.

K. Musgrave, "The Synthesis and Rendering of Eroded Fractal Terrains," *Computer Graphics (ACM-SIGGRAPH)* 23 (3) (July 1989): 41-50.

A. Norton, "Generation and Display of Geometric Fractals in 3-D," *Computer Graphics (ACM-SIGGRAPH)* 16 (1982): 61-67.

I. Peterson, "Ants in the Labyrinth and Other Fractal Excursions," *Science News* 125 (1984): 42-43.

T. Phipps, "Enhanced Fractals," *Byte* (March 1985): 21-23.

C. Pickover, "Symmetry, Beauty and Chaos in Chebyshev's Paradise," *The Visual Computer: An International Journal of Computer Graphics* 4 (1988): 142-147.

C. Pickover, "Biomorphs: Computer Displays of Biological Forms Generated from Mathematical Feedback Loops," *Computer Graphics Forum* 5 (4) (1987): 313-316.

C. Pickover, "Synthesizing Extraterrestrial Terrain," *IEEE Computer Graphics and Applications* 15 (2) (March 1995): 18-21.

A. Robinson, "Fractal Fingers in Viscous Fluids," *Science* 228 (1985): 1077-1080.

P. Schroeder, "Plotting the Mandelbrot Set," *Byte* (December 1986): 207-211.

M. Schroeder, "A Simple Function," *Mathematical Intelligencer* 4 (1982): 158-161.

P. Sorenson, "Fractals," *Byte* 9 (September 1984): 157-172 (A fascinating introduction to the subject.)

"Symmetries and Asymmetries," *Mosaic* 16 (1) (January/February 1985). (An entire issue on the subject of fractals, symmetry, and chaos. *Mosaic* is published six times a year as a source of information for scientific and educational communities served by the National Science Foundation, Washington DC 20550.)

D. Thomsen, "A Place in the Sun for Fractals," *Science News* 121 (1982): 28-32.

D. Thomsen, "Making Music—Fractally," *Science News* 117 (March 1980): 189-190.

S. Ushiki, "Phoenix," *IEEE Transactions on Circuits and Systems* 35 (7) (July 1988): 788-789.

R. Voss, "Random Fractal Forgeries," in *Fundamental Algorithms in Computer Graphics* ed., R. Earnshaw (Berlin: Springer-Verlag, 1985). S. West, "The New Realism," *Science* 84 (5) (1984): 31-39.

B. West and A. Goldberger, "Physiology in Fractal Dimensions," *American Scientist* 75 (1987): 354-365. (This article describes the fractal characterization of the lungs' bronchial tree, the Weierstrass function, the fractal geometry of the heart, and "fractal time.")

Fractal Horizons

I

Fractals in Education

1

Conquering the Math Bogeyman

William P. Beaumont

Computers have revolutionized the practice of mathematics, allowing it to become more visual and more experimental. Fractals are one of the more spectacular results of this revolution. A more visual and experimental mathematics has important implications for the teaching of the subject, but so far the mathematics classroom has seen little of the change. This chapter shows how fractals can be used in the teaching of mathematics, indicates the potential benefits of such use, and discusses some of the viewpoints that might prevent such benefits from being realized.

As I write this, I am lying on a sun-drenched beach on the southern coast of Australia, watching the surf break just a few meters away. Surf breaking is a chaotic thing. Billions of water particles respond separately to the conflict between gravity, wind, inertia and cohesion. Each droplet is like a computer processor, measuring its own local forces from moment to moment and calculating its own path through the chaos. The result is a thing of visual and aural beauty.

Different people will react in different ways to this reverie. Some of these reactions have an important bearing on the topic of this chapter, and will be considered later. For the present, let us consider the mathematics involved. Ten or fifteen years ago, it would have been difficult to think realistically about the mathematical modeling of a surf break. The forces at work were well understood, but there was too much complexity, too much calculation required for the process as a whole to be properly modeled. The mathematics ran out where the chaos started.

Computers have changed the situation. The influence of computers on mathematics has barely begun, and it has already been profound. For centuries, mathematicians suffered from the difficulty of not being able adequately to visualize the objects they studied. Mathematicians were often restricted to objects that could be described easily using conventional formulas, and that were "well behaved" in some analytic sense. Many interesting natural phenomena are not well behaved and so could not be easily studied.

These mathematical "monsters" cannot be described using conventional formulas, but they can be studied using numerical methods that yield not a formula but a sequence of numbers. Numerical methods were among the first applications of computers and are still one of the most important. The resulting mass of numbers would be overwhelming, but computers can help again by rendering the numbers as visible images whose properties can be grasped more easily. The importance of images is demonstrated by the lavish illustrations of many contemporary mathematics books. The images not only depict the phenomena being modeled; they are often also beautiful, and they are often fractals—objects and patterns that exhibit intricate structural detail at different size scales.

John Hubbard of Cornell University discusses the teaching of differential equations using computer techniques. He asserts, ". . . it should be possible to *show* students the solutions to a differential equation and how they behave, by using computer graphics and numerical methods to produce pictures for qualitative study."[1] Such an activity was possible before computers, but now "we no longer need to consume huge amounts of time making graphs or tables by hand. The students can be exposed to ever so many more examples, easily, and interactive programs . . . provide ample opportunity and inducement for experimentation any time by student (and instructor)."

Such application of computers to mathematics education is not common. Pictorial mathematics is somehow seen, even at the school level, as inferior to the more abstract formulas. Many teachers learned their mathematics before the pictorial revolution took place and are only dimly aware of its implications. Formal assessment of experimental mathematics is difficult, so it is seen to be irrelevant in an assessment-directed curriculum. Nevertheless, the pictures and experiment of modern mathematics are potentially appealing to students, and it seems a shame not to exploit them.

How can we make mathematics more interesting, more experimental, more inspiring by teaching something about fractals? I see fractals being used not as a topic in their own right but as an incentive to inspire interest in the more traditional parts of the curriculum. In our book, Paul Scott and I present them as an umbrella covering the other topics in a motivating and inspirational role.[2] The beauty, interest, and natural realism of fractal images may inspire interest in the underlying mathematics, at least in children whose curiosity has not yet been killed by schooling. Fractals also show, in concrete form, many mathematical concepts that are otherwise difficult to illustrate.

A *fractal* is usually described using an algorithm, a set of rules that, if followed, generate the fractal. Following an algorithm is a concept familiar even to quite young children, because it is just like following the rules of a game.

One such game requires us to imagine that we have three aunts, Agatha, Beatrice, and Cynthia. They are triplets; they live in separate houses (see figure 1.1) and celebrate their birthdays separately. This year they have all invited us to their birthday parties, and we cannot decide which to attend. We set off from a starting point (which can be anywhere, but for simplicity, make it one of the aunts' houses). We choose an aunt at random and start walking toward her house. When we are exactly halfway there, we are again overcome by indecision. We choose again (possibly the same aunt, but possibly a different one), and set off toward her house. The whole process is then repeated. Figure 1.1 shows the path we would take if, starting from Beatrice's house, we choose, in sequence, Agatha, Agatha again, Cynthia, Agatha, Cynthia twice more, Beatrice, then Agatha again.

The point of the game is not to decide which aunt wins. In fact, none of them can win. Even if we consistently choose the same aunt every time, we never reach that aunt's house (although we can go, as mathematicians say, "as close as we please"). The point of the game is to see if there is anything interesting about the pattern made by our decision points.

An attempt to examine this question manually, by continuing the process shown in the figure, is tedious. The computer comes to our rescue. The rules of the game can be easily programmed. (See program code 1.1.) The plotting of

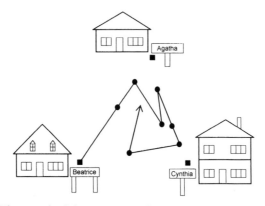

Figure 1.1 The aunts birthday game. At each step, we choose an aunt at random and move halfway to her house. The points we visit make up a fractal shape.

thousands of decision points results in a well-known fractal, the Sierpinski sieve. (See figure 1.2.)

In writing the program, we need to codify the rules very strictly. We need to define exact positions for the houses (the small black squares in the figure). We also need to know how to express, in mathematical terms, ideas such as "move halfway to Agatha's house." The approach to mathematics being advocated here does not eliminate the need to use and understand mathematical notation, but it does give the notation an immediate concrete application.

The educational benefits of the game go further. The Sierpinski sieve is itself a gallery of mathematical illustration. Here are just some of the concepts it illustrates:

Program code 1.1   The Rules for the Aunt's Birthday Game

```
Inputs:
   xa and ya are arrays containing screen coordinates of
      the aunts' houses.
Variables:
   x any y are the player's current coordinates.
   r holds the number of the chosen aunt (between 1 and 3).
Procedures:
   dot(x,y) draws a dot at screen coordinates (x,y).
   random(n) gives a random number between 1 and n.
```

```
x=xa[2]; y=ya[2];      /* Start from the second house */
do n = 1 to 10000      /* Plot 10,000 points */
   r=random(3);        /* Choose an aunt */
   x=(x+x[r])/2;       /* Move half way to her house */
   y=(y+y[r])/2;
   dot(x,y);           /* Draw a dot in this position */
end;
```

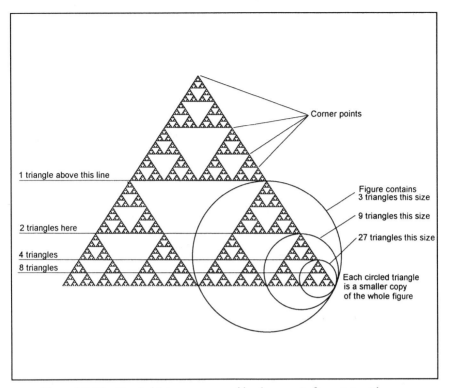

Figure 1.2 The Sierpinski sieve constructed by the game in figure 1.1, with some of its features indicated.

- "Powers of two (the number of triangles of successively smaller size as the eye moves down the shape is 1, 2, 4, . . . See the annotations on the left of figure 1.2).
- Powers of three (the number of triangles of successively smaller size in the total shape is 1, 3, 9, 27, . . . See the circled triangles in figure 1.2).
- Division of a triangle into four similar triangles.
- Perimeter and area. (Try calculating these quantities for the sieve.)
- Infinite sequences (the infinite sequence of smaller and smaller triangles as we approach a corner. See the circled triangles in figure 1.2).
- Convergence and limits. (An infinite number of triangles can still fit into a finite space.)
- Recursion. (The triangle is built up from smaller copies of itself. Each of the small circled triangles in figure 1.2 is a smaller copy of the whole figure.)

Figure 1.3 Constructing a fractal mountain chain using elastic cord.

- The relationship between rational and real numbers. (Along the sides of the triangle, there is a point for every real number and a "corner point" for every rational number within a given interval. In figure 1.2, some of the "corner points" are indicated.)

The exposition need not stop there. The algorithm of program code 1.1 is an example of an *iterated function system.* Within the same general algorithm, variation of the rules that determine the new position produces an infinite variety of interesting fractals. Many are discussed by Barnsley.[3] Exploration in this direction exposes students to the algebraic representation of rotations, translations, and scalings.

Another game illustrates the power of fractals to mimic natural landscapes. Suppose we wish to generate a mountain skyline. A suitable method is the midpoint displacement algorithm described by Saupe.[4] I like to demonstrate it this way. Two volunteers hold the ends of a length of elastic cord, at about waist height, on opposite sides of the room. Our "skyline" is now a straight line, like the horizon at sea. A third volunteer grasps the midpoint of the cord and raises or lowers it by any amount up to about a meter. This converts the skyline into a hill or a valley comprising two straight lines. Two more volunteers grasp the midpoints of these lines and raise or lower them by any amount up to about half a meter. The situation at this stage is shown in figure 1.3. The process continues until volunteers run out, or until there is no more room to fit them along the cord.

Here again, the computer comes to the rescue. It can go on subdividing the skyline to the limits of its graphics resolution (and even further, if we don't wish to see the results). A skyline requiring 1,025 "volunteers" is shown in figure 1.4. Several skylines can be arranged to form a picture. With removal of hidden lines and some simple rendering, an interesting piece of scenery can be produced. (See figure 1.5.) Within each mountain range, the shading starts with black at the

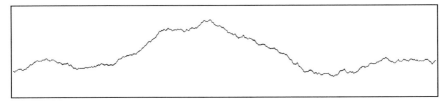

Figure 1.4 A mountain skyline made up of 1,024 segments.

Figure 1.5 Mountain scenery made from rendered skylines.

highest peak and becomes lighter with decreasing height. Horizontal stripes are avoided by using random variation to blur the boundaries between areas with the same shade. Programming the rules requires clarification of ideas that were vague in the game, such as "an amount up to about a meter." Clarifying this particular point leads to an exploration of Gaussian number distributions, because the displacements should be chosen randomly from such a distribution.

The rules of these games allow us to detect foul play. In the aunt's birthday game, for example, it is not uncommon for a group to try to test the rules by repeatedly choosing the same aunt. The algorithm allows this, and it just makes the result (in the short term) less interesting. In the mountain building, it is perfectly allowable for a volunteer not to move the cord at all. If everyone does this, we finish as we began, with a flat skyline. This is still a valid skyline, even if it is not mountainous.

Does this kind of play really evoke an interest in mathematics? My experience here is limited, but the signs are hopeful. On a few occasions, I have set up a "fractal classroom," usually with 12- or 13-year-olds. The results have been gratifying. Here, in a mathematics lesson, students had the opportunity to experiment creatively as they drew a fractal tree. (See figure 1.6.) How could they make the colors realistic? How could they change the shape of the tree? How could they make the tree grow exactly to a certain height on the screen? These are artistic

Figure 1.6 The tree constructed by program code 1.2.

```
Program code 1.2   Rules for the Two-Dimensional Tree in Figure 1.6

    Inputs:
      size: the total height of the tree.
      limit: puts a limit on the depth of recursion.
    Procedures:
      forward(d) moves the turtle forward by distance d,
          drawing as it goes.
      back(d) moves the turtle backwards by distance d without
          drawing.
      left(degrees) and right(degrees) turn the turtle left
          or right by the given angle.

    tree(height)
      if height>limit then do
        forward(size/4);    /* draw the trunk */
        left(40);           /* draw a left branch */
        tree(size/2)
        right(40);
        forward(size/12);   /* draw a little more trunk */
        right(40);          /* draw a right branch */
        tree (size*0.45);
        left(40);
        tree (size*2/3);    /* draw the top part of the tree */
        back(size/3);
      end;
```

questions, but in this context they have mathematical answers because the tree is defined mathematically, by an algorithm (see program code 1.2) to control a drawing turtle and implemented in Logo.[5] Most such questions have answers that are within, or not much beyond, the children's regular mathematics curriculum. But what incentive did they have previously to learn, for example, about the addition of fractions or about the sum of an infinite series?

Students who get serious about fractal mimicry of nature soon realize that two dimensions are not enough. Natural shapes are three-dimensional. Fractals

Program code 1.3   Rules for the Three-Dimensional Tree in Figure 1.8

```
Inputs:
  size: a measure of the size of the tree.
  limit: controls the depth of recursion.
Procedures:
  forward(d,w) moves turtle forward by the distance d,
      drawing a line of thickness w as it goes.
  back(d) moves turtle back by distance d, without
      drawing.
  pitch(degrees), yaw(degrees) and roll(degrees) turn the
      turtle around the appropriate axis by the given
      angle (see Figure 7).
```

```
ThreeDTree(size)
  if size>limit then do
    forward(size/5,size/75);   /* draw trunk */
    do i=1 to 3                 /* draw 3 branches */
      pitch(30);
      ThreeDTree(size*0.7);
      pitch(-30);
      roll(120);
    end;
    back(size/5);                /* return to starting point */
  end;
```

in three dimensions can be generated by a reasonably straightforward extension of the algorithms already presented. The problem is that our visualisation medium, the computer screen, is still only two-dimensional. We must render a three-dimensional object in two dimensions. Many methods are available, including simple isometric projection (squashing the three-dimensional object back into two dimensions), perspective projection (geometrically more accurate), stereo images (which gives true three-dimensional effect, at the cost of realistic color) and ray tracing (which gives perspective with correct rendering of shading, shadows, reflections, and surface textures). These must suffice until a truly three-dimensional medium becomes available. In the future, techniques such as virtual reality and stereolithography (a process that can render abstract mathematical shapes as three-dimensional plastic sculptures) will provide profound, additional stimulation for students.

Program code 1.3 shows the algorithm for a three-dimensional version of the tree. In three dimensions, the rotations are more complex, and are so unfamiliar that everyday English has no words for them. One solution is to turn to aviation for the necessary terms. As well as turning left and right (yaw), the turtle can now turn up and down (pitch) and twist around the axis of its body (roll). These rotations are shown in figure 1.7. The turns are interdependent. A turn around one axis affects the result of a subsequent turn around either of the others.

A three-dimensional turtle following the algorithm of program code 1.3 can provide instructions to a ray tracing program, which can in turn produce a reasonably realistic image of the tree. (See figure 1.8.) The addition of leaves and

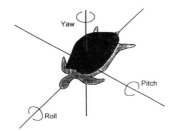

Figure 1.7 Turtle rotations in three dimensions.

flowers (figure 1.9) and modification of the original shape (figure 1.10) can produce a whole garden of botanical specimens. For more examples of images of this nature, see the work of Lindenmayer and Prusinkiewicz.[6]

Behind most fractal images lurks a *dynamical system*. The iterative function systems introduced earlier (the aunt's birthday game) are simple examples. The fractals associated with more general dynamical systems can be especially spectacular. The best known is the Mandelbrot set, the set of values c for which successive applications of the rule $z \rightarrow z^2 + c$ (starting from z equals 0) do not converge to infinity. Felsager and Vedelsby[7] have introduced dynamical systems to the Danish school curriculum. In students' journeys toward an understanding of these systems and their images, they move through many more basic mathematical ideas, illustrated with applications in many fields, including accounting, anatomy, environment, physics, and demographics.

By approaching to mathematics through fractals, mathematics can relate easily to other areas of the curriculum. The relationship to art is obvious. The mimicry of nature ties mathematics to the natural sciences. Many economic indicators show fractal behavior. The most valuable link set up by fractals is the link to the real world that students see every day. Fractal mathematics describes

Figure 1.8 The three-dimensional tree produced by program code 1.3, as rendered by a ray tracer.

the shapes that they see in the natural world: clouds, trees, mountains, even surf breaking on a beach.

This brings us back to my introductory reverie as I watch the chaos of the surf and wonder about the possible reactions of people to it. In education, there is not much agreement about how students learn (this awaits the development of a mathematical model for learning), or even what students should learn. Different educators would react to my reverie in different ways, and their reactions to this whole chapter would similarly differ.

Traditional aesthetes like poets and artists might share Walt Whitman's dismay at the "learn'd astronomer" destroying the beauty of the stars with his numbers "ranged in columns."[8] Keep mathematics out of nature, they cry, not realizing that mathematics is the language in which nature seems to be written. To these people, I would point to the fascinating objects, previously unsuspected, that have been discovered by mathematicians in recent years. No poet, no artist ever imagined the Mandelbrot set. No poet, no artist *could* imagine the Mandelbrot set. Without mathematics, without computers, we would not know it now.

Figure 1.9 A three-dimensional flower of uncertain species.

Figure 1.10 A three-dimensional flower of a different species.

Whitman at least believes that beauty is important. Not so, say the pragmatists, the philistines, for whom economic management and bureaucratic control reign supreme. They would condemn my thoughts on the beach as a waste of time and direct me back to administrative trivia. To these people I would point to the applications of mathematical modeling—the weather, the spread of forest fires, the flow of gases in the furnace of a power station, the dispersal of pollutants in a river. All these are studies of chaotic systems and all are economically important. Yet all involve the modeling, display, and interpretation of fractal images not unlike those I am watching in the surf.

A similar argument could be used for another group, the social engineers. They would regard my thoughts on the beach as not only wasteful but elitist as well. My time would be better spent investigating the gender balance of the people surfing, or worrying about whether the buffeting of their bodies by the waves constitutes some obscure kind of sexual abuse from which they should be saved. These people partake of the comforts of technology but dismiss as elitist the mathematics that led to these comforts and continues to sustain them.

Finally, what of mathematicians themselves? Some of them express concern that fractals would be trivialized in the classroom. The underlying math can be very difficult. The proof of the connectedness of the Mandelbrot set, for example, is something that Mandelbrot himself confesses he could not have devised.[9] We do not say that Mandelbrot should therefore not be interested in fractals. Number theorists may similarly worry that the elementary school syllabus trivializes the integers. Of course it does, but integers are so omnipresent and useful that everyone needs to know something about them, even if not the whole story. In the future, fractals may well have the same status.

Looking into the future, we can be almost certain that more and more powerful computers will continue to become available to schools. Education may remain at the bottom end of the computer market, but the quality of the bottom end is rising all the time. Today's supercomputers may eventually find a place in the classroom.

What use will education make of this technology? Each of the viewpoints just mentioned has its own answer. Traditional aesthetes may accept a computer as a convenient tool but nothing more. Pragmatists see it as a way of saving money, since the remotely accessed electronic class will need no classrooms, no physical plant at all, except what students provide for themselves. Social engineers rightly hail the potential of computers to serve the disabled but also welcome their perceived potential to make basic skills unnecessary, leaving more room in the curriculum for social conditioning. (Calculators make arithmetic unnecessary; word processors make spelling and grammar unnecessary; voice processing will make reading and writing unnecessary.)

It would be tragic if new generations of school computers were still used only for the same old collection of office tools and drill programs, ported from less powerful computers. It would be a pity if we missed the potential of computers to be partners in creativity, to allow experiment in areas where it was previously impossible, and to connect students with the mathematics of the world around them.

As I look again at the surf, I see the surfers riding the waves, reading the mathematical cacophony with varying degrees of skill, interpreting the result of the billions of droplet processors, reading the waves. Many of them are probably children playing truant from school. They have deserted the ideologically sound, philistine pragmatism of the classroom for a more involving and aesthetic experience. They are not aware that in reading the waves, they are, in a real sense, "doing mathematics." Perhaps, in the fractal classroom of the future, their awareness will awaken.

Notes

1. J. Hubbard and B. West, *Differential Equations: A Dynamical Systems Approach* (New York: Springer, 1991), vii-x.

2. B. Beaumont and P. Scott, *The Fractal Umbrella* (Adelaide: Australian Association of Mathematics Teachers, 1994). (Available from AAMT Inc., G.P.O. Box 1729, Adelaide S.A. 5001, Australia).

3. M. Barnsley, *Fractals Everywhere* (Boston: Academic Press, 1993).

4. H. Peitgen and D. Saupe, eds., *The Science of Fractal Images* (New York: Springer, 1988) 71-94.

5. H. Abelson and A. di Sessa, *Turtle Geometry* (Cambridge: MIT Press, 1982).

6. P. Prusinkiewicz and A. Lindenmayer, *The Algorithmic Beauty of Plants* (New York: Springer, 1990).

7. B. Felsager and M. Vedelsby, *Geometry and the Imagination: Chaos and Fractals* (Minneapolis: The Geometry Centre, 1992).

8. W. Whitman, "The Learnd Astronomer," *Leaves of Grass* (New York: Modern Library, 1950).

9. H. Peitgen, H. Jürgens, and D. Saupe, *Fractals for the Classroom* (New York: Springer, 1992), 1-16.

2

The Fractal Curriculum

David Fowler

A new curriculum is emerging that will replace the old list of school mathematics topics. Not only is the content of the curriculum rich in fractal concepts, but the curriculum itself will possess a fractal-like branching structure, replacing the linear model now in use. Students will explore this curriculum in collaborative study groups linked through a globally connected network.

INTRODUCTION

If mathematics is "The Music of Reason," as Jean Dieudonné has called it,[1] then we are entering an age in which everyone can join the orchestra. The new instruments are the members of the computer family with which one can play not only the classics—Archimedes, Newton, and Gauss—but also can improvise on the themes created by the new composers—Poincaré, Julia, and Mandelbrot.

The themes are simple—iteration of a rational function or repeated application of a small set of affine transformations to a geometric object. The results are complex and visually compelling. Mathematics has once again become an experimental subject, with experimental results that draw the attention of the mathematical bystander. Robert Devaney, a mathematician at Boston University, has produced a series of computer-generated animations illustrating the spectacular visual images associated with the dynamics of fractal sets.[2] Few people can walk past a video of Devaney's "Exploding Julia Sets" without pausing to watch at least one cycle of action.

Mathematical concepts once presented as static topics—symmetry, congruence, similarity—now have dynamic interpretations. Furthermore, the new mathematical themes generate models of the natural world that include the patterns and rhythms traditional mathematics has failed to describe. With these models comes a set of opportunities for introducing children to mathematics—mathematics that is connected to what real mathematicians do in the world. Watching teachers take advantage of these opportunities, particularly in the early grades, and working with fourth- and fifth-grade students myself, I'm convinced that a new mathematics curriculum—a fractal curriculum—is emerging and will flourish early in the next century.

FRACTALS FOR KIDS

At the earliest ages, children are being introduced to the concepts of iteration and pattern. They are using rubber stamps with single fern leaves to produce a fern collage, squeezing out dendritic patterns of tempera paint between pieces of clear acrylic plastic, constructing toothpick tetrahedra and then assembling them into a Sierpinski tetrahedron.[3] Following these simple tactile experiments, they are replicating the same patterns with computer simulations. These simulations are based on cellular automata systems,[4] multiple-turtle or three-dimensional turtle Logo systems [5] or higher-level object-manipulation programs.[6] Students are also videotaping natural processes—for example, multiple periodicity in children playing "Double Dutch" jump rope—and constructing side-by-side comparisons on the computer screen of the natural and simulated processes.[7] All these activities are now being tried out in pilot projects and will be appearing in regular school classrooms in the next few years.

Even when fractals are not being taught explicitly, the underlying concepts of iteration and pattern are present in the classroom. A series of videotapes made as part of a National Science Foundation (NSF) project to promote the use of calculators in elementary schools shows teachers using calculators in ways that go beyond simply verifying paper-and-pencil calculation—the transitional agreement under which calculators were first allowed into the classroom.[8]

In one example, a teacher is introducing her first-grade students to the calculator for the first time. The teacher begins with a discussion of the different colors of the keys, helping the children grasp the basic distinction among numbers, operations, and relations. Then she asks how many numbers can be displayed in the window. For these calculators, the answer is 8. The teacher directs the class's attention to a student who answered by entering consecutively the first eight positive integers, an efficient way to solve the problem. An interesting difficulty comes up immediately—one student begins with zero and is able to display integers only up to 7. The teacher tries to repeat the experiment and finds that when she tries to enter the sequence 0,1,2, . . . , the 1 overwrites the 0. How did the student do it? After a moment's thought, the teacher realizes the student first entered the decimal point.

Two lessons here: Zero is clearly a special kind of number, but the "." symbol can make zero "behave" and hold its place in a sequence of digits.[9] The teacher and students agree that there are still eight numbers displayed in the window. The principle of one-to-one correspondence and the "pigeon-hole principle,"[10] governing the placing of N objects into N locations, have made their appearance in the first grade.

Now the teacher asks the students to add 1 plus 1 and press the equals key. She then encourages them to repeatedly press the equals key to iteratively generate more numbers. They do this with enthusiasm, and the room is filled with the whir of clicking equals keys. A student observes that it's "like a game." The teacher encourages them to see who can get to 100 first. Then she asks the students to put repeating number patterns into the display. Some students put in two-digit repeating patterns and a few put in three-digit patterns. After the first two occurrences of a three-digit pattern, only the next two digits can be entered. A simple game emerges from this situation; that is, to guess the missing third digit. The children break up into small groups to quiz each other. Using a small electronic device for playing games is very natural to them.

Here are the origins of the fractal curriculum—playful use of technology to explore numerical iteration and basic concepts and processes of mathematics:

number, operations, relations, patterns, and experimentation and prediction. The children's first impression of the calculator is that it is a pattern-generating device and a source of interesting problems. When they break up into small groups to challenge each other with missing-digit patterns and other games, they are learning that mathematics is a process of human communication.

The children involved in these activities are approaching mathematics in a sense described by Seymour Papert in the early 1970s.[11] In a paper titled "Teaching Children to be Mathematicians Versus Teaching About Mathematics," Papert argued that children in elementary mathematics classes could be meaningfully engaged as mathematicians, just as children in music or art classes could be meaningfully engaged as musicians or artists. The NSF videos show students experimenting and then conversing, writing, and drawing pictures based on their mathematical discoveries. Their teachers have created contexts for mathematical exploration.

Further evidence of the potential for a fractal curriculum is based on my own experiments with a CAM-6 (Cellular Automata Machine) computer in a fifth-grade classroom. The *cellular automata* machine was developed at MIT to create a new environment for modeling complex phenomena.[12] The "machine" is actually a parallel processing card inserted in a basic 8086 IBM PC. The computer, thus transformed, displays a field 256 by 256 pixels wide. Rules for pixel behavior—"turn off, turn on, change color"—are written so that the state of a pixel depends on the state of its neighbors. The CAM-6 updates the state of each pixel 60 times per second, producing high-level animation effects.

The "game" of *LIFE* is the most familiar of the cellular automata models,[13] but there are many other interesting cellular automata phenomena.[14] Cellular automata are particularly good at modeling phenomena in which a simple set of rules applies to every element in an area being studied. For example, each molecule of air moving across an airplane wing will bump against the wing and against neighboring molecules according to the set of rules governing elastic collisions. Each molecule of oil moving through a porous metal filter, or each point along the edge of a prairie fire, can be described by a particular rule set. Even the "Wave" effect at American athletic events can be described by a cellular automata rule: "Stand if the person on your left stands, else sit." When initiated by a column of people standing at once and then sitting, this rule generates a wave form moving to the right across the stadium. In each of these cases, a cellular automata simulation can be constructed so that each pixel models the behavior at a particular point. From the local actions at each point, an overall interesting and often unpredictable behavior may result. The CAM-6 and its associated software make the programming a relatively simple task.

Two years ago I installed a CAM-6 machine in an elementary school media center and taught a small group of students how to use it. The following year I installed a second machine in a regular classroom, so that the first group of students

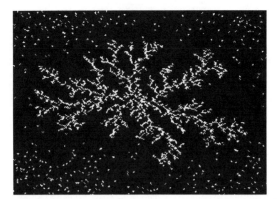

Figure 2.1 The "Dendrite." Taken from a videotape of a simulation of diffusion-limited aggregation running on a CAM-6 computer.

could teach their classmates how to use the machine. The students and I concentrated on a cellular automata model that simulates a process known as diffusion-limited aggregation,[15] although we always referred to the program as "the dendrite." This model simulates the motion of "warm" cells that move as green random dots on the screen until they encounter a motionless, blue "frozen" cell. When a warm cell meets a cold cell, the warm cell changes color and becomes part of a growing branched, hence "dendritic" structure. On the CAM-6 screen the warm cells resemble a furious swarm of green gnats, while the icy blue dendrite emerges against the black background as a strange apparition in space. Figure 2.1 shows a black-and-white rendering of a frame taken from a videotape of the terminal while the dendrite program was running.

Initially, my students used the CAM-6 system as a microworld for various experiments. The system allows students to set parameters and draw starting configurations without going into the actual programming language. For example, one student drew his initials as a set of blue frozen cells, then watched them "sprout"—his term—when the program was run. We didn't realize at the time that a computer scientist had recently written a chapter on "growing your own font" using a closely related process.[16] Figure 2.2 shows a replication of the experiment with the initials C-A-M. Notice the relatively sparse branching on the interior of the letters. As the randomly moving warm cells encounter the frozen cells, the resulting aggregation seals off the interiors, thus limiting any further growth.

The students suggested most of the experiments on their own. Two girls set out to time the growth of the dendrite from one initial seed until all cells were frozen—a time period of about 15 minutes. I showed them how to vary the initial concentration of warm cells, and we drew graphs that plotted initial concentration against growth time, starting from a single frozen cell. We then varied the initial

Figure 2.2 "Sprouted initials," formed by drawing three letters and then running
the diffusion-limited aggregation simulation on the CAM-6 computer.

number of frozen cells to produce multiple dendrites on the screen. Following this
experiment, we compared computer-generated dendrites with video images of
natural phenomena. Figure 2.3 is digitized from a still video frame of frost crystals
on an automobile window.

Figure 2.4 shows a close-up of an image generated by the CAM-6, starting
with an initial line of cold cells and a few adjacent cells. From this set of seeds, the
program generates a pattern with the same features as the natural crystals.

Our classroom discussion of how crystals could form from random motions
of molecules led to a general discussion of random processes. We watched a video
recording of a robin hopping across a lawn and discussed the differences between
random and purposeful motion. We also watched videos of sandhill cranes as they
took off from a lake, flying at first in apparently chaotic swarms and then
organizing into formations, and we tried to find the point at which randomness

Figure 2.3 Enlargement of a still video frame of frost crystals on an automobile
window.

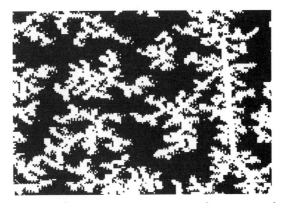

Figure 2.4 Portion of the CAM-6 computer screen showing a simulation of the frost crystals in figure 2.3.

became order. As our final activity, each student wrote out a set of commands on three-by-five cards: "(F)orward, (L)eft, (R)ight, (B)ack." The students were advised to make these as "random as possible." We then went outdoors, and the students simulated the random motions of molecules by moving one step at a time according to their own directions. Figure 2.5 shows the students "being molecules" on the playground.

An obvious extension of these experiments is to investigate the frequent occurrence of dendritic patterns in the natural world. A recent article in *National Wildlife* contains excellent illustrated examples, including deer antlers, tree roots, coral stalks, lightning bolts, and the veins in a jackrabbit ear.[17] Students can

Figure 2.5 Children simulating the motion of random molecules. Each child is following instructions for "random" movements written during class.

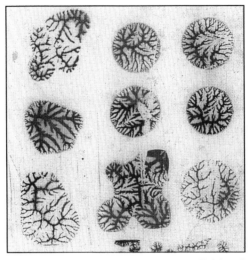

Figure 2.6 Dendritic images formed by squeezing small dabs of tempera paint between two pieces of clear acrylic plastic.

discover that not all dendritic structures have the same "topology." In other words, their fundamental connectedness is different. The branching antlers of a deer, for example, do not form closed loops, while the branches of the frost crystals can reconnect. Students also can study actual frost crystals in greater depth. A recent article in *The Science Teacher* contains a series of frost crystal experiments that could be paired with cellular automata simulations.[18] Students also could experiment with analog as well as digital techniques for creating dendritic patterns. Figure 2.6 shows a set of dendritic images formed by squeezing small dabs of tempera paint between two pieces of clear acrylic plastic. This series was produced by a group of mathematics teachers attending a summer workshop on fractal geometry.[19]

The deeper lesson in the CAM-6 experiments is that a pattern can form as the result of a set of simple rules of motion. No individual cell as it moves about on the screen has the "purpose" of creating a branching structure—the cell's rules are simply to move at random until it encounters a motionless cell. A primary feature of cellular automata models is that they provide intuitive demonstrations of emergent phenomena. Gaining an understanding of how a general form can emerge from simple local rules is an example of what one researcher calls "overcoming the centralized mind set."[20]

These examples represent contexts for exploration now being developed in elementary and secondary schools. The collection of such contexts constitutes a landscape in which the fractal curriculum will evolve. Fractal examples are already percolating into the regular curriculum. Heinz-Otto Peitgen and colleagues have laid out a topic-by-topic connection chart for linking fractal mathematics to the traditional school curriculum.[21] Algebra texts are including chapters devoted to fractals,[22] and schools are considering one-semester courses in fractal geometry.

The important changes, however, will be driven by the elementary curriculum. First of all, elementary teachers have a head start in technology-based mathematics. Although they rarely have the best equipment, these teachers use what they have in imaginative ways.[23] Elementary teachers are making calculators available to all students in all grade levels. They use estimation, experimentation, and writing as problem-solving tools. Children use glass calculators on overhead projectors to demonstrate their solutions to their classmates. All this at a time when many American college professors are still trying to prohibit use of graphing calculators on calculus exams.

A second reason that the fractal curriculum will be driven by the elementary schools lies in the fact that the most interesting cognitive research is being done at this level. New modes of thought are associated with fractal geometry and chaotic dynamics.[24] Although Piagetian models for cognitive development have proved robust with respect to conservative systems, researchers are now exploring the existence of a distinct set of abilities necessary for understanding dynamic systems.[25] Furthermore, the studies of Piaget and Inhelder[26] that showed a progression in a child's conception of space from topological, to projective, to affine and eventually to Euclidean reflect a "pre-fractal" bias, coming from a time when fractional dimension and self-affinity were considered counterintuitive.[27]

In a classic experiment, Piaget had children measure strips of paper folded into zigzag paths, using small cards as measuring units. The children described their measurements as "steps taken by a little man." Although the same length of paper was used each time, and the paper strips were folded in plain sight of the children, the very young children always thought that if the endpoints were farther apart, then the "little man" would have to walk a longer distance along the crooked path. Older children understood that the length of the path is conserved under the operation of folding.

It is interesting to reconsider these conclusions in light of Benoit Mandelbrot's paper "How Long Is the Coast of Britain?"[28] Mandelbrot argues that geographers walking a pair of dividers along a map will attain variable lengths between two points of a coastline. The length will depend not only on the setting of the dividers but the relative jaggedness, or fractal dimension, of the coast. Mandelbrot argues further that fractal dimension is not a mathematical curiosity but regularly found in nature. The curriculum, then, should include experiences that develop intuition for fractal dimension. The nature of these experiences is a new territory for cognitive research.

Finally, the fractal curriculum will be driven by the elementary curriculum because elementary teachers are good at making connections—a feature that has been observed internationally, especially in subjects such as science.[29] The study of fractals is inherently cross disciplinary, linking art, mathematics, and natural and physical science. Since Mandelbrot's seminal *Fractal Geometry of Nature*,[30] an efflorescence of science and mathematics writers have demonstrated the fractal connectedness of the world. As fractal experiments and lessons are developed at the elementary level, the elementary classroom can mirror this connectedness and, in doing so, become the foundation for a thematic-based curriculum extending into the upper grades.

AN IMAGE OF THE FRACTAL CURRICULUM

The word "curriculum" is derived from "running track," which conveys an inherently linear image as shown in figure 2.7.

In the past the mathematics curriculum was represented by a linear series of topics, broken down into subsets of objectives. In contrast, the fractal curriculum can be visualized as a series of thematic connections. The starting point for this concept is a report published by the National Research Council (NRC).[31] Entitled *On the Shoulders of Giants,* after Isaac Newton's famous phrase, the report describes an approach to mathematics education based on five mathematical themes: dimension, quantity, uncertainty, shape, and change.

These themes provide a context for the development of mathematical thinking. Each theme extends from the earliest grades through high school. Dimension begins in the lower grades with active experimentation with volumes and decomposition of solid figures, and advances to concepts of fractal dimension in high school. Quantity develops basic quantitative reasoning making strong use of the calculator and computer. Uncertainty provides children opportunities to experiment with the concepts of data and chance. Change lets children experiment with patterns of growth and compare types of change in the real world, including nonlinear models as well as the traditional "ballistic" models of calculus. Shape provides a rich set of hands-on opportunities for naming, analyzing, and representing spatial structures.

The NRC authors are not proposing a rigid structure of topics. In fact, there are many more than five possible themes. Symmetry could be a theme, as well as discourse—the ways people communicate mathematical ideas to each other. The essential idea is to conceive of the curriculum as a collection of lines of thought that students can pursue through time, rather than as a fixed list of topics ("objectives") to be studied ("mastered").

Throughout the NRC curriculum fractal concepts appear with regularity. The Sierpinski triangle provides a good example. This figure can be constructed by "recursive removal," that is, removing an equilateral triangle from the interior of an equilateral triangle, then continuing to remove equilateral triangles from the interior of each remaining triangle as shown in figure 2.8.

The Sierpinski triangle as an example of self-similarity; each subsection is a replica of the whole figure. With respect to this property, the Sierpinski triangle

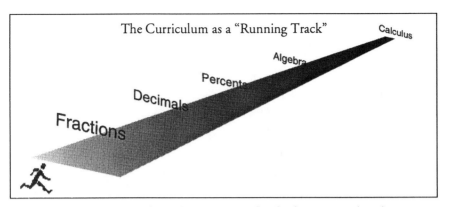

Figure 2.7 In the past, the mathematics curriculum has been imagined as a linear series of topics. The original meaning of curriculum is a "running track."

falls naturally within the Shape theme as an extension of the concept of similarity. An alternative construction of the Sierpinski triangle can be obtained by starting with the *Pascal triangle* and replacing each odd number with a 1 and each even number with a 0. The Pascal triangle describes the number of ways of selecting a group of items from a larger set; thus, it will be encountered in discrete probability theory. This property links the Sierpinski triangle to the Uncertainty theme.

The Sierpinski triangle emerges in surprising fashion from the *Chaos game,* originally described by mathematician Michael Barnsley.[32] In this "game," a player selects an arbitrary point within an equilateral triangle, then chooses a vertex of the triangle at random and selects a new point halfway between the initial point and the selected vertex. From this new point, the player again selects a vertex at random and moves halfway toward the selected vertex. Given this description, one would guess that the result of repeating the procedure many times would be to fill

Figure 2.8 The eighth stage of "recursive removal" in the creation of a Sierpinski triangle.

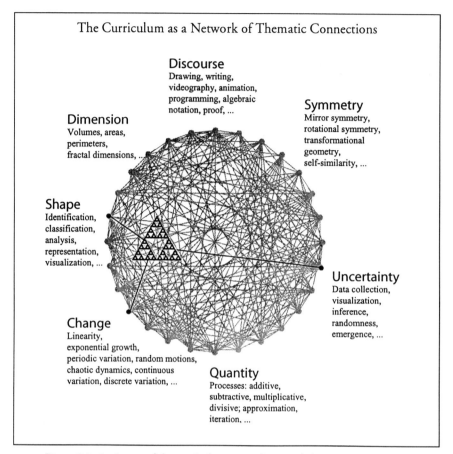

The Curriculum as a Network of Thematic Connections

Discourse
Drawing, writing,
videography, animation,
programming, algebraic
notation, proof, ...

Symmetry
Mirror symmetry,
rotational symmetry,
transformational
geometry,
self-similarity, ...

Dimension
Volumes, areas,
perimeters,
fractal dimensions, ...

Shape
Identification,
classification,
analysis,
representation,
visualization, ...

Uncertainty
Data collection,
visualization,
inference,
randomness,
emergence, ...

Change
Linearity,
exponential growth,
periodic variation, random motions,
chaotic dynamics, continuous
variation, discrete variation, ...

Quantity
Processes: additive,
subtractive, multiplicative,
divisive; approximation,
iteration, ...

Figure 2.9 An image of the curriculum as mathematical themes intersecting at many different topics. Here the themes of Shape, Change and Uncertainty are shown intersecting in the Sierpinski triangle.

the interior of the triangle with a random collection of points. In fact, after several hundred iterations a pattern begins to emerge—and that pattern is a Sierpinski triangle. More precisely, the pattern converges to an underlying Sierpinski triangle, even if the initial point chosen is outside the equilateral triangle. The Sierpinski triangle is an "attractor" for the sets of points generated by the Chaos game. Once again, this time as an example within the theme of Change, the Sierpinski triangle makes an appearance.

In a fully realized fractal curriculum there are many themes and many connections. The "running track" image can then be replaced with an image such as figure 2.9.

An exhaustive list of themes would be as pointless an undertaking as a database of all known musical themes in the world. Instead, general descriptions of these themes are essential. Details are important but should be left for students to work out in their own ways. In addition to the themes described by the NRC authors, the best of the mathematical writers, including Gardner,[33] Stewart,[34] Pickover,[35] Davis,[36] and others, have already suggested many fascinating curricular themes. Teachers can use this material to set up a new kind of environment that implements the thematic structure of the fractal curriculum. This environment, taking advantage of emerging technologies, will connect students in a worldwide network.

THE SUPERCONNECTING SUPER CURRICULUM

Imagine the curriculum as consisting of many themes and associated with each theme is a set of platforms. A platform consists of a set of tools, suggestions of what can be built with these tools, and pointers to related platforms. A platform is analogous to an activity area in a regular classroom, where the tools may be traditional hard-copy materials, such as pencils, rulers, paste, and scissors, and the suggestions for working with the tools provided by teachers and other students.

In their advanced form, platforms will be full-immersion environments linked through subnets of the Internet.[37] Tools will include "virtual scissors" for cutting apart and reassembling geometric objects.[38] Other Internet users will suggest ideas for connecting to other platforms. In the present jargon of the Internet, sites for collaborative exploration are called *multi-user domains* and are supervised by "wizards" or, as at the MIT Virtual Media Lab, by "janitors."[39] The supervisors of the mathematics platforms could be called "teachers." They would earn their "certification" by constructing the mathematical framework for a particular platform.

The technology is already in place for such a system. Multi-user domains for children are under way.[40] A networked environment for collaborative problem solving between students has recently been established.[41] New utilities make it possible to transfer fractal visualization tools from one site to another in a few seconds. For example, a student connected to the Internet can use the command ftp.forum.swarthmore.edu to access a list of programs. The command get, followed by the name of the program, brings the software to the student's computer. An example is the program "Julia's Dream,"[42] a program that simultaneously projects a Julia set when the student clicks on any point in the Mandelbrot set, as shown in figure 2.10.

Within a few years such programs will take the form of three-dimensional graphic user interfaces.[43] Students using three-dimensional viewing systems with multiple-input controls will use such programs to travel along thematic connections through a curricular space. Researchers are already designing virtual reality interfaces for mathematics.[44] The technological problems, while not trivial by any means, are all tractable.

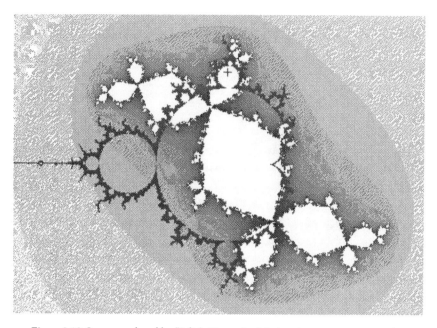

Figure 2.10 Image produced by "Julia's Dream," a Macintosh computer program by
Reinoud Lamberts, that associates a Julia set with each point of the Mandelbrot set.

In curricular space, platforms will be represented by three-dimensional icons.
Once a student selects an icon and moves "into" the platform, the display will change
to the environment for the specific platform. For example, the platform
"Mandelbutte" would allow a student to explore a mesa shaped like an extruded
Mandelbrot set. (See figure 2.11.) Students could meet at a particular location, perhaps
one symbolized by a Julia set, for seminars related to iteration, fixed points, periodic
points, and general and local connectivity. At some point a teacher might suggest that
one group of students visit the Dragon platform, to study this particular class of fractals,
and a second group visit the Spiral platform, to review some concepts of basic calculus.
When the students reconvene, the teacher challenges the students to construct a fractal
dragon that contains a logarithmic spiral.[45] Novice students visiting the platforms
could be young children or senior citizens wanting to learn about the new mathematics.
They can get compressed, digital video "take-home" instructions for hands-on activities. They also can practice with the virtual tools available on the platforms.

CONCLUSION

In this chapter I have outlined a rather conservative view of the possibilities for
mathematics education in the early twenty-first century. I haven't mentioned

Figure 2.11 "Mandelbutte" display screen.

neural network–based systems for student record keeping or advanced artificial intelligence systems for constructing new thematic strands through curricular space. All the interactions described in this chapter are simple extensions of human communication taking place at this moment.

A skeptical reader may notice that I haven't mentioned the cost of the fractal curriculum. In fact, the relative costs may be rather small, compared with curriculum projects driven by expensive textbook adoptions and massive in-service training.[46] The infrastructure is being built, and people are using it. Computing power continues to become cheaper and more accessible. Although some colleges have based their resistance to multi-user domains on fears of heavy additional loads on their systems, recent calculations suggest that interactive conversation is a relatively small portion of the Internet traffic.[47]

In a real sense, this curriculum will be constructed by the students as they communicate their ideas across the networks. One of the main tasks of teachers today should be to insure the right of every person to her or his share of the Internet bandwidth. If teachers help their students make the initial connections, the fractal curriculum will emerge as a self-organizing entity. Perhaps then the Music of Reason will be heard by the larger audience that mathematics deserves.

Notes

1. J. Dieudonné, *Mathematics, the Music of Reason* (New York: Springer Verlag, 1993).

2. R. L. Devaney, "Film and Video as a Tool in Mathematical Research," *The Mathematical Intelligencer* 11 (2) (1989): 33-37.

3. D. Fowler, *Fractals for Kids: Summer Workshop Report.* Teachers College, University of Nebraska–Lincoln (1994).

4. K. Karakotsios, *CASim User's Guide: Exploring Art, Simulation, Artificial Life Through Cellular Automata* [Macintosh software] (San Francisco: Algorithmic Arts, 1990).

5. S. Wagon, "Flying a Space Turtle Along a Peano Curve," in *Mathematica in Action* (New York: W. H. Freeman, 1991).

6. B. Lindow, *LS Man: A Program for Lindenmeyer [L-system] Constructions* (Livermore, CA: National Energy Research Supercomputer Center, 1993).

7. A. Rubin, "VIEW: Video for Exploring the World: National Science Foundation Directorate for Education and Human Resources," in *Applications of Advanced Technologies: Abstracts of Current Projects* (Washington, DC: NSF publication, 1992).

8. D. Fowler, *The Calculator in the Elementary School* [Videotape series] (Lincoln, NE: The Nebraska Systemic Initiative for Mathematics and Science Education, 1993).

9. B. Rotman, *Signifying Nothing: The Semiotics of Zero* (Stanford, CA: Stanford University Press, 1987).

10. D. Wood, *Theory of Computation* (New York: Harper & Row, 1987).

11. S. Papert, "Teaching Children to Be Mathematicians versus Teaching about Mathematics," *International Journal of Mathematics, Education, Science and Technology* (3) (1972): 263-268.

12. T. Toffoli and N. Margolis, *Cellular Automata Machines: A New Environment for Modeling* (Cambridge, MA: MIT Press, 1987).

13. W. Poundstone, *The Recursive Universe* (Oxford: Oxford University Press, 1987).

14. R. Rucker, *CA Lab* [MS DOS software] (Sausalito, CA: Autodesk, 1989).

15. P. Meaking and S. Tolman, "Diffusion-Limited Aggregation," in *Fractals in the Natural Sciences,* eds. M. Fleischmann, D. J. Tildesley, and R. C. Ball (Princeton, NJ: Princeton University Press, 1989).

16. C. A. Pickover, *Computers and the Imaginations: Visual Adventures Beyond the Edge* (New York: St. Martin's Press, 1991).

17. L. Gemery, "When Nature Branches Out: The Biological Functions of Branching" *National Wildlife,* 31 (6) (1993): 52-60.

18. R. Crump, "Snowflake science," *The Science Teacher* 61 (1) (1994): 45-48.

19. D. Fowler and C. Friesen, "Fractal Connections: A Report on a Summer Workshop for Nebraska Mathematics and Science Teachers" (Lincoln, NE: The Nebraska Coordinating Commission for Postsecondary Education, 1994).

20. M. Resnick, *Overcoming the Centralized Mindset* (Cambridge, MA: MIT Media Laboratory, 1992).

21. H-O Peitgen, J. Saupe, and L. E. Yunker, *Fractals for the Classroom: Strategic Activities* (New York: Springer Verlag, 1991).

22. L. Yunker, G. D. Vannatta, and F. J. Crosswhite, *Advanced Mathematical Concepts* (Columbus, OH: Merrill, 1992).

23. G. H. Wheatley and R. Shumway, "The Potential for Calculators to Transform Elementary School Mathematics," in *Calculators in Mathematics Education*, eds. J. T. Fey and C. R. Firsch (Washington, DC: National Council of Teachers of Mathematics, 1992).

24. M. Resnick and F. Martin, *Children and Artificial Life* (Cambridge, MA: MIT Media Laboratory, 1990).

25. M. M. Chandler and R. G. Boutiller, "The Development of Dynamic System Reasoning," *Human Development* 35 (1992): 121-137.

26. J. Piaget, B. Inhelder, and A. Szeminska, *The Child's Conception of Geometry* (New York: Harper & Row, 1964).

27. H. Hahn, "Geometry and Intuition," in *Space, Intuition and Geometry*, ed. W. L. Schaaf (Stanford, CA: SMSG Press, 1967).

28. B. B. Mandelbrot, "How Long is the Coast of Britain? Statistical Self-Similarity and Fractional Dimension," *Science* 156 (1967): 636-639.

29. A. Blum, "Integrated Science Studies," in *The International Encyclopedia of Curriculum*, ed. A. Lewy (Oxford: Pergamon Press, 1991).

30. B. B. Mandelbrot, *The Fractal Geometry of Nature* (San Francisco: W. H. Freeman, 1992).

31. L. A. Steen, ed., *On the Shoulders of Giants* (Washington, DC: National Academy Press, 1990).

32. M. Barnsley, *Fractals Everywhere* (San Diego: Academic Press, 1988).

33. M. Gardner, *The New Ambidextrous Universe: Symmetry and Assymmetry from Mirror Reflections to Superstrings* (New York: W. H. Freeman, 1990).

34. I. Stewart and M. Golubitsky, *Fearful Symmetry: Is God a Geometer?* (Oxford: Blackwell, 1992).

35. C. A. Pickover, *Computers, Pattern, Chaos, and Beauty: Graphics from an Unseen World* (New York: St. Martin's Press, 1990).

36. P. J. Davis, *Spirals: From Theodorus to Chaos* (Wellesley, MA: A. K. Peters, 1993).

37. M. Benedikt, ed., *Cyberspace: First Steps* (Cambridge, MA: MIT Press, 1992).

38. J. Doenias and R. E. Crandall, *Topology Lab: Visualization of Mathematically Defined Surfaces via PostScript Polygonal Rendering* [NeXT Computer software], (1989).

39. See, for example, Usenet discussions moderated by Greg Swan at `cbnvee@`
    `mcmuse.mc.maricopa.edu`.

40. MOOSE Crossing, an Internet context for kids. Amy Bruckman, `asb@`
    `media.mit.edu`.

41. Mathmagic, a collaborative problem-solving environment. Annie Fetter,
    `annie@forum.swarthmore.edu`.

42. R. Lamberts, *Julia's Dream* [Macintosh software]. Available by ftp from
    `forum.swarthmore.edu` (1991).

43. D. Mandelkern, "Graphical User Interfaces: The Next Generation," *Communi-
    cations of the ACM* 36 (4) (April 1993): 36-39.

44. W. Winn and W. Bricken, "Designing Virtual Worlds for Use in Math Educa-
    tion," *Educational Technology* 32 (12) (December 1992): 12-19.

45. D. Fowler and C. Friesen, "The Spiral and Dragon: Connecting Continuous and
    Discrete Mathematics," Ms.

46. M. Lieberman, "The Real Cost of Education," in *Public Education: An Autopsy*
    (Cambridge, MA: Harvard University Press, 1993).

47. B. Kort, re: Commercialization and Education on the Internet. `cbnvee@`
    `mcmuse.mc.maricopa.edu` (1993).

3

Fractals and Education: Helping Liberal Arts Students to See Science

Michael Frame

———

I describe an introductory fractals and chaos course directed toward liberal arts students. The course has been offered for six years and in that time has settled into a reasonably constant pattern that seems successful with this audience of students. After outlining the content, I give a detailed picture of fractals and iterated function systems, the first section of material presented in the class. The general pedagogic context of the course is explored, together with some assessment of its success and impact on students. Finally, I suggest some future directions for this material.

———

WHAT'S WRONG WITH THIS PICTURE?

Here is an all-too-familiar picture: a class of bright liberal arts students fulfilling their college mathematics requirement by learning the mechanics of computing derivatives and applying this to determine how many coat hangers a small business must produce to maximize profit. Calculus is beautiful mathematics but is best appreciated in a broader context. Moreover, its level of abstraction makes calculus a formidable challenge for more visually oriented students. As the sole exposure to college mathematics, an experience of this sort is not very useful. Students see mathematics as sterile, abstract, the work of people long dead. "Applications" involve either such fantastic oversimplifications that they become the subject of parody, or such detailed information that an entire class period is involved setting up a single problem. Why do we so often fail to communicate to our students the excitement we feel for our field?

Not every student we encounter has the good taste to be interested in those subjects we love. Students majoring in mathematics, science, computer science, or engineering usually develop such interests early, and our teaching is made easier by their appreciating the relevance of many topics. Liberal arts students, on the other hand, often have little inherent interest in mathematics. Because some degree of technology literacy is universally recognized as important for the educated person, increasing numbers of liberal arts students find themselves facing requirements for mathematics and laboratory science courses. All too often we offer unappetizing choices: a watered-down calculus course with "business applications" or the standard introductory biology class awash with beginning premed students. It is little wonder that such students rarely have fond memories of how they fulfilled their mathematics and science requirements. These attitudes contribute to the marginalization of the worth of doing science and mathematics. The majority of our fellow citizens thus are excluded from understanding that science is driven by the same passion firing all creative activities, that intuition and feeling play an active role in our work. Instead they often see a dry, deductive presentation devoid of any spark of humanity—the results of our work, in compact fashion, with little regard for why or how anyone would think in such an abstract way. Science and mathematics are the work of other people—people with thick glasses and stringy hair who are unable to utter a sentence devoid of technical terms.

The widespread availability of computer graphics has opened up another possibility: courses centered on visualization, on thinking in pictures. Building such courses from the standard offerings is challenging because we must overcome prejudices about the content of these courses. Some topics no longer hold their relevance in a world dominated by computers—for example, I am not concerned that my students cannot use a slide rule or a table of logarithms. More important, our ideas remain constrained by our own student experiences in the precomputer age of education. Thinking outside these bounds is challenging, though sure to be important. A more direct route to visualization-based courses is to explore fields in which pictures have played an essential role. Fractals, chaos, and complexity science are such fields. Since the computer and visualization are essential elements in the birth and continuing growth of these fields, courses on these topics are natural candidates for the incorporation of these tools. This is not to say that programs used in research can be transported directly to the introductory classroom. Some care must be given to pedagogy—programs must be accessible and at the same time foster learning by discovery. When used only as a classroom demonstration tool, the computer is little better than a videotape. Used creatively, the computer can empower even timid liberal arts students and stimulate them to explore and discover fascinating aspects of dynamics.

A central lesson of fractal geometry is that many complicated shapes of nature become easier to understand when viewed dynamically, as a process rather than as a static figure. In this sense, fractal geometry is not only an instance of, but also a metaphor for, presenting science and mathematics as a process. In this form, the subject seems to speak more directly to the liberal arts student and so offers a more effective introduction to science.

In this chapter I describe an introductory fractals and chaos course for liberal arts students, developed over the last six years with David Peak of the physics department of Union College. We have taught this course at Union College, the University of Richmond, and Yale University. This material and approach have evolved through trials with many different students, so we are reasonably confident of the robustness of the course. This is only one out many possible expressions of such material. I present it as an example of incorporating visualization and computing into introductory science courses for liberal arts students. The future of such courses is filled with great promise.

What's right with this picture? Using our approach, students can see some pieces of how science and mathematics are done and understand it as a human activity. They use discovery at even this introductory level of instruction and thus must become engaged, participating in the process—they see some of the excitement of doing science. We believe this is an important lesson in the context of liberal education.

ONE APPROACH TO TEACHING MATHEMATICS

By now the fields of fractals and chaos have grown to such an extent that they encompass enough material for many different courses. We have selected topics arranged in several categories, organized roughly thematically, that we present in outline form. For students to seek interconnections between topics as the course unfolds, we start the course with a short survey of all the topics.

Fractal Constructions and Iterative Geometry

We begin by describing self-similarity, emphasizing from the first the differences between mathematical fractals, for which the scaling is over an infinite range, and physical fractals, with scaling over only a limited range. Fractals are constructed by iterative geometrical tools, beginning with the *Chaos game,* (described in detail later) and proceeding through *iterated function systems* (IFS). The latter involves the mechanics of transformations of the plane—translation, scaling, rotation, and reflection. We investigate two methods of generating images by IFS: *deterministic,* in which all the transformations are applied simultaneously, and *random,* in which the transformations are applied one at a time in random order. Arguments for the convergence of the deterministic method motivate us to introduce the notion of limits. Showing that the deterministic and random methods produce the same pictures requires the notion of IFS addresses, a symbolic encoding of locations. Students gain familiarity with these concepts through homework using interactive graphics programs and through calculations done by hand. In particular, they have some practice with the "inverse problem"—finding the affine transformations to produce a given fractal. Finally, we introduce an approach to driving IFS by a nonrandom sequence, for example, Jeffrey's analysis of the base sequence of certain DNA strings.[1] The lesson of this section is that simple dynamic rules can give rise to geometric shapes that appear complex when viewed as static objects.

Quantifying Fractals: Dimension Computations

We began thinking of dimension as an exponent by computing the mass dimension of familiar objects such as line segments, filled-in squares, and cubes. Taking successively smaller pieces of the object, we record the mass and linear size (length for the line segment, side length for the square and cube) of the pieces and do a curve fit on the mass vs. size graph. This yields a functional dependence

$$M = K \cdot r^d$$

where M denotes the mass of the piece, r the size of the piece, K is a constant, and d is the mass dimension. Not surprisingly, we obtain $d = 1$, 2, and 3 for the line segment, square, and cube, respectively.

Such arguments are not needed to show us a cube is three-dimensional, of course, but to analyze fractal shapes. One of the simplest examples of a fractal is the

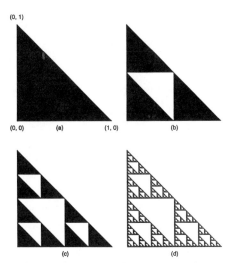

Figure 3.1(a) A right isosceles triangle. (b) and (c) The first and second iterations of the construction of a right isosceles Sierpinski gasket. (d) A right isosceles Sierpinski gasket—the limiting shape of this iterative process.

Sierpinski gasket formed by starting with a filled-in triangle, connecting the midpoint of each edge, removing the smaller triangle formed by these connections, then repeating the process on each of the three remaining smaller triangles, and so on. (See figure 3.1d.) Since the Sierpinski gasket consists of three copies half the size of the original, and this pattern persists (nine copies one-fourth the size, 27 copies one-eighth the size, . . .), we obtain points $(M, r) = (1, 1), (\frac{1}{3}, \frac{1}{2}), (\frac{1}{9}, \frac{1}{4}), (\frac{1}{27}, \frac{1}{8})$, . . . (We have taken both the mass and size of the original gasket to be 1. This amounts to choosing the units of our scale and ruler.) The curve-fitting program gives a value of d equals 1.585, the mass-dimension of the Sierpinski gasket.

To obtain analytic values in such symmetrical cases, we introduce the logarithm as a function to extract the value of d from the mass-size relationship:

$$d = \lim_{r \to 0} \frac{\log(M)}{\log(r)}$$

For the Sierpinski gasket we find $d = \log(3)/\log(2)$.

To understand the sense in which noninteger dimensions are possible, we compute the "measure" of the filled-in unit square in the wrong dimensions. Computed below its correct dimension (two), the square has infinite measure. The "one-dimensional measure" of the square is its length, computed by covering the square with infinitely thin thread. More complete coverings take more thread, increasing without bound, so we say the square has infinite length. On the other hand, computed above its correct dimension, the square has zero measure. The "three-dimensional measure" of the square is its volume, computed by covering

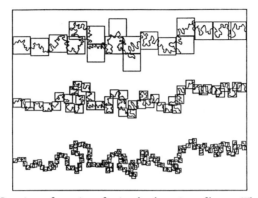

Figure 3.2 Coverings of a section of a river by three sizes of boxes. The (number of boxes, side length of each box) have these values: (18, ¼), (45, ⅛), and (143, ¹⁄₁₆).

the square with cubes. Using small cubes, the covering is a box with length and width close that of the square and height that of the cubes. Smaller cubes form a box of smaller total volume, decreasing without bound, so we say the square has zero volume. For the gasket, measuring in any dimension greater than $\log(3)/\log(2)$ gives zero, while measuring in any dimension less than $\log(3)/\log(2)$ gives infinity.

For shapes without exact symmetries—in particular, for natural fractals— this method of covering is generalized to the *box-counting dimension*. Here the shape is covered by boxes of a given size, then by boxes of a smaller size, then by boxes of still smaller size, and so on. The box-counting dimension is the exponent giving how the number of boxes in the cover depends on the size of the boxes. Examples include computing the dimension of a section of a river (see figure 3.2), a crumpled paper ball, and the structures formed by the intrusion of air into viscous substances. The paper crumpling experiment gives a good interpretation of the significance of dimension. A sheet of paper is approximately two-dimensional. By crumpling the paper, we are forcing it to try to be three-dimensional, but at the expense of introducing folds and cavities over a range of size scales. This is the expression of the fractal nature of the crumpled paper. Dimension as an indicator of structural complexity is the lesson of this section.

The Geometry of Complicated Dynamics

The method of graphical iteration is introduced through linear (compound interest) and piecewise linear *(Tent Map)* examples. Typical dynamics computations are approached not through algebraic manipulations, but through the geometry of *graphical iteration*.

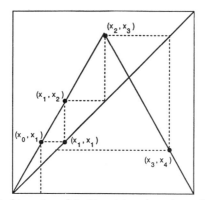

Figure 3.3 Graphical iteration of the Tent Map: the point x_0 is taken to x_1, that to x_2, that to x_3, and so on.

Figure 3.3 illustrates this process. The box containing the picture is $0 \le x \le 1$ and $0 \le y \le 1$. The Tent Map $T_r(x) = (r/2) - r \left| x - \frac{1}{2} \right|$, plotted for $r = 7/4$, is the inverted V in the picture, and the auxiliary line $y = x$ (called the diagonal) also is shown. As its name suggests, graphical iteration is a visual process for tracing the sequence of points x_0, $x_1 = T_r(x_0)$, $x_2 = T_r(x_1)$, ... , the *orbit* of x_0 under iteration of the Tent Map. (Of course, graphical iteration can be applied to any real-valued function of a real variable, not just to the Tent Map.) For an initial point x_0, the point (x_0, x_1) lies on the graph of the Tent Map. The horizontal line through (x_0, x_1) intersects the diagonal at (x_1, x_1), and the vertical line through this point intersects the graph of the Tent Map at (x_1, x_2). Graphical iteration is this process: going vertically to the diagonal, then horizontally to the graph of the function being iterated.

Though graphical iteration reveals it to be a convincing source of complicated behavior from a very simple function, the Tent Map does not exhibit the variety of stable orbits observed in many physical systems. Connecting with real applications, dynamics is tied to biology through the *Logistic Map* $L_r(x) = rx(1 - x)$, a cartoon model of population growth with competition for limited resources. We develop graphical methods of detecting periodic points and of determining their stability by the slope of the tangent line. We introduce the technical notion of chaos, characterized by *sensitive dependence on initial conditions,* the presence of unstable order, and thorough mixing. The intermingling of chaos and order appears in the bifurcation diagram of the logistic map, and we explore its fractal nature, manifested through the presence of small copies of the bifurcation diagram within the periodic windows of the main diagram. (See figure 3.4.) To show we have not spent this time studying a single function, we end with a discussion of the universality of quadratic maps. The lesson of this section is the surprisingly intricate relation of chaos and order, the distinction between deterministic and predictable.

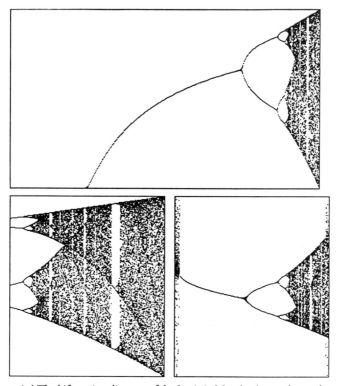

Figure 3.4 The bifurcation diagram of the Logistic Map (top), together with a pair of magnifications: $3.5 \leq r \leq 4$, $0 \leq x \leq 1$ on the left; $3.8275 \leq r \leq 3.8575$, $0.4 \leq x \leq 0.6$ on the right. The right picture magnifies the middle of the conspicuous "window" on the left, revealing a shape similar to the entire diagram (after a reflection).

Seeing Chaos

Suppose we have a sequence of values $\{z_1, z_2, z_3, \ldots z_N\}$ generated from some dynamical process such as the Logistic Map or by measurements made (at equal time intervals) from the real world. Plotting the points $(1, z_1)$, $(2, z_2)$, $\ldots$, (N, z_N) gives the *time series graph* of the sequence. This is a good way to detect stable periodic behavior—the graph eventually cycles between points lying on horizontal lines representing the periodic values. In many cases, however, the time series graph takes on a jagged form without apparent order. How can we tell if this is random or chaotic?

One visual method of distinguishing chaos from randomness is through *close-return plot* that detects the unstable order of chaos.[2] We select some tolerance, ε, and say two values z_i and z_j of the sequence are "close" if $|z_i - z_j| < \varepsilon$. The close-return plot is constructed on a grid $\{(i, k): 1 \leq i \leq N, 1 \leq k \leq N - i\}$ by the

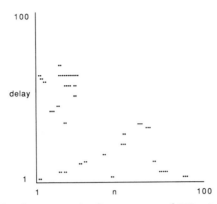

Figure 3.5 The close-return plot for a sequence of 100 points generated by the Logistic Map $x_{n+1} = 3.95x_n(1-x_n)$. The tolerance is $\varepsilon = 0.01$.

following rule: Plot a dot at (i, k) if $|z_i - z_{i+k}| < \varepsilon$. The number k is called the delay. If some value z_n falls close to an unstable k-cycle, then z_{n+k} will be close to z_n, z_{n+k+1} will be close to z_{n+1}, and so on for a few more points. Because the k-cycle is unstable, the sequence eventually will wander away from the cycle, but as a footprint it will leave a horizontal band of dots (n, k), $(n+1, k)$, . . . in the close-return plot. Figure 3.5 is the close-return plot for a sequence of 100 values generated by the Logistic Map for $r = 3.95$ and $\varepsilon = 0.01$. (To make the picture cleaner, isolated dots have been removed. That is, only those with horizontal neighbors have been plotted.) The horizontal bands suggest underlying unstable order—for example, the lower right band of dots are (88, 3), (89, 3), and (90, 3), indicating that z_{88} has gotten close to an unstable three-cycle. Such patterns are very uncommon in random sequences.

Another visual method is *driven IFS*, introduced in our study of IFS as a way to analyze patterns in DNA sequences. First, we find the minimum and maximum values, z_* and z^*, of the sequence, so $d = z^* - z_*$ is the range of values of the sequence. Starting from $(x_0, y_0) = (0, 0)$, a sequence of points (x_1, y_1), (x_2, y_2), . . . is generated by

$$(x_i, y_i) = D_k(x_{i-1}, y_{i-1}),$$

where

$$k = 1 \text{ if } z_* \leq z_i < z_* + d/4,$$
$$k = 2 \text{ if } z_* + d/4 \leq z_i < z_* + d/2,$$
$$k = 3 \text{ if } z_* + d/2 \leq z_i < z_* + 3d/4,$$
$$k = 4 \text{ if } z_* + 3d/4 \leq z_i < z^*,$$

and where

$$D_1(x, y) = (x/2 + 1/2, y/2+ 1/2),$$
$$D_2(x, y) = (x/2 -1/2, y/2 + 1/2),$$
$$D_3(x, y) = (x/2 -1/2, y/2 -1/2),$$
$$D_4(x, y) = (x/2 + 1/2, y/2 -1/2).$$

If the sequence of z_i is uniformly randomly distributed (so each of the functions $D_1(x, y)$, $D_2(x, y)$, $D_3(x, y)$, and $D_4(x, y)$ has the same probability of being applied), then the points (x_i, y_i) uniformly fill up the inside of the square with vertices $(1, 1)$, $(-1, 1)$, $(-1, -1)$, and $(-1, 1)$. On the other hand, figure 3.6 shows an IFS picture driven by the Logistic Map with $r = 3.95$. This is a chaotic sequence and the picture has structure, clearly distinct from filling in the square uniformly. Though not as striking as figure 3.6, figure 3.7 also appears different from filling in the square uniformly. In fact, figure 3.7 was made by driving the IFS with a random sequence, but in this case $D_1(x, y)$ and $D_4(x, y)$ are applied with probability 0.1 each, while $D_2(x, y)$ and $D_3(x, y)$ are applied with probability 0.4 each. (Do you see how these unequal probabilities give rise to the vertical banding of the picture?) How can we distinguish between chaotic and nonuniform random driving?

An answer lies in recognizing how structures arise in these two pictures. In figure 3.6, successive values of the driving sequence are correlated by the Logistic Map—that is, $z_{i+1} = L_r(z_i)$. For example, since the graph of the Logistic Map is a parabola opening downward, graphical iteration shows us that z_i near the maximum are taken to z_{i+1} near the minimum. On the other hand, since the driving sequence of figure 3.7 is random, there is no correlation between z_i and z_{i+1}. Any structure must come from the relative frequency with which the $D_j(x, y)$ are applied. This suggests a test for distinguishing structures arising from deterministic correlations and those arising from random selection of functions with different relative frequencies. Take the original sequence $\{z_1, z_2, z_3, \ldots z_N\}$, scramble the order of its terms,

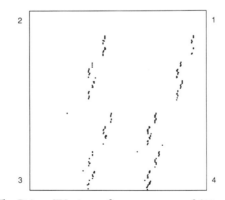

Figure 3.6 The Driven IFS picture from a sequence of 400 points generated by the Logistic Map with $r = 3.95$.

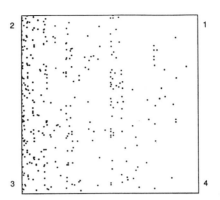

Figure 3.7 The Driven IFS picture from a random sequence of 400 points, but here D_1 and D_4 are applied with probability 0.1, and D_2 and D_3 with probability 0.4.

and drive the IFS with this scrambled sequence. Figure 3.8 shows the IFS driven by the scrambled Logistic Map sequence producing figure 3.6, and figure 3.9 shows the IFS driven by the scrambled nonuniform random sequence producing figure 3.7. Comparing figures 3.6 and 3.8, we see scrambling destroys much of the order, whereas figures 3.7 and 3.9 are quite similar. From this example, we deduce that if scrambling destroys much of the structure of a driven IFS picture, then the original sequence is produced at least in part by chaotic, not random, behavior.

As a real-world example, figure 3.10 shows the IFS picture driven by 20 years of DEC stock return data (after being detrended to remove the effects of long-term market growth). The scrambled data produce figure 3.11. Comparing these suggests DEC stock has little chaotic component but rather seems to reflect a nonuniform random process. In addition to financial data, students have used this method to seek out chaos in daily high-temperature data for New York City (with the seasonal variations removed) and daily milk production of a dairy farm, for

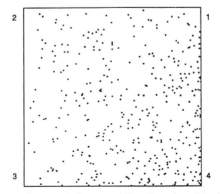

Figure 3.8 Replotting figure 3.6 after scrambling the order of the driving sequence.

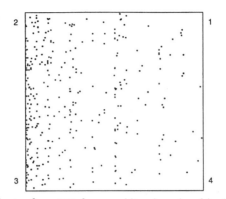

Figure 3.9 Replotting figure 3.7 after scrambling the order of the driving sequence.

example. Part of the appeal of these methods is their use of the eye, which is too often marginalized in this age of digital data, as a tool for analysis.

Having learned to recognize chaos, what do we do with this knowledge? We approach this question by describing recent work on controlling chaos, exploring a simple example of the method by controlling the Tent Map (in the chaotic range) to an unstable fixed point and to an unstable two-cycle.[3] After this, we study some of the physical experiments in controlling chaos: magnetic ribbons, cardiac chaos, and oscillating chemical reactions.

The Mandelbrot Set and Julia Sets

After introducing the arithmetic and geometry of complex numbers, we present the Mandelbrot set and its simple iteration scheme, $z_{n+1} = z_n 2 + c$, where the z_n and c are complex numbers. (Recall that c belongs to the Mandelbrot set if the

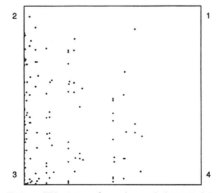

Figure 3.10 The Driven IFS picture from detrended DEC data.

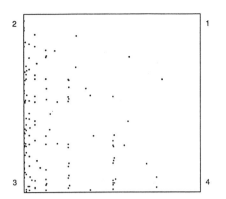

Figure 3.11 Replotting figure 3.10 after scrambling the order of the driving sequence.

sequence of z_i , starting from $z_0 = 0$, does not diverge to infinity as n becomes large.) The amazing visual complexity of the Mandelbrot set is illustrated with several animated video zooms into the Mandelbrot set.[4] Students see its self-similarity revealed through the halo of small Mandelbrot sets surrounding the boundary of the original set. In addition, these zooms exhibit patterns of features that we catalog through some of the combinatorics of the decorations around the Mandelbrot set. For example, every amateur Mandelbrot set explorer knows taking c from the main heart-shaped body of the set gives a sequence of z_i converging to a single point, taking c from the large disc attached on the left of the heart gives z_i converging to alternate between two points (a two-cycle), taking c from the prominent disc attached at the top of the heart gives a three-cycle, taking c from the next largest disc to the right gives a four-cycle, and so on. Moreover, between the two-cycle and the three-cycle discs, the largest disc gives a five-cycle, between the three- and the four-, the largest disc is a seven-, and so on. Understanding these patterns helps the students to navigate through the Mandelbrot set.

Next, Julia sets are introduced and their shapes are related to the local dynamics of the orbit of 0 under the Mandelbrot set iteration. This section's underlying lesson is that we must rethink our ideas about simplicity and complexity: The Mandelbrot set appears amazingly complicated, filled with baroque filigrees and spirals, revealed in ever-changing detail as we magnify the boundary ever more closely, and yet the entire set is generated by iterating a simple quadratic formula. How can such a simple process generate such a complicated picture?

The Fragility of Order

We begin this section with a geometric formulation of *Newton's method* for finding roots of real equations. Even in this simple case, we see that if an equation has multiple

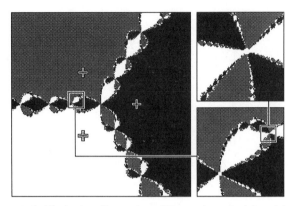

Figure 3.12 Left: The basins of attraction of Newton's method for solving $z^3 = 1$. The three solutions are indicated by crosses, and the basin of attraction of each solution consists of all points colored the same as the immediate neighborhood of the solution. Right: two successive magnifications.

roots, a small change in the initial guess can cause Newton's method to converge to a different root. This observation leads to the problem of determining the *basins of attraction* of the different roots and to the question of the nature of the basin boundaries. We show how Cayley extended Newton's method to the complex case, present a pictorial version of his solution for the basins of the roots of $z^2 = 1$ (Newton's method converges to 1 or -1 accordingly as the initial point z_0 has a positive or negative real part), and ask his question about the basins for the roots of $z^3 = 1$. Computer-generated pictures of these basins explain Cayley's inability to solve the problem by hand, and give a good picture of fractal basin boundaries.[5] (See figure 3.12.)

We contrast the lack of predictability in this case, where the final outcome is usually a stable fixed point, with the lack of predictability of chaos. Here the fractal nature of the boundary between the basins of attraction makes small uncertainties in the knowledge of the initial point more likely to affect in which basin the point lies. In a sense, the loss of predictability associated with fractal basin boundaries is more pernicious than that associated with the sensitivity to initial conditions of chaos—we don't expect to be able to make predictions for chaotic systems but perhaps do expect this for systems having stable outcomes.

We end this section with experiments on a parameterized family of cubic polynomials $f_c(z) = z^3 + (c-1)z - c$, z and c complex. For any c, observe that $z = 1$ is a root of $f_c(z)$. The original experiments began by testing for which c Newton's method for $f_c(z)$, starting at $z = 0$, converges to $z = 1$.[6] The top of figure 3.12 paints these c values black. In refining the experiment, another color was added: Paint a point c gray if Newton's method starting at $z = 0$ does not converge to any fixed point but rather to a stable (nonfixed) cycle. This is illustrated in the bottom of figure 3.13. Surprisingly, the Mandelbrot set reappears. A benefit of this analysis

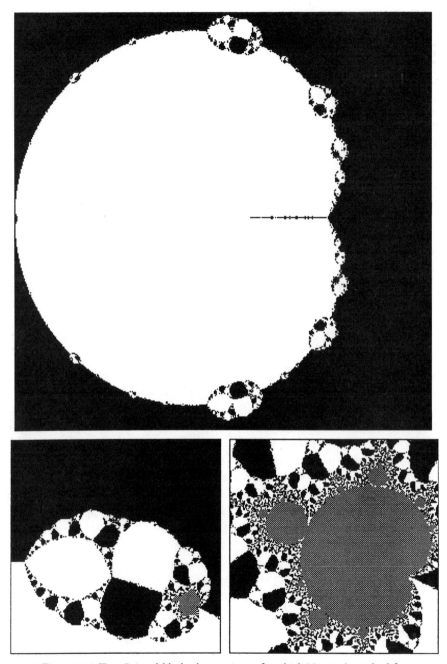

Figure 3.13 Top: Painted black, those points c for which Newton's method for $f_c(z) = z^3 + (c-1)z - c$, starting at $z = 0$, converges to $z = 1$. Bottom left: A magnification of the "top scarab" of the top figure, painting gray those c for which Newton's method starting from $z = 0$ converges to a stable cycle other than a fixed point. Bottom right: A magnification of the left, showing more clearly the familiar shape of the gray region.

is the realization that the time spent in the last section is relevant to a much wider class of equations than had originally been thought, that the Mandelbrot set is a sort of universal object, appearing in many different settings.

The topics of this section can be taken as a metaphor for the observation that even in the absence of chaos, small changes can have large effects. Students are encouraged to speculate on this phenomenon in their own life histories.

Cellular Automata, Neural Nets, and Artificial Life

Starting with von Neumann's solution to the problem of whether a machine can make a copy of itself, we abstract to cellular automata. Through computer experiments, we observe different types of behavior arising in both one- and two-dimensional automata, in particular noting that one-dimensional automata often give patterns similar to the Sierpinski gasket (described in another context in the next section). Conway's *LIFE* game is one of our main examples of a two-dimensional automaton, and we spend some time studying some specific *LIFE* configurations; for example, gliders, glider guns, and the r-pentomino. We observe various kinds of self-organizing behavior and simulate a biological model of how leopards get their spots.[7] The appearance of self-organization leads into a survey of recent work in artificial life, including Langton's parameterization of different classes of automaton behavior, Kauffman's self-catalysis of the origin of life, Holland's genetic algorithms, Ray's digital ecosystem, and Arthur's digital stock market.[8] In this section we focus on the important issue of levels in a model, that some large-scale features, perhaps including the emergence of order, result from "formal" properties of the system and are independent of detailed mechanics. This leads to speculation, sometimes passionately argued by students, about the origin of consciousness and the identity of individuals.

The course concludes with speculations of future developments and with critical analyses of the validity of some models.

SOME SOFTWARE

We developed Macintosh software for the course that enhanced class demonstrations and student exploration. We used five principal packages:

- TreenessEmerging: Provides the Chaos game, deterministic IFS, and random IFS.
- IterateItAgainSam: Provides graphical iteration and histograms, time series, and bifurcation diagrams for the Tent and Logistic maps.

- DesparatelySeekingChaos: Provides driven IFS and close returns for time series generated by the Tent and Logistic maps and by imported data files.
- WaitingForMandelbro: Magnifies selected regions of the Mandelbrot set, plots the Julia set or the orbit of 0 for a selected point in or near the Mandelbrot set.
- Cellebration: Plots successive generations of one- or two-dimensional cellular automata.

Program code 3.1 presents the program code for these packages.

A SKETCH OF THE FIRST PART OF THE COURSE— FRACTAL CONSTRUCTIONS AND ITERATIVE GEOMETRY

In this section we present a fairly complete outline of the topics covered in the first part of the course. Class presentation is supplemented by many examples, in every case beginning with the particular and only later abstracting to the general. We structured the course this way in part because liberal arts students often have more success with new concepts introduced by numerical examples, and also because in some instances students can discover versions of the general principles from the particular examples.

What Are Fractals?

The shapes of Euclidean geometry—for example, lines, circles, cones, and paraboloids—are described by equations. Euclidean geometry does not provide simple descriptions of many shapes from nature because natural objects are irregular over a range of scales, and so the corresponding equations should have the same form over this range of scales. This is not the case for most Euclidean shapes—for example, on smaller and smaller scales a circle looks more like a straight line (the tangent line at the point about which it is being magnified), and yet a circle and a straight line have different equations. Benoit Mandelbrot approached this problem from an entirely different perspective by denying the centrality of equations for describing nature. Rather, he looked more closely at how natural shapes arise and asserted the primacy of process over form. His fractal geometry is successful at describing nature because it emphasizes the dynamic rather than the static.[9] Before considering fractals mimicking nature, we use simpler examples to illustrate the methods.

Program code 3.1 Chaos & Fractal Courseware

TreenessEmerging
  Chaos Game
      select the corner points (a_1, b_1), ..., (a_N, b_N) with the mouse
      select the starting point (x_0, y_0) with the mouse
      repeat
            select i from $\{1, ..., N\}$ uniformly randomly
            $(x_{next}, y_{next}) = ((x_0 + a_i)/2, (y_0 + b_i)/2)$
            Plot(x_{next}, y_{next})
            $(x_0, y_0) = (x_{next}, y_{next})$
      until the quit command is entered
  Deterministic IFS
      enter parameters for the affine contractions T_1, ..., T_N
      with the mouse, enter the set $P = \{(x_1, y_1), ..., (x_M, y_M)\}$ of
      pixels of the starting picture
      repeat
            apply each of T_1, ..., T_N to each pixel of P, producing a
            new set S of pixels
            plot each pixel of S
            set P = S
      until the quit command is entered
  Random IFS
      enter parameters for the affine contractions T_1, ..., T_N
      set $(x_0, y_0) = (0, 0)$
      repeat
            select i from $\{1, ..., N\}$ randomly by the assigned
                  probabilities
            $(x_{next}, y_{next}) = T_i(x_0, y_0)$
            Plot(x_{next}, y_{next})
            $(x_0, y_0) = (x_{next}, y_{next})$
      until the quit command is entered

IterateAgainSam
  select the function $f(x)$ = Tent Map or Logistic Map
  Time series
      select the parameter, r, for the function
      select the starting value of x_0
      select N, the number of points to be generated
      i = 0
      repeat
            $x_{i+1} = f(x_i)$
            plot(i, x_{i+1})
            i = i+1
      until i=N
  Graphical iteration
      initialize the histogram counters: $h_1 = 0$, ..., $h_M = 0$
      plot the graph of the function, $(x, f(x))$ for $0 \leq x \leq 1$
      plot the diagonal line (x, x) for $0 \leq x \leq 1$
      select the parameter, r, for the function
      select the starting value of x_0
      repeat

```
            x_next = f(x_0)
            draw the line from (x_0, x_0) to (x_0, x_new)
            draw the line from (x_0, x_new) to (x_new, x_new)
            k = Trunc(M*x_new)
            increment the histogram counter: h_k = h_k + 1
            plot the histogram point (k, h_k)
            x_0 = x_new
        until the quit command is entered
    Bifurcation diagram
        enter window bounds (r_low, x_low) and (r_high, x_high)
        rRange = r_high - r_low
        xRange = x_high - x_low
        enter the number N_d of iterations dropped
        enter the number N_p of iterations plotted
        L = drawing window length (in pixels)
        H = drawing window height (in pixels)
        i = 0
        repeat
            r = r_low = i*rRange/L
            x_0 = 0.5
            set count = 0
            repeat
                    x_new = f(x_0)
                    x_0 = x_new
            until count = N_d
            set count = 0
            repeat
                    x_new = f(x_0)
                    x_0 = x_new
                    if ((x_low ≤ x_0) and (x_0 ≤ x_high)
                            then plot(i,Trunc(H*(x_0 - x_low)/xRange))
            until count = N_p
        until i = L

DesparatelySeekingChaos
    enter the data set {x_1, ..., x_N}
    Driven IFS
```
$$D_1(x, y) = (\tfrac{1}{2}x + \tfrac{1}{2}, \tfrac{1}{2}y + \tfrac{1}{2})$$
$$D_2(x, y) = (\tfrac{1}{2}x - \tfrac{1}{2}, \tfrac{1}{2}y + \tfrac{1}{2})$$
$$D_3(x, y) = (\tfrac{1}{2}x - \tfrac{1}{2}, \tfrac{1}{2}y - \tfrac{1}{2})$$
$$D_4(x, y) = (\tfrac{1}{2}x + \tfrac{1}{2}, \tfrac{1}{2}y - \tfrac{1}{2})$$
```
    min = minimum{x_1, ..., x_N}
    max = maximum{x_1, ..., x_N}
    range = max - min
    (x_0, y_0) = (0, 0)
```

```
        i = 0
        repeat
            i = i+1
            if min ≤ x_i < min + (range/4)
                    then (x_new, y_new) = D_1(x_0, y_0)
            if min + (range/4) ≤ x_i < min + (range/2)
                    then (x_new, y_new) = D_2(x_0, y_0)
            if min + (range/2) ≤ x_i < min + (3*range/4)
                    then (x_new, y_new) = D_3(x_0, y_0)
            if min + (3*range/4) ≤ x_i < max
                    then (x_new, y_new) = D_4(x_0, y_0)
            plot(x_new, y_new)
            (x_0, y_0) = (x_new, y_new)
        until i = N
    Close returns
        ·enter the tolerance e
        i = 1
        repeat
            j = 1
            repeat
                    if |x_i - x_{i+j}| < e then plot(i, j)
                    j = j + 1
            until j = N - i
            i = i + 1
        until i = N - 1

    WaitingForMandelbro
        plot a stored image of the Mandelbrot set
    Magnify
        enter the maximum dwell, MD
        enter window bounds (a_low, b_low) and (a_high, b_high)
        aRange = a_high - a_low
        bRange = b_high - b_low
        L = drawing window length (in pixels)
        H = drawing window height (in pixels)
        i = 1, j = 1
        repeat
            b = b_low + (j/bRange)
            repeat
                a = a_low + (i/aRange)
                k = 0
                x_0 = 0, y_0 = 0
                repeat
                    k = k + 1
                    x_new = x_0^2 - y_0^2 + a
                    y_0 = 2*x_0*y_0 + b
                    x_0 = x_new
                    dist = x_0^2 + y_0^2
                until (k = MD) or (dist > 4)
                if not(dist > 4) then plot(i, j)
```

```
                    i = i+1
                    until i = L
            j=j+1
     until j = H
 Julia
     with the mouse, select a point (a, b) on the plotted stored image
     of the Mandelbrot set
     enter the maximum dwell, MD
     L = drawing window length (in pixels)
     H = drawing window height (in pixels)
     i = 1,  j = 1
     repeat
            y = -2 + (4*j/H)
            repeat
                   x = -2 + (4*i/L)
                   k = 0
                   x_0 = x,  y_0 = y
                   repeat
                          k = k + 1
                          x_new = x_0^2 - y_0^2 + a
                          y_0 = 2*x_0*y_0 + b
                          x_0 = x_new
                          dist = x_0^2 + y_0^2
                   until (k = MD) or (dist > 4)
                   if not(dist > 4) then plot(i, j)
                   i = i + 1
            until i = L
            j = j+1
     until j = H
  orbit
     with the mouse, select a point (a, b) on the plotted stored image
     of the Mandelbrot set
     L = drawing window length (in pixels)
     H = drawing window height (in pixels)
     x_0 = 0,  y_0 = 0
     repeat
            x_new = x_0^2 - y_0^2 + a
            y_0 = 2*x_0*y_0 + b
            x_0 = x_new
            dist = x_0^2 + y_0^2
            plot(-2 + (4*x_0/L), -2 + (4*y_0/H))
     until (the quit command is entered) or (dist > 4)

Cellebration
  1-dim  automaton
     H = drawing window height (in pixels)
     initialize the cells: c_1 = dead, ..., c_N = dead
     enter the automaton rule (the collection of neighborhood
         configurations  giving rise to a live cell)
     enter the initial configuration of live cells
     j = 1
     repeat
            j = j + 1
```

```
            i = 1
            repeat
                    if cell_i = live then plot(i,1)
            until i = N
            i = 1
            repeat
                    if the neighborhood configuration of cell_i appears in
                            the automaton rule, then set newcell_i = live
                    else set newcell_i = dead
                    i = i + 1
            until i = N
            i = 1
            repeat
                    cell_i = newcell_i
                    if cell_i = live then plot(i,j)
                    i = i + 1
            until i = N
        until j = H
2-dim automaton
    initialize the cells:  c_ij = dead for 1 ≤ i ≤ N and 1 ≤ j ≤ N
    enter the automaton rule (the collection of neighborhood
        configurations  giving rise to a live cell)
    enter the initial configuration of live cells
    repeat
            i = 1, j = 1
            repeat
                    repeat
                            if cell_ij = live then plot(i,j)
                            i = i + 1
                    until i = N
                    j = j + 1
            until j = N
            i = 1, j = 1
            repeat
                    repeat
                            if the neighborhood configuration of cell_ij
                                appears in the automaton rule, then set
                                newcell_ij = live
                            else set newcell_ij = dead
                            i = i + 1
                    until i = N
                    j = j + 1
            until j = N
    until the quit command is entered
```

Figure 3.1a (p. 39) is a right isosceles triangle. For definiteness, take the vertices to be (0, 0), (1, 0), and (0, 1). Now define a process for changing the triangle: Find the midpoint of each side, connect these midpoints, and remove the middle triangle formed in this way. Figure 3.1b illustrates this process. Note that three triangles remain, each similar to the original triangle. We can apply the same process to these three, producing nine still smaller triangles, each similar to the original. (See figure 3.1c.) Figure 3.1d shows the limiting shape after this process has been carried out infinitely many times. Call this limiting shape G. Note the part of the shape contained within the vertices $(0, 0)$, $(\frac{1}{2}, 0)$, $(0, \frac{1}{2})$ is just a copy of the entire picture but scaled down by a factor of $\frac{1}{2}$ in the x- and y-directions. Call this smaller copy G_1. Similarly, the part with vertices $(\frac{1}{2}, 0)$, $(1, 0)$, $(\frac{1}{2}, \frac{1}{2})$ is a copy (call it G_2) of the entire picture scaled down by a factor of $\frac{1}{2}$ in the x- and y-directions and translated by $\frac{1}{2}$ in the x-direction; and the part with vertices $(0, \frac{1}{2})$, $(\frac{1}{2}, \frac{1}{2})$, $(0, 1)$ is a copy (call it G_3) of the entire picture scaled down by a factor of $\frac{1}{2}$ in the x- and y-directions and translated by $\frac{1}{2}$ in the y-direction. This shape, called the Sierpinski gasket, is self-similar, that is, the Sierpinski gasket is made up of pieces, each similar to the whole. We shall give the name *fractal* to any shape made up of subsets, each similar in some way to the whole. Writing a description of the gasket in the language of Euclidean geometry is an overwhelming task—infinitely many lines are needed to describe all the triangles in the gasket. We shall see there is a much more efficient way to describe the gasket, a way that emphasizes the recursive process by which we generated the gasket, rather than the static snapshot of the final picture.

The Chaos Game

The Chaos game is simple: Consider four points $P_1 = (0, 0)$, $P_2 = (1, 0)$, $P_3 = (1, 0)$, and $P_4 = (1, 1)$, corners of the unit square S. Select a starting point (x_0, y_0) somewhere in the plane. Now suppose we have a random number generator producing a uniformly distributed sequence of $1s$, $2s$, $3s$, and $4s$. (By "uniformly distributed" we mean approximately the same number of $1s$, $2s$, $3s$, and $4s$ occur in a long sequence.) Suppose the first random number produced is 1. Then we define the point (x_1, y_1) by going halfway between (x_0, y_0) and P_1. Suppose the next random number produced is 3. Then (x_2, y_2) is halfway between (x_1, y_1) and P_3, and so on. In general, after producing points (x_0, y_0), (x_1, y_1), (x_2, y_2), ... (x_n, y_n), the point (x_{n+1}, y_{n+1}) is halfway between (x_n, y_n) and P_k, where k is the $(n+1)$st random number produced. What sort of picture will this make?

One thing is clear: If some (x_n, y_n) falls inside the unit square S, then all the later points (x_{n+1}, y_{n+1}), (x_{n+2}, y_{n+2}), ... lie in the square because the next point is halfway between the current point and one of the corners of the square. Also, though seeing this is more difficult, the randomness of the sequence of numbers guarantees the points eventually will fall inside the square S. Once points are inside the square, because their positions are determined by a uniform random process, we guess they will fill up the square uniformly. Figure 3.14 shows the result of generating several thousand points in this way.

Now suppose instead of playing the Chaos game with the corners of the unit square, we use the corners $P_1 = (0, 0)$, $P_2 = (1, 0)$, and $P_3 = (1, 0)$ of a right isosceles

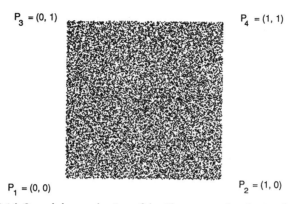

$P_3 = (0, 1)$ $P_4 = (1, 1)$

$P_1 = (0, 0)$ $P_2 = (1, 0)$

Figure 3.14 Several thousand points of the Chaos game played using the corners of the unit square.

triangle. Of course, now the random number generator produces only 1s, 2s, and 3s. As before, once a point (x_n, y_n) falls inside the triangle, all the following points (x_{n+1}, y_{n+1}), (x_{n+2}, y_{n+2}), . . . lie inside the triangle. Encouraged by the correctness of our guess for the square Chaos game, we suspect the randomness of the point-generating process will fill up the inside of the triangle uniformly. Figure 3.15 shows several thousand points generated by this triangular Chaos game. Our guess is wrong. (In playing this trick on my students last year, when the triangular Chaos game image began to appear on the screen, one of them shouted out, "It's the Sierpinski gadget.")

We shall return to why the triangle Chaos game produces the gasket, but for now notice this: In playing this Chaos game, we used only three rules: Move halfway toward (0, 0), move halfway toward (1, 0), and move halfway toward (1, 0). This observation is central to understanding how the dynamical geometry of

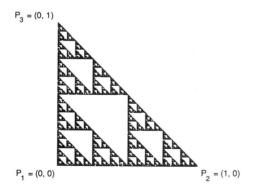

$P_3 = (0, 1)$

$P_1 = (0, 0)$ $P_2 = (1, 0)$

Figure 3.15 Several thousand points of the Chaos game played using the corners of a right isosceles triangle.

fractals gives rise to a simple description of these shapes. To see this, first we consider another version of this game.

Deterministic Iterated Function Systems

First, we reformulate the rules of the triangular Chaos game. Scaling by $\frac{1}{2}$ in the x- and y-directions can be viewed as applying the function $T_1(x, y) = (x/2, y/2)$, scaling by $\frac{1}{2}$ in the x- and y-directions and translating by $\frac{1}{2}$ in the x-direction as applying $T_2(x,y) = (x/2 + \frac{1}{2}, y/2)$, and scaling by $\frac{1}{2}$ in the x- and y-directions and translating by $\frac{1}{2}$ in the y-direction as applying $T_3(x,y) = (x/2, y/2 + \frac{1}{2})$. Recall G denotes the Sierpinski gasket of figure 3.1d, and by its self-similar nature, we divide G into three pieces: G_1, G_2, and G_3. Our first observation is so simple it hardly seems worth noting separately. Nevertheless, to draw a parallel between the development of ideas in Euclidean geometry and fractal geometry, we give three propositions establishing the connection between the shape G and the functions T_1, T_2, and T_3.

For i = 1, 2, and 3, denote by $T_i(G)$ the set obtained by applying T_i to each point of G. Then a simple calculation shows:

Proposition 1: For the gasket G, $G_1 = T_1(G)$, $G_2 = T_2(G)$, and $G_3 = T_3(G)$.

Now define a composite function T on sets in the plane:

$$T(S) = T_1(S) \cup T_2(S) \cup T_3(S).$$

(Here the symbol $\cup$ denotes the union of sets. For example, $T_1(S) \cup T_2(S)$ denotes the collection of all points of $T_1(S)$, together with all points of $T_2(S)$.) Along with the self-similar decomposition of G, Proposition 1 implies a second proposition.

Proposition 2: The gasket G is left invariant by T: T(G) = G.

Proof: From the definition of T, we see $T(G) = T_1(G) \cup T_2(G) \cup T_3(G)$. Applying Proposition 1, $T_1(G) \cup T_2(G) \cup T_3(G) = G_1 \cup G_2 \cup G_3$. Combining these with the self-similar decomposition of the gasket $G_1 \cup G_2 \cup G_3$ = G, we see T(G) = G, completing the proof.

Having noticed this, a natural question is What happens if we apply T to some shape other than the gasket? Figure 3.16 shows the results of applying G to three different starting shapes, the unit square, a flower, and a cat. In the top we see the result of applying T to the square is three smaller squares—S_1, S_2, and S_3—arranged

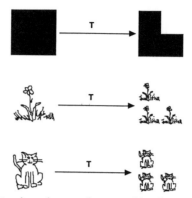

Figure 3.16 Applying the gasket transformation T to the unit square, to a flower, and to a cat.

in an L-shaped pattern, roughly in the locations of G_1, G_2, and G_3. Together, S_1, S_2, and S_3 are not the original square, so right away we expect the gasket is some sort of "special shape" for T. This result is reinforced by the middle and the bottom of figure 3.16, showing the results of applying T to a flower and to a cat.

Suppose we continue this process by applying T to S_1, S_2, and S_3. The third picture of figure 3.17 shows the result: nine still smaller squares. Repeated application of T generates the other pictures of figure 3.17, a sequence of pictures converging to the gasket. (This rough notion of convergence can be made precise using the Hausdorff metric [see p. 114 of Falconer, *Fractal Geometry*[10]], but this rigorous development exceeds the prerequisites for the course. We mention it to let the students know the subject can be pursued at a much more mathematical level.) Figure 3.18 presents evidence that these pictures converge to the gasket, regardless of the starting shape. So at least we have evidence for Proposition 3. The transformations T_1, T_2, and T_3 characterize the gasket G in this sense:

$$T(G) = G, \text{ and}$$

$$T^{on}(S) \text{ converges to G for any shape S.}$$

Here T^{on} means apply the composite function T n times in succession, and the implication is to observe limiting behavior as *n* becomes large.

A collection of functions such as $T_1 \cup T_2 \cup T_3$ generating a picture by repeated application is called an *Iterated function system (IFS)*. This process of applying all the transformations at each step is referred to as a *Deterministic IFS*, in contrast to the *random IFS*, applying the transformations one at a time in random order, reminiscent of the Chaos game. Early work along these lines was done by Dekking, and the proof of convergence of this method—for functions T_i

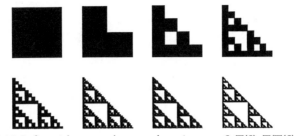

Figure 3.17 Left to right, top to bottom: the unit square S, T(S), T(T(S)), ...

called *affine contractions* (defined in a later section)—was established by Hutchinson.[11] Popularization and application to data compression and communications are due mostly to Barnsley.[12] Both chapter 9 of Falconer's book and a recent *Scientific American* article contain other good expositions of IFS.[13] In essence, the IFS method is an efficient way to encode images of many natural objects because, as Mandelbrot recognized, natural objects often exhibit the same structures on many scales of length. IFS is a functional way of unpacking this scale independence, a way of recursively finding smaller pictures within larger and hence of elucidating the fractal nature of objects.

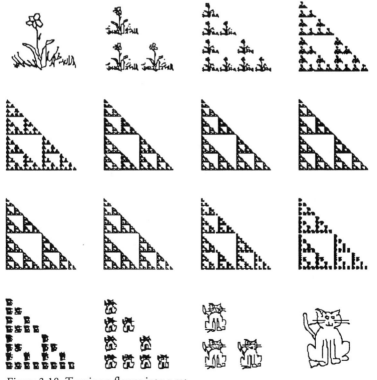

Figure 3.18 Turning a flower into a cat.

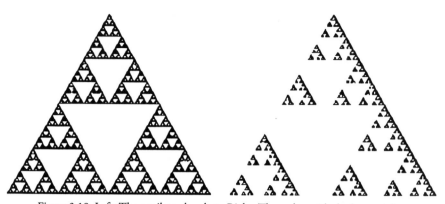

Figure 3.19 Left: The equilateral gasket. Right: The gasket with the lower right shrinking factor ⅓ instead of ½.

Magnified sufficiently, the last picture of figure 3.17 really is 3^7 tiny copies of the square, but the limited resolution of the computer screen renders it indistinguishable from the gasket. This "loss of information" is not surprising, since any starting picture produces a sequence the limit of which is the gasket. Together with a touch of whimsy, this convergence can be used to produce some interesting fake transformations of one picture into another. Figure 3.18 gives an example.

The main idea so far is that at least for some shapes, S, there are sets of transformations that, when applied together, leave S invariant and when repeatedly applied together to any other shape yield a sequence of shapes converging to S. Questions that arise naturally now are: If we change the transformations a small amount, will the limiting picture change by a small amount? and Given a shape S, how can we find the transformations which have S as a limiting picture? (This second question is called the "inverse problem.") A "yes" answer to the first gives some reason to be hopeful of finding an answer to the second: If small changes in the functions produce small changes in the picture, then perhaps we need not find the right functions exactly, but only get close to them. So we are lucky that the answer to the first question is yes, for in a sense that can be made precise, the limiting process depends continuously on the transformations. Consequently, a small change in the transformations gives rise to a small change in the limiting shape. This should not suggest that small changes are uninteresting, as we shall see in the next section.

Changing the Rules

It is not difficult to see that an equilateral gasket (see the left side of figure 3.19) can be produced, by making slight changes to the original transformations:

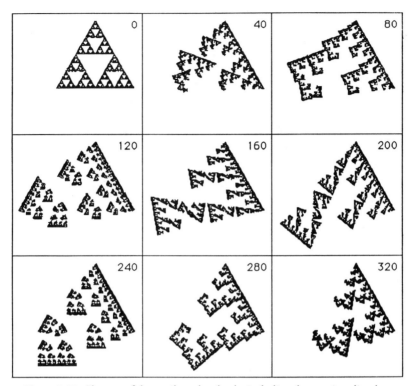

Figure 3.20 Changes of the equilateral gasket by including the rotations listed to the transformation W_1. Note that the lower right triangle of the lower right triangle rotates twice in the 360°.

$W_1(x, y) = (x/2, y/2),$
$W_2(x, y) = (x/2 + 1/2, y/2),$ and
$W_2(x, y) = (x/2 + 1/4, y/2 + \sqrt{3}/4).$

The right side of figure 3.19 shows the results of changing the shrinking factors of W_1 from 1/2 to 1/3. We might expect that the lower left part of the equilateral gasket is a bit smaller than in the original equilateral gasket, but we see there is a complication. Specifically, the lower left part of every piece of the gasket is shrunk relative to the part in which it lies. This points out an interesting consequence of the self-similar nature of fractals: that changing one part of the shape changes the corresponding parts of all subshapes.

Another example is shown in figure 3.20: a sequence of nine modifications of the equilateral gasket, each adding a 40 degree rotation to W_1 of the previous picture. Of course, many other modifications are possible, and exploration usually produces surprises.

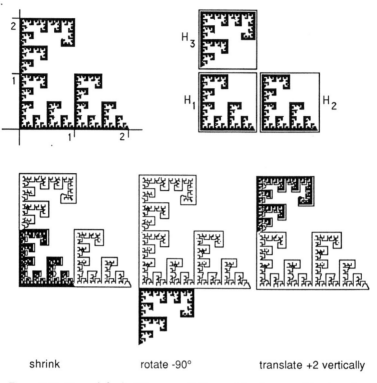

shrink   rotate -90°   translate +2 vertically

Figure 3.21 Upper left, the "Hangman." Upper right, decomposition into three pieces. Bottom, shrink, rotate -90°, translate +2 vertically: the transformation giving H_3.

Finding the Rules to Fit the Picture

As for the Inverse Problem of finding the transformations to produce a given picture, consider for a moment the reason behind Proposition 1. We recognized G could be decomposed into three pieces—G_1, G_2, and G_3—and for each of these it was a simple matter to find the transformations T_1, T_2, and T_3 taking G to each of G_1, G_2, and G_3. So a rough guideline for finding the transformations to encode a shape, S, is to split S into pieces "looking like" S in some sense and then find the transformations that realize this similarity. This method of finding decompositions is called the *Collage Theorem,* by Barnsley. Figure 3.21 gives an example of this process for another fractal, H, the "Hangman." Transformations R_1 and R_2, producing H_1 and H_2, are clear: $R_1(x, y) = (x/2, y/2)$ and $R_2(x, y) = (x/2 + 1, y/2)$. Figure 3.21 illustrates how R_3 is found.

Finding the transformations for more complicated shapes is made easier by giving a geometric interpretation to the components of the transformations. First, we shall consider only transformations of the form

Figure 3.22 A fractal tree.

$$w(x,y) = (a \cdot x + b \cdot y + e, \ c \cdot x + d \cdot y + f)$$

where a, b, c, d, e, and f are constants. Such a function w is called an *affine transformation*. (To guarantee a unique picture is associated with an IFS, we must further restrict our attention to *affine contractions*—that is, for any pair of points (x_1, y_1) and (x_2, y_2), the distance between (x_1, y_1) and (x_2, y_2) must be more than the distance between $w(x_1, y_1)$ and $w(x_2, y_2)$. To use these affine transformations, we need to understand the geometric effects of the coefficients a, b, c, d, e, and f. Observe that e and f are simply translations in the x- and y-directions. For a, b, c, and d, some more effort is required. Write $a = r \cdot \cos(\theta)$, $b = -s \cdot \sin(\varphi)$, $c = r \cdot \sin(\theta)$, and $d = s \cdot \cos(\varphi)$. Why these are good choices will become clear in a moment. Trigonometry allows us to solve for r, s, θ, and φ if we are given a, b, c, and d, but r, s, θ, and φ have direct geometric meaning and so these are the quantities with which we shall work.

To understand this interpretation, set $e = f = 0$ and apply w to the vector $(1,0)$, obtaining the vector $(r \cdot \cos(\theta), r \cdot \sin(\theta))$. So r is the amount by which the unit vector in the x-direction is shrunk or stretched, and θ is the angle (measured counterclockwise from the x-axis) through which this vector is rotated. Similarly, applying w to the vector $(0,1)$ shows that s is the amount of shrinking or stretching of the unit vector in the y-direction and φ is the angle (measured counterclockwise from the y-axis) through which this vector is rotated. By allowing θ and φ to take on different values and allowing r and s different values, w can produce skewings as well as rotations and changes of scale. Taking r negative reflects the image across the y-axis, while taking s negative reflects across the x-axis. We shall represent transformations by a 6-tuple $(r, s, \theta, \varphi, e, f)$. (See table 3.1.)

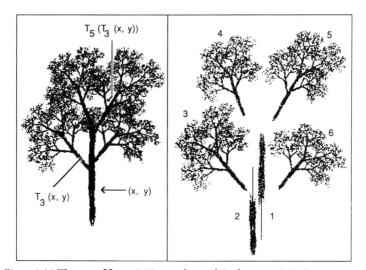

Figure 3.23 The tree of figure 3.22, together with its decomposition into six pieces. The left panel exhibits a point (x, y) of the trunk, its image $T_3(x, y)$ in the lower left branch, and the image $T_5(T_3(x, y))$ in the upper right branch.

Of course, the Sierpinski gasket and the Hangman only look complicated for a moment, until we notice the exact self-similarity, the precise replication of the shape at all magnifications. Viewed in this fashion, it is not such a surprise that these shapes can be described using only a few functions. Can we do a similar encoding for a more complicated natural shape, a shape like the tree in figure 3.22?

A Natural Fractal

The answer, of course, is yes. Figure 3.23 illustrates the decomposition of this shape into the relevant pieces. A moment's study of the tree reveals most of the decomposition: the four main branches are shrunken, rotated, and translated copies of the entire tree. The trunk posed more of a challenge: a solution, shown in the right side of figure 3.23, uses two copies of the tree (reduced significantly in the x-direction and by a factor of one-half in the y-direction) to cover the trunk. Since each of these functions is determined by six parameters, and since only six functions are used to generate figure 3.22, we can say the tree picture is described by 36 numbers. Viewed in this way, the information content of figure 3.22 is quite small. Again, there is a rigidity to the picture, each piece exactly reflecting the whole, but now the image is so complicated that one may have some difficulty perceiving this rigidity. Of course, this picture is too regular to fool anyone who has carefully studied trees. For example, a botanist criticized

			Table 3.1			
r	s	θ	φ	e	f	prob
0.050	0.600	0.000	0.000	0.000	0.000	0.100
0.050	-0.500	0.000	0.000	0.000	1.000	0.100
0.600	0.500	40.000	40.000	0.000	0.600	0.200
0.500	0.450	20.000	20.000	0.000	1.100	0.200
0.500	0.550	-30.000	-30.000	0.000	1.000	0.200
0.550	0.400	-40.000	-40.000	0.000	0.700	0.200

the picture for the fact that the branches do not exhibit heliotropy. This could be remedied, but at the expense of including more functions.

Table 3.1 gives the parameters for the transformations to grow the tree of figure 3.22. The additional parameter, "prob," represents the relative frequency with which each transformation is applied. Each row corresponds to a transformation: the first row to the copy labeled 1 on the right side of Figure 3.23, the second to the copy labeled 2, and so on.

By varying the parameters of the tree functions, in principle a sufficiently diligent person could make the pictures for an animation of the tree "swaying in the breeze." Unhappily, the result of these efforts cannot be seen until all the pictures have been made and "pasted" into an animation package. Several years ago a student of mine wanted to make such an animation. He guessed at the appropriate parameter changes, ran the program again and again, pasted each picture into *Mathematica,* and found he had made a movie of . . . a tree doing jumping-jack exercises. So beware—think carefully before you animate.

Some shapes are more naturally amenable to fractal encoding than are others. Familiar botanical examples include ferns, Queen Anne's lace, and some types of leaves. A particularly successful lab exercise involves scanning a leaf and using a Paint program to shrink, rotate, and translate copies to cover the original image. From this, students can try to deduce the IFS parameters and test their deductions with the TreenessEmerging program: If the resulting picture looks like the leaf, the IFS parameters are close to correct. More than listening to hours of lectures, this lab exercise gives students a grasp of the Collage process and of how the parameters are related to the shapes of the small copies. With this practice, students often go exploring on their own, occasionally producing surprising results. One of the most interesting was a nine-function IFS a student found to make a recognizable caricature of his face.

Some Extensions

Having produced a reasonable forgery of a natural structure, two more questions are suggested: Can these methods be extended to produce three-dimensional pictures? and Can we produce color images in a similar fashion?

The answer to the first is straightforward: all the arguments and results for two-dimensional pictures extend easily to three dimensions, though of course more parameters are needed to specify a transformation, and visualizing the three-dimensional images presents its own set of problems. While we have used *Mathematica* to display three-dimensional images of fractals similar to Sierpinski gaskets, the rigors of *Mathematica*'s syntax make it unsuitable for students. Nevertheless, our students have approached the visualization problem by building models from plaster of paris, wood, and Legos.

The second question has (at least) two answers. Chapter 8 of Barnsley's *Fractals Everywhere* describes a method of assigning a measure (probability distribution) to the picture and determining colors by this measure. The mathematical techniques guaranteeing the IFS method will produce a picture can be adapted directly for measures, and the color Inverse Problem can be solved by extending the Collage Theorem to measures. This approach is very versatile, though visualizing colors in terms of measures requires some practice.

Another way to color IFS images is described in Frame and Erdman.[14] To describe this method, we need to understand the notion of "address" of a region of a set generated by IFS. To introduce addresses, we sketch the proof that the Random and Deterministic IFS produce the same picture.

As an illustration, consider the tree again. IFS images can be rendered as we have described, by applying all the functions simultaneously (Deterministic IFS), or by following the path of a single point as the functions are applied one at a time in random order (random IFS). To see why this random IFS produces the same picture as the deterministic IFS, consider figure 3.24. Since the gasket $G = W_1(G) \cup W_2(G) \cup W_3(G)$, we can assign to regions of the gasket *addresses* by which combinations of transformations cover that region. For example, the three regions outlined in the left of figure 3.24 can be given addresses 1, 2, and 3. Now consider the region with address 3. Applying W_1 to this region gives a part of the region with address 1—in fact, just the region with address 13 shown in the right of the figure. Continuing in this way, we see that longer addresses determine ever smaller regions of the gasket. Now pick a starting point in the gasket—the origin, say—and begin applying the transformations in random order, for instance, 12232113132123211112323. . . . This generates a point in region 1, then in region 21, then 221, then 3221, then 23221, ... Since every sequence of length n eventually will appear as the first n symbols of the address string of the point whose orbit we follow, the orbit will visit all regions determined

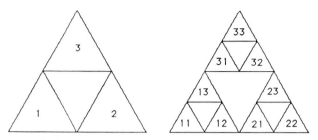

Figure 3.24 The first two steps in assigning addresses to regions of the gasket. The left panel shows the addresses of length 1; the right panel, addresses of length 2.

by addresses of length n. By making n as large as we want, arbitrarily small regions of the gasket will be visited by the orbit. The gasket is the closure of this orbit.

The coloring method of Frame and Erdman is described using random IFS. Denote by $T_1, \ldots, T_6$ the IFS transformations used to generate the tree. Following the coding of table 3.1, let T_1 and T_2 cover the trunk, T_3 the lower left branch, T_4 the upper left branch, T_5 the upper right branch, and T_6 the lower right branch. In the left side of figure 3.23 we have given the label (x, y) to a point of the trunk. We know $T_3(x, y)$ will lie in the lower left branch, but in fact we know more about its position: $T_3(x, y)$ lies in the image in the lower left branch of the trunk. Similarly, $T_5(T_3(x, y))$ lies in the image in the upper right branch. So an elementary way to color the picture is to keep track of the last several transformations applied and paint the point brown if any of these transformations has been T_1 or T_2; otherwise, paint the point green. Frame and Erdman provide an illustration of this coloring method.

As a final extension, consider the question: What happens if the random IFS selects transformations by some nonrandom method? Some results are easy to see. For instance, play the Chaos game using the corners of the unit square, and instead of selecting the corners randomly, suppose we always selected the corner $(1, 1)$. Each point is halfway between the previous point and $(1, 1)$, so the sequence converges rapidly to the single point $(1, 1)$. What would happen if we alternately moved between the corners $(1, 1)$ and $(1, 0)$?

As mentioned previously, an interesting application of this idea has been made by Jeffrey. Recall gene sequences are strings of four characters, C, G, T, and A, encoding instructions for building components of organisms. Jeffrey labeled the corners of the unit square C, G, T, and A and used various gene sequences to select the order of the transformations. Figure 3.25 shows an example of this method, using the approximately 4,000 characters describing the production of the protein amylase. (The sequence begins TGAATTCAAGTTTGGTGCAAAACTTGGCACAGTTATC. . . .) Comparing this with figure 3.14, we see this genetic sequence is certainly not random. One characterization of a random sequence is that it cannot be described more

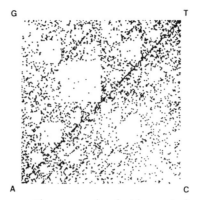

Figure 3.25 The square Chaos game played with genetic data.

succinctly than by listing all its elements. Consequently, departures from randomness imply the sequence contains less than the maximal amount of information for strings of its length. The nonrandomness might represent redundancy for error correction or perhaps the presence of spacers between active parts of the code. This type of assay still is quite young but seems potentially useful.

Natural objects such as trees, ferns, and flowers manifest fractal structures over a range of length scales, and over these scales their formation is governed by a complex nonlinear dynamical process. Large-scale structures in the universe probably are determined by forces just as complex, and so a fractal interpretation of these structures might be appropriate. As the differences between randomness and chaos are better understood, there is more hope of reading dynamical features from fractal appearances. For example, wispy, interwoven ribbons now can be read (sometimes) for the underlying stretching and folding characterizing chaos. By looking for the functional dependencies generating a shape, rather than simply considering the shape as frozen in time, we open up a new way to understand the world. Fractals certainly cannot describe everything, but they provide a new perspective, a more direct reading of form and process. Their visual aspect makes the practice of fractal image encoding accessible to liberal arts students.

BUT DO THEY LEARN ANYTHING?

As with any other course, students take away knowledge in proportion to their level of engagement. Thus we ask two natural questions: What sort of things does a typical student learn, and To what extent is a typical student engaged?

Students in our class gain some mastery of basic mathematics skills including limits, the geometry of affine maps of the plane, logarithms, the derivative as the slope of the tangent line, the computation of derivatives of polynomial functions, complex arithmetic, iteration, and Newton's method. Examples of manual (noncomputer) computations are given in class, while students' mastery is checked by both homework and examinations. In all cases, these topics were introduced in the context of the course material, as tools to solve particular problems in fractals and chaos. Course-specific skills include geometric construction of fractals, IFS (including the inverse problem), dimension computations of mathematical and physical objects, graphical iteration and the interpretation of chaotic dynamics, construction of bifurcation diagrams, generation of Mandelbrot and Julia sets, recognizing final state sensitivity resulting from fractal basin boundaries, and mechanics of cellular automata (including the appearance of self-organization).

Emphasizing the visual aspect of fractals and chaos, in-class demonstrations made considerable use of computer graphics, and the homework included problems requiring the use of these programs. This provided the students with the opportunity to explore actively, not just to listen. Based on students' remarks made throughout the course, the use of these programs was a major factor in their grasping the more intricate aspects of the subject.

In all courses, students have the right to ask about the relevance of the material, but such questions are asked only rarely in traditional mathematics and science classes. Because fractals and chaos have been developed to address real-world situations, we find students are much more likely to ask such questions in this course. The questions are asked not in a confrontational fashion, but seeking elaboration and often, with some excitement, wondering about how these concepts are relevant to their own worlds. Many have commented that the course changed the way they look at nature, since sometimes to the annoyance of their roommates, they see fractals everywhere around them: "frost on our window," "cracks in the sidewalk," "the outline of the Heldeburg mountain range against the sky," and "the vines covering the Arts building." Other common remarks echo the sentiments of these examples: "I am excited about science now," "When I told my mother that math is my favorite class this year, she asked if she had called the right person," and "If high school math had been so interesting, I wouldn't have avoided taking math again until my senior year [of college]."

The frequency with which comments such as these have been repeated over the years gives us some confidence in the success of courses designed along these lines.[15] We believe fractals and chaos, together with interactive computer graphics, are effective in energizing liberal arts students and in convincing them of the relevance of science and mathematics to their lives.

WHERE NOW?

The early 1990s has seen the proliferation of courses developed to introduce fractals and chaos. Reflecting the interests and training of the individual instructors, most of these are directed toward students with some experience in mathematics and science. To be sure, these are valuable contributions. More than many standard courses, these provide a clear picture of science as a process, and they introduce students to important new developments in mathematics and science. Representing natural applications of computer graphics, these are very important additions to standard curricula and deserve to be extended on a wider scale.

The course outlined here is of a different kind. Using the visual appeal of fractals and chaos, together with modeling realistic dynamics, we introduce liberal arts students to a lively science, approachable and relevant to their lives. These students can complete what may be their only exposure to science or mathematics in a more engaged fashion, with more positive impressions of science and mathematics. By expanding the offering of these courses, and ultimately placing them in the high school curriculum, we have an opportunity to increase general scientific literacy. In a global, technological world, such opportunities cannot be ignored.

The increasing availability of powerful graphics systems in the near future will only increase the visual appeal of this material. Here are two instances, though many others are possible. In terms of involving students, one of the most successful aspects of our course is experimenting with finding IFS parameters to produce particular pictures. Equipment limitations restrict these to relatively simple explorations, but even at that, students often invest a lot of time and cleverness in producing "their fractal." How much more work will they do to produce an entire fractal landscape or world, especially if they can explore their world in a virtual reality simulation? The intricacy of such a task makes it appropriate for a collaborative project.

The second instance benefitting from increased computing power is an extension of cellular automata to the digital ecosystems common in Artificial Life. In our course, students often are enthusiastic about exploring the evolution of cellular automata and trying to predict this evolution from the rule structure. By using parallel computers, students will be able to perform reasonably quick experiments in evolution and emergence. They will design their own automaton "animals," watch them compete for resources and avoid predators, and evolve through mutation and crossover. As interesting as is designing their own static world (the earlier instance), watching the evolution of their own dynamical world is sure to be a powerful, almost hypnotic, motivator.

Fractals and chaos have proven to be especially effective introductions to science for liberal arts students. The added dimension of virtual reality simulations and of parallel processing may well make these the most effective, seductive windows that science has ever shown to general audiences. Teaching students skills—how to do things with these new tools—is amazingly more effective than simply lecturing about the underlying ideas. Benoit Mandelbrot has said the combination of fractals, chaos, and easy computer graphics may bring about a real revolution in science education. Let me repeat, in a global, technological world, such opportunities cannot be ignored.

Notes

1. H. Jeffrey, "Chaos Game Visualization of Sequences," *Computers & Graphics* 16 (1992): 25-33.

2. C. Gilmore, "A New Approach to Testing for Chaos, with Applications in Finance and Economics," *International Journal of Bifurcations and Chaos* 3 (1993): 583-587.

3. E. Ott, C. Grebogi, and J. Yorke, "Controlling Chaos," *Physical Review Letters* 64 (1990): 2296-2299, 2837. W. Ditto and L. Pecora, "Mastering Chaos," *Scientific American* 269 (August 1993): 78-84.

4. H. Smith, "Mandelbrot Sets and Julia Sets," *Art Matrix* (1990), PO 880 Ithaca, NY, 14851-0880.

5. H.-O. Peitgen, D. Saupe, and F. von Haeseler, "Cayley's Problem and Julia Sets," *Mathematical Intelligencer* 6 (1984): 11-20.

6. J. Curry, L. Garnett, and D. Sullivan, "On the Iteration of Rational Functions: Computer Experiments with Newton's Method," *Communications in Mathematical Physics* 91 (1983): 267-277.

7. D. Young, "A Local Activator-Inhibitor Model of Vertebrate Skin Patterns," *Mathematical Biosciences* 72 (1984): 51-58.

8. C. Langton, "Studying Artificial Life with Cellular Automata," *Physica* D 10 (1986): 120-149. S. Kauffman, *The Origins of Order: Self-Organization and Selection in Evolution* (Oxford: Oxford University Press, 1993). J. Holland, *Adaptation in Natural and Artificial Systems* (Ann Arbor: University of Michigan Press, 1975). T. Ray, "An Approach to the Synthesis of Life," in *Artificial Life II,* eds. C. Langton, C. Taylor, J. D. Farmer, and S. Rasmussen (Redwood City, CA: Addison-Wesley, 1992). M. Waldrop, *Complexity: The Emerging Science at the Edge of Order and Chaos* (New York: Simon & Schuster, 1992).

9. B. Mandelbrot, *The Fractal Geometry of Nature* (New York: W. H. Freeman, 1983).

10. K. Falconer, *Fractal Geometry: Mathematical Foundations and Applications* (Chichester: John Wiley & Sons, 1990).

11. F. Dekking, "Recurrent Sets," *Advances in Mathematics* 44 (1982): 78-104. J. Hutchinson, "Fractals and Self-Similarity," *Indiana University Mathematics Journal* 30 (1981): 713-747.

12. M. Barnsley, *Fractals Everywhere,* (Boston: Academic Press, 1988). M. Barnsley and L. Hurd, *Fractal Image Compression* (Wellesley, MA: AK Peters, 1993).

13. H. Jurgens, H.-O. Peitgen, and D. Saupe, "The Language of Fractals," *Scientific American* 263 (1990): 60-67.

14. M. Frame and L. Erdman, "Coloring Schemes and the Dynamical Structure of Iterated Function Systems," *Computers in Physics* 4 (1990): 500-505.

15. D. Peak and M. Frame, *Chaos Under Control: The Art and Science of Complexity* (New York: W. H. Freeman, 1994).

II

Fractals in Art

4

The Computer Artist and Art Critic

J. C. Sprott

For decades, the generation of art by machines has fascinated both scientists and artists. In this chapter, I describe how modern computers can generate and evaluate fractal patterns. The patterns are the products of mathematical feedback loops, better known to mathematicians as iterated maps and iterated function systems. The computer solves the equations with random choices of parameters and thereby produces an unlimited variety of patterns. These patterns can be characterized by numerical quantities known as the Lyapunov exponent and the fractal dimension. In studies with human subjects, I have found a correlation between these quantities and the aesthetic quality of the patterns. This suggests that the computer can be taught to generate fractal patterns that appeal to humans. I provide computer code and examples of the patterns produced by this technique. I also discuss the related problem of using a computer to evaluate art produced by humans. The future of these methods holds unlimited promise.

INTRODUCTION

In the last decade or two it has become widely recognized that simple mathematical equations can have solutions of such complexity as to preclude long-term prediction. Such *deterministic chaos* has been observed in many natural processes as well as in mathematical models.[1] The widespread interest in chaos is a recent phenomenon fostered by the proliferation of powerful desktop computers. These computers have become the primary tool for the study of chaos.

One advantage of using personal computers and workstations to study chaotic systems is that you can display the solutions graphically on the computer screen and manipulate them in real time. These solutions are generally fractals, intricate objects that exhibit structures at different size scales.[2] Since chaotic processes, like nature, combine unpredictability with determinism, these fractals often resemble natural forms and have considerable aesthetic appeal. Chaotic equations thus provide a powerful new generator of computer art.[3]

A curious feature of chaotic systems is that a small change in a single parameter of the equations can radically change the appearance of the solution. Thus a single equation or system of equations with several parameters can be used to produce an endless number of patterns, like snowflakes, no two of which are alike. An interesting computational task is to program the computer to produce a succession of such patterns and to select automatically those that are likely to have aesthetic appeal. A human can then intervene at the last stage to evaluate those cases selected by the computer.

An infinite number of monkeys with an infinite number of word processors will eventually replicate all the works of Shakespeare. Perhaps a sufficiently powerful computer could do the same for Picasso. The challenge is to do it quickly and to automate the elimination of those patterns of low quality and appeal. In this chapter I describe a few primitive steps in this direction. I suggest some simple equations suitable for producing a multitude of interesting fractal patterns. I show you how to program your computer to select those that are likely to have aesthetic appeal. Finally, I speculate on what you might do in the future to improve these techniques and to apply them to art produced by humans. The potential of these approaches is limitless. I look forward to hearing from those of you who have used these ideas to produce interesting fractal art.

CHAOTIC ITERATED MAPS

For a system to exhibit chaos, the generating equations must be *nonlinear*. A quadratic equation is perhaps the simplest such example. However, a single equation in a single variable has solutions that lie along segments of a curve and thus tend to be uninteresting. (Recall that the graph of a quadratic equation is a parabola.)

With a pair of equations involving x and y, the solutions are more interesting and are well suited for display on a computer monitor or sheet of paper. The simplest such example is the two-dimensional iterated quadratic map given in its most general form by

$$x_{n+1} = a_1 + a_2 x_n + a_3 x_n^2 + a_4 x_n y_n + a_5 y_n + a_6 y_n^2$$
$$y_{n+1} = a_7 + a_8 x_n + a_9 x_n^2 + a_{10} x_n y_n + a_{11} y_n + a_{12} y_n^2$$

These equations are an example of a mathematical feedback loop. They predict the next ($n+1$) value of x and y in terms of the present (n) value of x and y. The new values are then put back into the equations to produce an endless sequence. The appearance of the solution is determined by the values of the 12 constant coefficients a_1 through a_{12} and the initial values x_0 and y_0 at $n = 0$.

With some initial x_0 and y_0, you can determine successive values of x and y by repeatedly iterating these equations. You can plot the iterates as points on a two-dimensional surface. After a number of iterations, the solution will do one of four things: (a) It will converge to a single *fixed point*; (b) it will take on a succession of values that eventually repeat, producing a *periodic cycle*; (c) it will be unbounded and diverge to infinity; or (d) it will exhibit chaos and gradually fill in some often complicated region of the *xy*-plane.

The visually interesting solutions are the chaotic ones, and they are usually *strange attractors*.[4] They are *attractors* because a range of starting values of x and y, within the *basin of attraction*, yield the same eventual pattern. They are *strange* because the attractor generally has fractal microstructure and a noninteger dimension. They are chaotic because the iterates jump unpredictably from place to place on the attractor and because the sequence of positions depends sensitively on the initial values of x and y. The successive iterates of a chaotic system produce an *orbital fractal*. The first few iterates should be discarded since they almost certainly lie off the attractor. The boundary of the basin of attraction of a strange attractor may itself be a fractal, but as we shall see shortly, much more computation is required to reveal the boundary.

SENSITIVITY TO INITIAL CONDITIONS

To search for chaotic solutions, we need a criterion for detecting chaos. One such criterion is the *sensitive dependence to initial conditions*. (First, I'll give you a theoretical explanation. Later in the chapter I give BASIC code so that you can carry out the necessary computations.) Imagine iterating the two-dimensional quadratic map in the previous section with two initial conditions that differ by a small amount. If successive iterates approach a fixed point or periodic cycle, the difference between the two points, will on average, decrease with each iteration. If the solution is unbounded or chaotic, the difference will increase with each iteration. Unbounded solutions can be eliminated by discarding cases in which x or y grows beyond some large value, such as a million.

The difference between the two solutions initially grows on average at an exponential rate for a chaotic system. The rate of divergence is characterized by the *Lyapunov exponent,*[5] which can be thought of as the power of 2 (or sometimes *e*) by which the separation increases on average for each iteration. Thus, if the separation doubles with each iteration, the Lyapunov exponent is 1 bit per iteration. The Lyapunov exponent is the rate at which information about the initial condition is lost or, equivalently, the rate at which the accuracy of a prediction declines as you project farther into the future.

A two-dimensional map has two Lyapunov exponents, since a cluster of nearby initial points may expand in one direction and contract in another, stretching out like a cigar. The more positive one signifies chaos, and it is the one that dominates after a few iterations.[6] Kaplan and Yorke have conjectured[7] that the fractal dimension F of a strange attractor embedded in a two-dimensional space is related to the two Lyapunov exponents by

$$F = 1 - L_1 / L_2$$

where L_1 is the more positive of the two exponents. The fractal dimension is a measure of the space-filling property of the object. If the two Lyapunov exponents are equal and opposite, there is as much expansion in one direction as there is contraction in the other, and a cluster of initial points scattered throughout the *xy*-plane maintains its two-dimensional shape (thus $F = 2$).

A further difficulty is that the two solutions eventually get far apart, approaching the size of the attractor, and the average rate of separation slows to zero. This problem can be remedied if, after each iteration, one of the points is moved back to its original small separation along the direction of the separation in a process called *renormalization*. The Lyapunov exponent is then determined by the average of the distance the point must be moved for each iteration to maintain a constant separation. If the two cases are separated by a distance d_n, after the *n*th iteration and by d_{n+1} after the next iteration, the largest Lyapunov exponent is determined by averaging the logarithm of the ratio d_{n+1} / d_n over many *(N)* iterations:

$$L = \frac{1}{N} \sum_{n=0}^{N-1} \log_2 (d_{n+1} / d_n)$$

After each iteration, the value of one of the iterates is changed to make $d_{n+1} = d_n$. (See program code 4.1.)

FRACTAL DIMENSION

Calculation of the Lyapunov exponent is only one way to identify chaos. An alternative is to look for solutions with a noninteger dimension. Although such a calculation is relatively time-consuming compared with calculating the Lyapunov exponent, especially if its value is very close to an integer, it is useful because the dimension influences the visual quality of the attractor, offering a measure of how filled-in it is. The fractal dimension is a measure of the strangeness of an attractor, whereas the Lyapunov exponent is a measure of its chaoticity.

It is easier and more efficient to calculate a related quantity called the *correlation dimension,* defined as the logarithmic slope of the *correlation sum* $C(r)$.[8] (See program code 4.1.) This correlation sum is the probability that the separation r_{ij} of two randomly chosen points *(i* and *j)* is less than *r,* where

$$r_{ij} = \sqrt{(x_i - x_j)^2 + (y_i - y_j)^2}$$

The correlation dimension is a lower bound on the fractal dimension, but in practice it closely approximates it. The two are identical if the points are uniformly distributed over the attractor.

You can calculate the correlation dimension in real time as the pattern develops by taking each new point and calculating its separation from one or more randomly chosen previous points.[9] Increment a counter N_1 if the separation is less than r_1 and N_2 if it is less than r_2; you can choose r_1 and r_2 arbitrarily except that they should be much smaller than the attractor, but large enough that N_1 and N_2 are statistically significant and $r_1 < r_2$. The correlation dimension is

$$F = \log(N_2/N_1) \,/\, \log(r_2/r_1)$$

Since F depends on the ratio of two logarithms, you can calculate them in any convenient identical base.

For a given r_2, you can choose r_1 to minimize the statistical uncertainty in F. Unfortunately, this value depends on F and requires that you solve the following transcendental equation:

$$F\log_e(r_2\,/\,r_1) = 2 - 2(r_1\,/\,r_2)^F$$

For the interesting cases of $1 < F < 2$, the optimum r_1 lies in the range $(0.20r_2 < r_1 < 0.45r_2)$, and the uncertainty depends only weakly on r_1. The statistical uncertainty in F can be estimated from

$$\delta F = [1 / N_1 - 1 / N_2]^{\frac{1}{2}} / \log_e(r_2 / r_1)$$

COMPUTER SEARCH PROCEDURE

The procedure for implementing a computer search for strange attractors is straightforward. Choose the coefficients a_1 through a_{12} randomly over some interval, choose initial conditions x_o and y_o, iterate the equations for the map while calculating the Lyapunov exponent and testing for boundedness, and keep only those solutions that are bounded and have a positive Lyapunov exponent.

A computer program that repetitively performs these operations is given in program code 4.1. It is written in BASIC so as to be widely accessible and easily understood. The program should run without modification under Microsoft QBASIC, QuickBASIC, or VisualBASIC for MS-DOS; PowerBASIC, Inc. PowerBASIC, or, with slight modification, under Microsoft VisualBASIC for Windows on an IBM PC or compatible. It assumes VGA (640 by 480 pixel) graphics. If your computer or BASIC compiler does not support this graphics mode, change the SCREEN 12 command to a lower number (*i.e.*, SCREEN 2 for CGA mode). I recommend a compiled BASIC and a computer with a math coprocessor.

Program code 4.1.   BASIC Program to Search for Strange Attractors from
Two-Dimensional Quadratic Maps

```
DEFDBL A-Z                            'Use double precision
DIM a(12)                             'Array of coefficients
RANDOMIZE TIMER                       'Reseed random numbers
SCREEN 12                             'Assume VGA graphics
n% = 0
WHILE INKEY$ = ""                     'Loop until a key is pressed
    IF n% = 0 THEN CALL setparams(x, y)
    CALL advancexy(x, y, n%)          'Advance the solution
    CALL display(x, y, n%)            'Display the results
    CALL testsoln(x, y, n%)           'Test the solution
WEND
END

SUB advancexy (x, y, n%)              'Advance (x, y) at step n%
SHARED a()
xnew = a(1) + x * (a(2) + a(3) * x + a(4) * y) + y * (a(5) + a(6) * y)
y = a(7) + x * (a(8) + a(9) * x + a(10) * y) + y * (a(11) + a(12) * y)
x = xnew
n% = n% + 1
END SUB

SUB display (x, y, n%)                'Display (x, y) at step n%
STATIC xmin, xmax, ymin, ymax, w%, h%
SELECT CASE n%
```

```
   CASE 1                              'Initialize min and max x and y
      xmin = 1000: ymin = xmin: xmax = -xmin: ymax = -ymin
   CASE 2 TO 99                        'Skip these
   CASE 100 TO 999                     'Update min and max x and y
      IF x < xmin THEN xmin = x
      IF x > xmax THEN xmax = x
      IF y < ymin THEN ymin = y
      IF y > ymax THEN ymax = y
   CASE 1000                           'Clear the screen and rescale
      CLS
      IF CSNG(xmax) = CSNG(xmin) THEN xmax = xmin + 1
      IF CSNG(ymax) = CSNG(ymin) THEN ymax = ymin + 1
      dx = (xmax - xmin) / 10: xmin = xmin - dx: xmax = xmax + dx
      dy = (ymax - ymin) / 10: ymin = ymin - dy: ymax = ymax + dy
      w% = 640 / (xmax - xmin): h% = 480 / (ymin - ymax)
   CASE ELSE                           'Plot the data
      xp% = w% * (x - xmin)
      yp% = h% * (y - ymax)
      PSET (xp%, yp%)                  'Illuminate screen pixel
END SELECT
END SUB

SUB lyapunov (x, y, n%, l)            'Calculate Lyapunov Exp (l)
STATIC xe, ye, lsum
IF n% = 1 THEN lsum = 0: xe = .000001#: ye = 0
xsave = x: ysave = y: x = xe: y = ye
CALL advancexy(x, y, n% - 1)          'Reiterate equations
dx = x - xsave: dy = y - ysave: d2 = dx * dx + dy * dy
df = 100000000000# * d2: rs = 1# / SQR(df)
xe = xsave + rs * (x - xsave): x = xsave
ye = ysave + rs * (y - ysave): y = ysave
lsum = lsum + LOG(df)
l = .721348 * lsum / n%                'Convert to bits per iteration
END SUB

SUB setparams (x, y)                   'Set a() and initialize (x, y)
SHARED a()
x = 0: y = 0
FOR i% = 1 TO 12: a(i%) = (INT(25 * RND) - 12) / 10#: NEXT i%
END SUB

SUB testsoln (x, y, n%)                'Test solution at (x, y, n%)
nmax% = 10000                          'Bailout value
CALL lyapunov(x, y, n%, l)             'Get Lyapunov exponent (l)
IF n% = nmax% THEN n% = 0              'Bailout value reached
IF n% > 100 AND l < .005 THEN n% = 0   'Solution is not chaotic
IF ABS(x) + ABS(y) > 1000000# THEN n% = 0   'Solution is unbounded
END SUB
```

The coefficients are chosen in increments of 0.1 over the range -1.2 to 1.2 (25 possible values). Smaller coefficients result in missing many chaotic solutions, and larger coefficients produce mostly unbounded solutions. The increment was chosen to make each attractor visibly different and so that the coefficients could be represented as letters of the alphabet (A = -1.2, B = -1.1, through Y = 1.2) for easy reference and replication. Thus each attractor is uniquely identified by a 12-letter name. The number of possible cases is 25^{12}, or about 6 times 10^{16}. Of these, approximately 1.6 percent are chaotic, or about 10^{15} cases.[10] Viewing them all at a rate of one per second would require over 30 million years! Thus it is very unlikely that any of the patterns you produce in this way will ever have been seen before. The coefficients can be considered as settings of a combination lock, each revealing a different pattern for you to admire. They open doorways to an infinite reservoir of magnificent shapes and forms.

Initial conditions are set arbitrarily to $x_0 = y_0 = 0$. Other small initial values produce the same result for most cases, as expected for an attractor. The Lyapunov exponent is calculated using two x_0-values whose initial conditions differ by 10^{-6}. The program performs 100 iterations before testing the Lyapunov exponent to insure that the iterates lie on the attractor. After 100 iterations, the program also begins saving the minimum and maximum values of x and y so that after 1,000 iterations the screen can be cleared and resized, allowing a 10 percent border around the attractor. If 10,000 iterations are reached with a positive Lyapunov exponent and a bounded solution, the result is assumed to be a strange attractor. Otherwise, the parameters are discarded because they likely produce an unaesthetic image. The search immediately resumes after each attractor is confirmed and continues until you press a key.

The search procedure is surprisingly fast. On my 33-MHz 486DX computer running PowerBASIC 3.0, the program finds about 1,400 strange attractors per hour. The listed program only displays the attractors on the screen. A more versatile program would call a subroutine to print the attractors, perhaps after user confirmation or evaluation, or would save the coded coefficients in a disk file for later analysis.

SAMPLE STRANGE ATTRACTORS

The ten pieces of art comprising figure 4.1 show samples of the shapes that arise from the iteration of such two-dimensional maps. These cases are all strange attractors that I selected for their beauty and diversity from a much larger collection. However, they are not atypical, and there are many others that I could have exhibited. It is remarkable that such a diversity of shapes comes from the same simple set of equations with only different values of the coefficients.

I produced these cases using a laser printer with 300 dots-per-inch resolution on an 8.5 by 11-inch page after 500,000 iterations. The program needs modification to output the plots to a printer at high resolution. However, you can obtain satisfactory results using any of the various utilities that print a screen image.

Also shown on each figure is the code name preceded by the letter E to denote a two-dimensional quadratic map, the Lyapunov exponent L (in bits per iteration), and the correlation dimension F. The dimension is slightly ill-defined because it varies somewhat with scale. The dimension is taken here at a scale of about 1 percent of the largest diameter of the attractor, and hence it might be called a visual dimension. Program code 4.2 is a subroutine that your program can call once per iteration to return a continually updated estimate of the correlation dimension F and its uncertainty δF.

EBCQAFMFVPXKQ   F = 1.32 L = 0.13

ECDJXIYLSYQUM   F = 1.43 L = 0.11

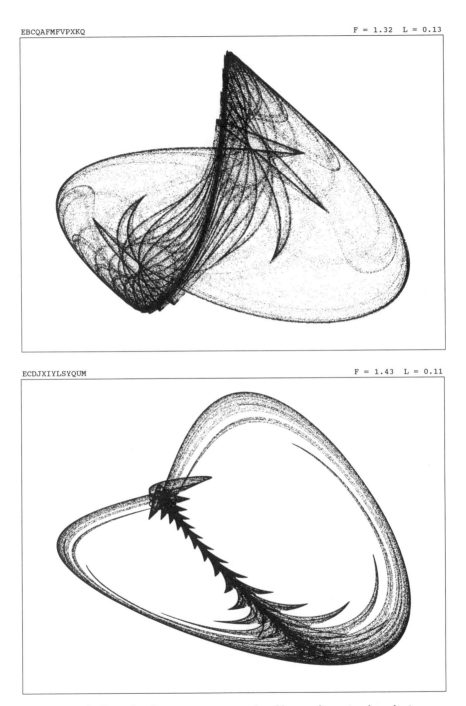

Figure 4.1 Examples of strange attractors produced by two-dimensional quadratic maps.

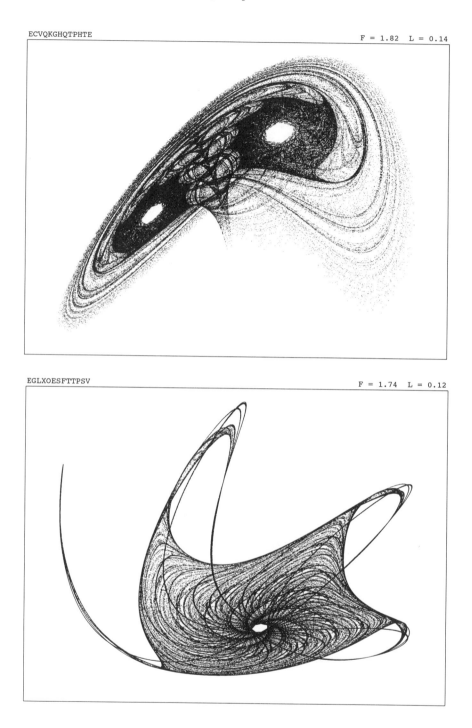

ECVQKGHQTPHTE   F = 1.82   L = 0.14

EGLXOESFTTPSV   F = 1.74   L = 0.12

EJXAICXIXFRHI   F = 1.37   L = 0.12

EMCRBIPOPHTBN   F = 1.39   L = 0.05

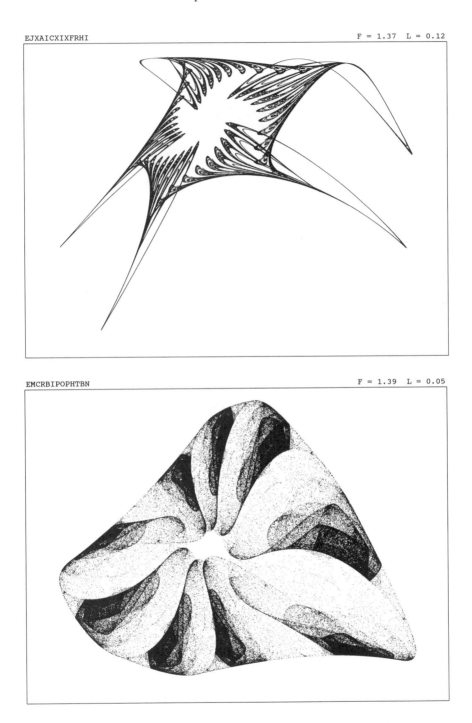

EMDVAIDOYHYEA   F = 1.57   L = 0.26

EMTGETXEJWCUR   F = 1.44   L = 0.07

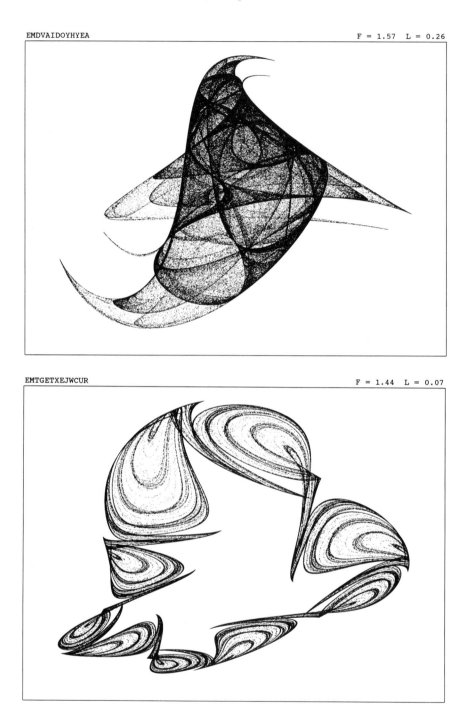

EUEBJLCDISIIQ   F = 1.24   L = 0.24

EUWACXDQIGKHF   F = 1.70   L = 0.15

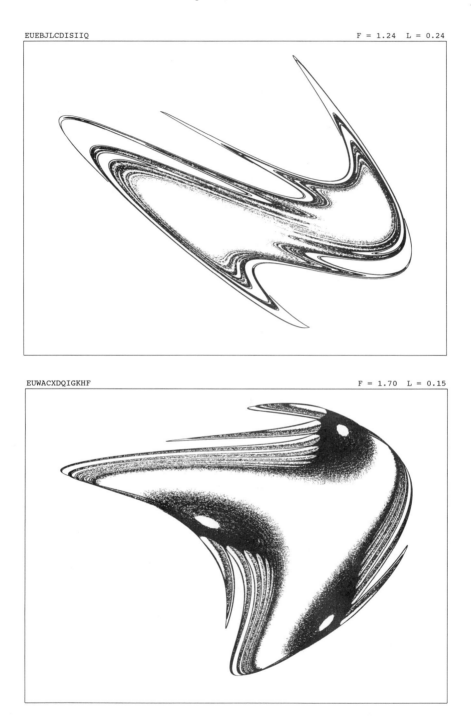

J. C. Sprott

EVDUOTLRBKTJD   F = 1.54   L = 0.04

EWLDREDHHWTTN   F = 1.54   L = 0.37

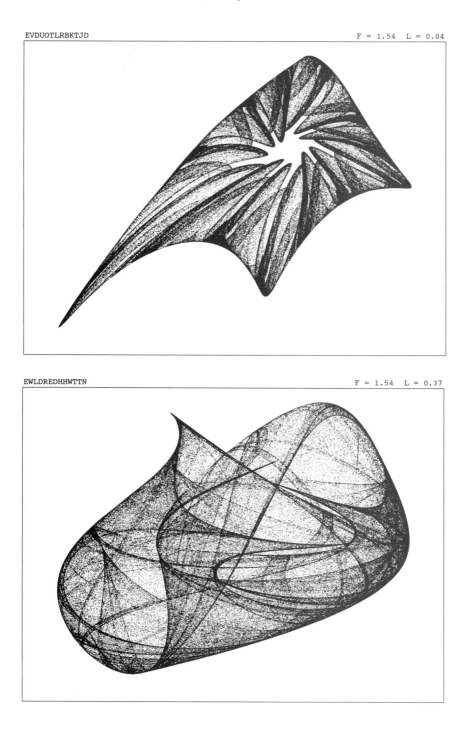

Program code 4.2   BASIC Subroutine to Estimate the Correlation
Dimension and Its Uncertainty

```
SUB corrdim (x, y, n%, f, df)          'Returns correlation dim (f±df)
STATIC n1, n2, xs(), ys(), rsqm, newf, newdf
ns% = 1000                             'Number of previous points saved
IF n% = 1 THEN                         'Initialize variables
    ERASE xs, ys: DIM xs(999), ys(999)
    n1 = 0: n2 = 0: rsqm = 0: newf = 0: newdf = 0
END IF
i% = n% MOD ns%
j% = (i% + INT(ns% * RND / 2)) MOD ns%  'Choose a random reference point
dx = x - xs(j%): dy = y - ys(j%)
rsq = dx * dx + dy * dy                'Calculate square of separation
IF n% < ns% THEN
    IF rsq > rsqm THEN rsqm = rsq      'Save maximum rsq in rsqm
ELSE
    IF rsq < .0003 * rsqm THEN         'Point was inside large sphere
        n2 = n2 + 1
        IF rsq < .00003 * rsqm THEN    'Point was inside small sphere
            n1 = n1 + 1
            newf = .868589 * LOG(n2 / n1)
            newdf = .868589 * SQR(1 / n1 - 1 / n2)
        END IF
    END IF
END IF
xs(i%) = x: ys(i%) = y                 'Replace oldest point with new
f = newf: df = newdf
END SUB
```

AESTHETIC EVALUATION

I evaluated 7,500 such attractors with the help of seven volunteers, including two graduate art students, a former art history major, three physics graduate students, and a former mathematics major.[11] All evaluators were born and raised in the United States. The evaluations were done by choosing attractors randomly from a collection of about 18,000 and displaying them sequentially on the computer screen without any indication of the quantities that characterize them. The volunteers were asked to evaluate each case on a scale of one to five according to its aesthetic appeal. It only took a few seconds for each evaluation.

At the end of each session I produced a graph similar to figure 4.2 in which the average rating is displayed using a gray scale on a plot in which the largest Lyapunov exponent (L) and the correlation dimension (F) are the axes. The darkness of each box increases with the average rating of those attractors whose values of L and F fall within the box. Figure 4.2 summarizes all the evaluations, although the cases examined by the various individuals are similar. In particular, all evaluators preferred attractors with dimensions between about 1.1 and 1.5 and Lyapunov exponents between zero and about 0.3. Some of the most interesting cases have Lyapunov exponents below about 0.1.

The dimension preference is not surprising since many natural objects have dimensions in this range. The Lyapunov exponent preference is harder to understand,

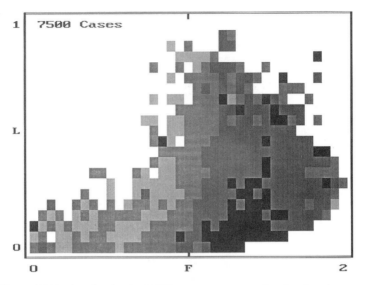

Figure 4.2 Results of evaluating 7,500 strange attractors, showing that the most visually appealing cases are those with small Lyapunov exponents (L) and with correlation dimensions (F) somewhat greater than one.

but it suggests that strongly chaotic systems are too unstructured to be appealing. The 443 cases that were rated five (best) by the evaluators had an average correlation dimension of $F = 1.30 \pm 0.20$ and an average Lyapunov exponent of $L = 0.21 \pm 0.13$ bits per iteration, where the errors represent plus or minus one standard deviation. About 28 percent of the cases evaluated fall within this range.

ESCAPE-TIME FRACTALS

Let me describe another class of patterns, generated with similar methods. I think you will find them to have considerable artistic appeal.

An attractor is defined as the set that a system approaches after many iterations. Some attractors attract initial conditions from the entire xy-plane. More often there is a finite basin of attraction, outside of which the solutions are unbounded and move ever farther from the attractor. It is also possible for two or more attractors to compete with one another. You might like to plot the basin boundary for various attractors. Points outside the basin escape from the vicinity of the attractor. Some of the most impressive computer fractal art has been produced by plotting with colors the number of iterations required for the orbit to escape beyond some given radius in the $x_0 y_0$-plane of initial conditions.[12] The Julia sets of the complex map $z_{n+1} = z_n^2 + c$ are perhaps the most celebrated examples of such escape-time fractals.[13] The objects are also sometimes called nonattracting chaotic sets.

Program code 4.3 BASIC Subroutine to Plot Escape-Time Contours

```
SUB basin (xmin, xmax, ymin, ymax)        'Plot escape-time contours
nmax% = 1040                              'Bailout value
w% = 640: h% = 480: nc% = 16             'Screen size and colors
dx = (xmax - xmin) / w%: dy = (ymax - ymin) / h%
FOR i% = 0 TO w% - 1
    FOR j% = 0 TO h% - 1
        x = xmin + dx * i%
        y = ymax - dy * j%
        n% = 0
        WHILE n% < nmax% AND x * x + y * y < 1000000!
            CALL advancexy(x, y, n%)
        WEND
        c% = n% mod (2 * nc%)
        if c% > nc% - 1 then c% = 2 * nc% - 1 - c%
        PSET (i%, j%), c%
    NEXT j%
NEXT i%
END SUB
```

The BASIC subroutine in program code 4.3 plots color escape contours for the equations in the subroutine *advancexy* in program code 4.1. You can call it from some appropriate place in the program to see the basin for each attractor that you find. You might expect that a complicated (strange) attractor would have a complicated basin, and sometimes this is the case. More often the basin boundary is relatively smooth and uninteresting. It seems that simple (fixed-point and cyclic) attractors tend to have complicated boundaries. Curiously, the basin boundary appears to touch many of the attractors, at least to the resolution of the computer screen.

These observations suggest that searching for strange attractors by the method previously described is not the optimal way to find interesting escape-time fractals. It is better to look for a set of coefficients for which the solution is unbounded but escapes very slowly. Many of these cases will be *repellers* rather than attractors. The illustrations in figure 4.3 show examples in which the orbit of a two-dimensional quadratic map (given in the beginning of this chapter) starting with $x_0 = y_0 = 0$ takes between 100 and 1,000 iterations to escape from a square that extends to $|x| = |y| = 10^6$. The figures all show a range of x_0 and y_0 between -1 and +1. The colors cycle from 0 to 15 and then from 15 back to 0. In these cases I have mapped the colors to a 16-level gray scale for printing, but the color versions that you can produce on your computer screen are stunning. The curves of constant color resemble elevations on a contour map, and you might try rendering them with various three-dimensional techniques. Dr. Clifford Pickover (IBM) and I have since pushed these methods to very high resolution and speed, and to three-dimensional representations, using a parallel graphics supercomputer called the IBM Power Visualization System.[14]

Figure 4.3 Examples of escape-time contours of various two-dimensional maps.

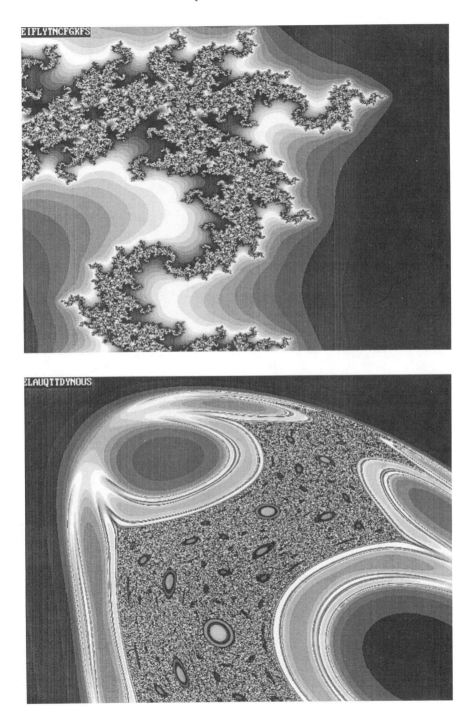

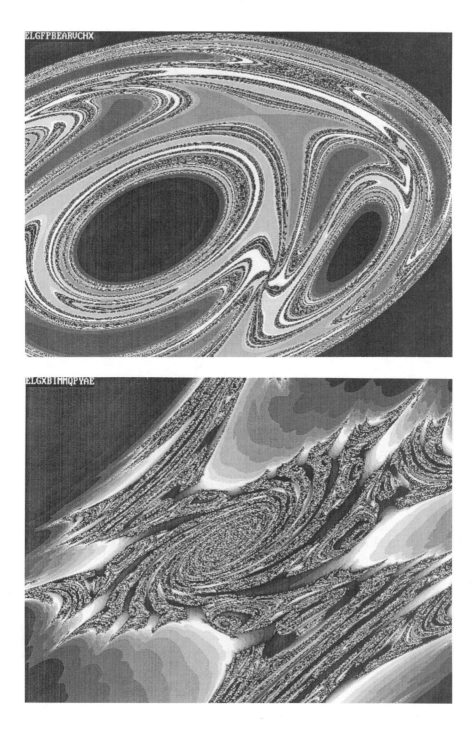

If you have a slow computer or if you are otherwise impatient, you can speed the production of the patterns by reducing the bailout value (see program code 4.3), increasing the range of initial values that you plot, or using a lower-resolution screen mode. The technique I use is to run the computer overnight, capture each screen image to a graphics file, and in a few minutes the next day view the hundred or so that were captured, printing any that were especially interesting and discarding the rest.

ITERATED FUNCTION SYSTEMS

An alternative to producing patterns with chaotic maps is to use *iterated function systems*.[15] Such systems were originally studied by Hutchinson and more recently applied to data compression and transmission by Barnsley.[16] Barnsley also introduced much of the terminology and the random iteration algorithm for their solution.[17] The simplest example of an iterated function system is a set of two-dimensional affine maps:

$$x_{n+1} = a_1 x_n + a_2 y_n + a_5$$
$$y_{n+1} = a_3 x + a_4 y + a_6$$

Such a mapping has a single fixed point (x^*, y^*) given by

$$x^* = [a_2 a_6 - a_5(a_4-1)] / [(a_1-1)(a_4-1) - a_2 a_3]$$
$$y^* = [a_3 a_5 - a_6(a_1-1)] / [(a_1-1)(a_4-1) - a_2 a_3]$$

(The fixed point is the solution of the equations $x_{n+1} = x_n$ and $y_{n+1} = y_n$.) This fixed point may be either stable (attracting) or unstable (repelling). A stable fixed point attracts initial conditions within its basin, which in this case is the entire xy-plane. With an unstable fixed point, successive iterates grow ever larger, and the system is unbounded. For simplicity, I consider saddle points (attracting in one direction and repelling in another) to be unstable and ignore periodic orbits, which seldom occur.

With each iteration, such an affine map takes a set of points in the xy-plane and moves it to a new location in the plane generally with scaling, translation, rotation, reflection, and shear. Stable solutions necessarily scale in such a way that the area of the new set is less than the area of the previous set, in which case we say that the set has contracted. Continually contracting mappings cause the set eventually to collapse into a region of negligible area surrounding the fixed point, and continually expanding mappings are unbounded. The amount of area contraction is determined by the determinant of the *Jacobian* matrix (hereafter simply Jacobian) given by

$$J = a_1 a_4 - a_2 a_3$$

which is the ratio of the area after a contraction to the area before. Thus the condition for contraction is $|J| < 1$. Note that area contraction does not guarantee boundedness, since a set can continually contract in one direction and expand in another, approaching a thin filament of zero area and infinite length. We are generalizing the usual meaning of a contraction mapping in which every pair of points is moved closer together by the mapping.

Now suppose there exists a second different contracting affine map,

$$x_{n+1} = a_7 x_n + a_8 y_n + a_{11}$$
$$y_{n+1} = a_9 x_n + a_{10} y_n + a_{12}$$

which is applied to the set after some finite number of iterations of the first map. The result is to displace the set of points away from the original fixed point toward which it was converging in the direction of a new fixed point in discrete steps of ever smaller size. However, the progression toward the new fixed point need not be along a straight line but rather can spiral around it. Now suppose that the two maps are applied in some arbitrary sequence. Then the two attracting fixed points compete for the set, producing a succession of points distributed over the xy-plane in some pattern. Each point can be transformed into another point in the pattern by some sequence of the two affine mappings. The collection of all such sequences of two or more maps is an iterated function system (IFS).

A practical method for producing iterated function systems is the *random iteration algorithm* in which a computer in essence repeatedly flips a coin and uses one map if it comes up heads and the other if it comes up tails. All possible sequences of heads and tails are obtained eventually, and long sequences of all heads or all tails, which would densely populate the regions near the fixed points, are rare. The locations of the fixed points usually are not obvious in such patterns. The patterns evolve much more quickly and uniformly if the coin is weighted so that the probability of applying each mapping is proportional to its Jacobian. Strongly contracting mappings don't need to be applied very often since they shrink so rapidly. The starting point can be chosen arbitrarily since the basin of attraction is the entire xy-plane. You can start at one of the fixed points, but it is easier just to start at the origin ($x_0 = y_0 = 0$) and discard the first few points to ensure that the points have collapsed onto the attractor.

The resulting pattern is a deterministic fractal, despite being produced by a random algorithm. Different sequences of random numbers will produce the same eventual pattern. The pattern may have an integer dimension, but in such cases the boundary of the pattern usually has a noninteger dimension. (The Mandelbrot set is an example of an object with integer dimension (2) and a fractal boundary.) Although the attractor is a fractal, it is not usually called a strange attractor. That term is reserved for chaotic dynamical systems. Affine mappings cannot exhibit chaos because they lack the requisite nonlinearity. They do not exhibit sensitivity to initial conditions.

Like strange attractors, iterated function systems can be categorized by their Lyapunov exponent. Imagine two bounded iterated function systems produced by the same pair of affine maps but with initial conditions that differ slightly. If the two are produced by a different sequence of coin flips, the sequences of (x, y) values would be completely different. Indeed, this would be true even if they had the same initial condition. Thus there is extreme sensitivity to initial conditions resulting from the underlying randomness used to produce the sequence. However, if the same sequence of coin flips was used in the two cases, successive iterates would approach one another, implying insensitivity to initial conditions and a negative Lyapunov exponent as expected for a deterministic nonchaotic process. A two-dimensional IFS has two Lyapunov exponents whose sum is related to the weighted average Jacobian $<J>$ of the maps by

$$L_1 + L_2 = \log_2 | <J> |$$

where the vertical bars denote the absolute value of the quantity they enclose.

The fractal patterns produced by iterated function systems arise from repeated affine mappings that make distorted copies of the pattern at successively smaller scales. Because the copies are generally sheared, they are not self-similar but rather are *self-affine*. Furthermore, the mappings may overlap one another. Therefore, calculation of the fractal dimension directly from the equations is generally not straightforward, but it is easy to calculate the correlation dimension numerically as previously described.

I've implemented a search procedure for visually interesting iterated function systems that is similar to the one used for strange attractors. The simplest way to produce the iterated function system is to replace the subroutine *advancexy* in program code 4.1 with the version in program code 4.4. In this case, the Lyapunov exponent test is not used, and you can remove it to speed the calculation. You also can save some computation time by choosing initial conditions at one of the fixed points to avoid having to discard the first few iterates. The illustrations in figure 4.4 are samples of the shapes that I produced with this method. The figures include the codes for the 12 coefficients preceded by the letter *a* to denote a two-dimensional IFS with two mappings.

Program code 4.4 2-D Iterated Function System

```
SUB advancexy (x, y, n%)                    'Advance (x, y, n%) for IFS
SHARED a()
r% = 6 * INT(2 * RND)
xnew = a(1 + r%) * x + a(2 + r%) * y + a(5 + r%)
y = a(3 + r%) * x + a(4 + r%) * y + a(6 + r%)
x = xnew
n% = n% + 1
END SUB
```

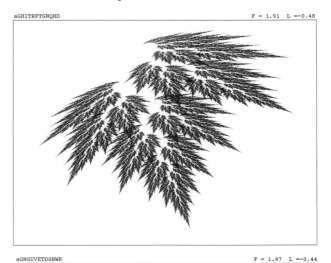

aGHITRFTGNQHD   F = 1.91 L =-0.48

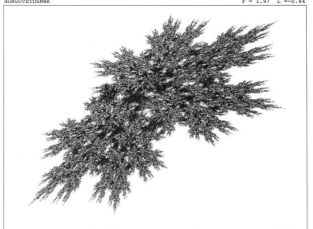

aGNGUVETDSNWK   F = 1.97 L =-0.44

aHTIMREXHRQNO   F = 1.89 L =-0.44

Figure 4.4 Examples of two-dimensional iterated function systems.

aKDOWQMMYEMWD F = 1.38 L =-0.03

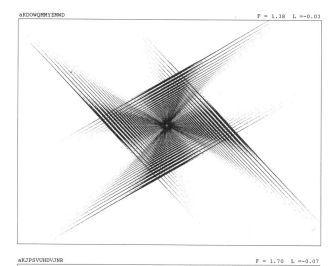

aKJPSVUHDVJNR F = 1.70 L =-0.07

aOHQNGUSFSRBR F = 1.69 L =-0.44

aPSBSHNPLFJAH   F = 1.86   L =-0.29

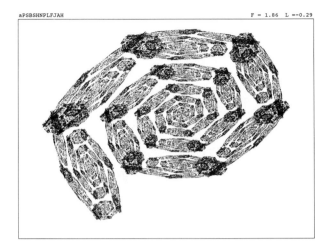

aRHTTQYGQHKXQ   F = 1.84   L =-0.40

aTLPVPLMQEIFC   F = 1.88   L =-0.47

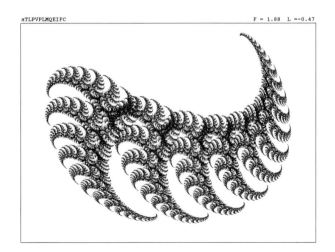

aTTGUJNMRLWBR   F = 1.79 L =-0.05

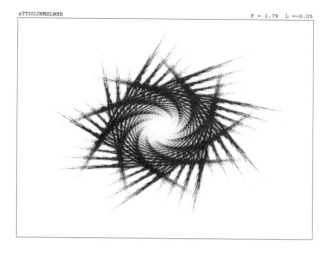

aULLMVQJWATTE   F = 1.74 L =-0.02

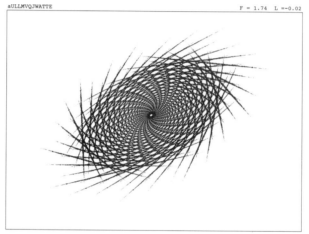

aUTDQOJLQOLBJ   F = 1.82 L =-0.15

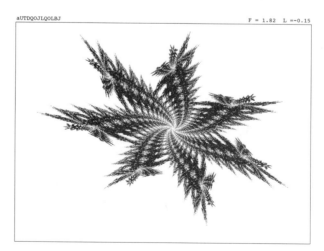

I evaluated 7,500 such patterns as I did for the strange attractors, and the result is shown in figure 4.5. There is a clear preference for patterns with correlation dimensions greater than 1 and for large negative Lyapunov exponents. For a given dimension, the largest negative values for the Lyapunov exponents correspond to cases in which the two exponents are equal, implying the same contraction in all directions. Thus the patterns preserve their shape and are exactly self-similar. The largest negative Lyapunov exponent and fractal dimension are bounded by a curve

$$-FL < \log O\,/\log D$$

where O is the number of mappings (two in this case) and D is the dimension of the system (two in this case). Points on the boundary correspond to exact self-similarity since they contract equally in all directions. For the 76 cases that were rated best (a 5), the average correlation dimension was $F = 1.51 \pm 0.43$, and the average Lyapunov exponent was $L = -0.24 \pm 0.15$ bits per iteration, where the errors represent plus or minus one standard deviation. About 31 percent of the cases evaluated fall within the range.

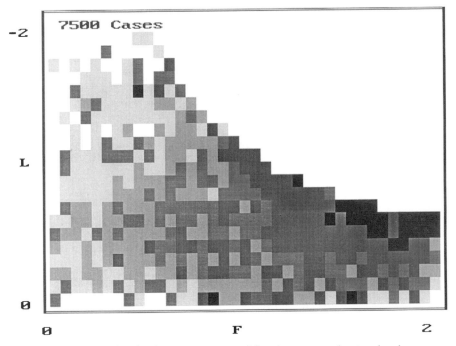

Figure 4.5 Results of evaluating 7,500 iterated function systems, showing that the most visually appealing cases are those with large negative Lyapunov exponents (L) and with correlation dimensions (F) close to 2.

These results suggest that the computer could be taught to select cases that have a high probability of aesthetic appeal. One such criterion that I devised by surrounding the best cases in the *LF*-plane with an ellipse is

$$[(2 - F) / 1.2]^2 + [(2 + L / \log O) / 1.6]^2 < 1$$

which eliminates about 98 percent of the two-dimensional cases. The 2 percent that remain are nearly all visually interesting.

EVALUATION OF HUMAN ART

The previous examples of computer fractal art were produced by *dynamical systems* in which the evolution of the patterns is determined by a system of equations. In such cases, it is simple to evaluate the patterns based on their Lyapunov exponent and correlation dimension because the computer has access to the equations. With art produced by humans, there are no equations (no "temporal" information); we are presented only with a finished static object. It is interesting to ask whether the computer can aid in evaluating this kind of art.

Suppose that the pattern has been digitized using a scanner and displayed on the computer screen, and that the piece employs only black-and-white pixels (no colors or shades of gray). (The method could be extended to include color and intensity by treating them as additional dimensions.) Since we don't know the sequence in which the picture was originally drawn, we cannot calculate a true Lyapunov exponent. Likewise, we cannot calculate a correlation dimension because many of the dots that make up the image may fall on the same screen pixel.

We can, however, calculate the capacity dimension (also called the *box-counting dimension*) by looking at the number of illuminated pixels *(N)* as a function of their linear size (*r*). For small *r*, this dimension[18] is

$$F = \log(N_2/N_1) / \log(r_2/r_1)$$

You can estimate this quantity by counting the total number of screen pixels that are illuminated (N_2). Next divide the screen up into squares each containing four pixels as shown in figure 4.6 and count the number of squares (N_1) that have at least one of their pixels illuminated. In such a case, r_2/r_1 is 2, and $F = \log_2(N_2/N_1)$. The capacity dimension is typically greater than the correlation dimension, but the two are often similar, and they are equal if the points that make up the image are uniformly distributed.

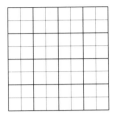

Figure 4.6 The capacity dimension is calculated by counting the number of boxes containing illuminated screen pixels on two different scales.

Whereas the dimension is inherently a static property of an object, the Lyapunov exponent is a dynamical property. However, imagine that the pattern was produced by an artist who worked systematically from the upper left to the lower right in raster fashion (the way one reads). The result is a *time series,* which in simplest form would consist of a string of zeros and ones to indicate whether each successive pixel is illuminated. (Something like this was probably done in digitizing the image and transferring it to the computer screen.) The time series is sometimes compressed to reduce disk storage requirements. One simple compression scheme generates a time series consisting only of the number of successive identical values of zeros or ones. You can imagine that the paintbrush is alternately raised and lowered for the prescribed number of pixels as the image is raster scanned.

From such a time series, you can calculate a quantity related to the Lyapunov exponent. Pick a window of, for example, five values representing a point and its four immediate predecessors, and move that window through the time series. You may allow the windows to overlap or not as you prefer. Each set of five values represents a point in a five-dimensional space (called the *embedding space*[19]). At each window position, search the entire time series for the point in this five-dimensional space that is nearest (a distance d_n away) but not at exactly the same position. Then find the next successive point corresponding to each of these points and determine their separation d_{n+1}. The average logarithm of the ratio d_{n+1}/d_n

$$U = \frac{1}{N} \sum_{n=0}^{N-1} \log_2 (d_{n+1} / d_n)$$

is a measure of the unpredictability of the pattern and resembles a Lyapunov exponent. It is crudely a measure of how unpredictable the pattern would be if a portion of it were obscured and you had to guess what was missing. BASIC subroutines for estimating the capacity dimension and unpredictability from a static screen display are given in program code 4.5.

Program code 4.5 BASIC Subroutines to Calculate Capacity Dimension
and Unpredictability from a Screen Display

```
SUB capdim (f, df)                       'Returns capacity dim f±df)
w% = 640: h% = 480                       'Screen width and height
n1 = 0: n2 = 0: n2old = 0
FOR j% = 0 TO h% - 2 STEP 2: FOR i% = 0 TO w% - 2 STEP 2
    FOR di% = 0 TO 1: FOR dj% = 0 TO 1
        IF POINT(i% + di%, j% + dj%) THEN n2 = n2 + 1
    NEXT dj%, di%
    IF n2 > n2old THEN n1 = n1 + 1: n2old = n2
NEXT i%, j%
f = 1.442695 * LOG(n2 / n1)
df = 1.442695 * SQR(1 / n1 - 1 / n2)
END SUB

SUB unpredict (u)                        'Returns unpredictability (u)
DIM a%(32766)                            'Big integer array
e% = 5                                   'Embedding dimension
w% = 640: h% = 480                       'Screen width and height
pold% = POINT(0, 0): count% = 0: k% = 0: usum = 0
FOR j% = 0 TO h% - 1: FOR i% = 0 TO w% - 1
    p% = POINT(i%, j%)
    IF p% = pold% THEN                   'Pixel is same as previous
        IF count% < 32766 THEN count% = count% + 1
    ELSE                                 'Pixel is different
        pold% = p%
        a%(k%) = count%
        IF k% < 32766 THEN k% = k% + 1
        count% = 0
    END IF
NEXT i%, j%                              'Data now stored in array a%(k%)
FOR i% = 1 TO k% - 1 - e%
    dsqm = 1E+37
    FOR j% = 1 TO k% - 1 - e%            'Find closest different point
        dsq = 0
        FOR n% = 0 TO e% - 1
            d = a%(i% + n%) - a%(j% + n%)
            dsq = dsq + d * d
        NEXT n%
        IF dsq > 0 AND dsq < dsqm THEN dsqm = dsq: im% = i%: jm% = j%
    NEXT j%
    dsq = 0
    FOR n% = 1 TO e%                     'Find separation of next points
        d = a%(im% + n%) - a%(jm% + n%)
        dsq = dsq + d * d
    NEXT n%
    IF dsqm * dsq > 0 THEN usum = usum + LOG(dsq / dsqm)
NEXT i%
u = .721348 * usum / i%                  'Convert to bits per iteration
END SUB
```

It is interesting to determine the extent to which the capacity dimension estimated in this way agrees with the correlation dimension and the unpredictability agrees with the Lyapunov exponent derived from the dynamics. Figure 4.7 shows such a comparison for 172 strange attractors generated randomly from two-dimensional quadratic maps as described earlier. The correlation dimension and capacity dimension are reasonably close, but the unpredictability correlates poorly with the Lyapunov exponent. There are other possible measures of the unpredict-ability of a static pattern, such as the LZ-measure of *algorithmic complexity*[20] or simply

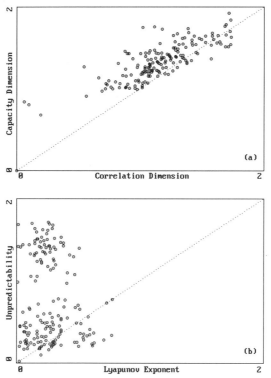

Figure 4.7 Comparison of capacity dimension with correlation dimension (a) and unpredictability with Lyapunov exponent (b) for 172 randomly chosen strange attractors produced by two-dimensional quadratic maps.

the factor by which a given file compression algorithm can reduce the size of a graphics file. For this purpose, IFS compression schemes may be especially appropriate.[21]

Nevertheless, the capacity dimension and unpredictability are experimentally accessible quantities that might relate to the aesthetics of the pattern. To test this proposition, I calculated the two quantities for 686 IFS patterns that I had previously evaluated as either very good (rating of 5) or very bad (rating of 1). The results are plotted in the usual way in figure 4.8. The visually appealing patterns have capacity dimensions slightly below two and intermediate values of unpredictability. In retrospect, it is reasonable that intermediate unpredictability is desirable since the perfect predictability of a wallpaper pattern is boring, and the perfect unpredictability of paint thrown randomly at a canvas is too disordered. For the 236 cases that were rated 5, the average capacity dimension was $F = 1.69 \pm 0.16$, and the average unpredictability was $U = 1.33 \pm 0.20$ bits per iteration, where the errors represent plus or minus one standard deviation. About 30 percent of the cases evaluated fall within the range.

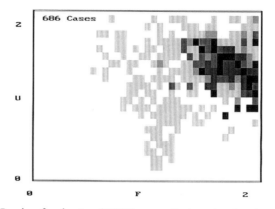

Figure 4.8 Results of evaluating 686 IFS screen displays, showing that the most visually appealing cases are those with intermediate unpredictability *(U)* and capacity dimensions *(F)* close to 2.

The ability of a computer to reduce a piece of art to a set of numbers raises many intriguing possibilities. Someone needs to determine which of these numbers correlate best with the aesthetic quality of the piece. It is likely that different artists tend toward different values of these quantities. If so, the method might provide a means to detect forgeries or a way to program the computer to produce art in a particular style. Of course, such an approach may be far too simplistic, needing very many numbers to capture these subtle differences. If the numbers contain as much information as the original picture, then little has been gained. Such studies are becoming practical because of the exponential growth of computer speeds and the increasing availability of high-quality digital images on the Internet.

THE FUTURE OF FRACTAL ART

If you want to pursue further the automatic generation of computer art, you should refer to my book, *Strange Attractors: Creating Patterns in Chaos*.[22] You will see about 350 examples of strange attractors produced by maps and ordinary differential equations in dimensions up to four with polynomials up to fifth order as well as an assortment of nonpolynomial functions. In dimensions higher than two, special techniques are required to exhibit the patterns. The book describes several such methods, including anaglyphs in which stereoscopic three-dimensional patterns are viewed through red/blue glasses, which are included.[23] The book comes with an IBM-PC disk that allows you to replicate all the patterns shown as well as an endless variety of others. It includes many suggestions for further studies, which fall into two categories—the first oriented toward the artistic aspects and the second toward the scientific aspects.

As an artistic tool, the technique could be extended by finding more artistic ways to display the solutions of the equations, especially in high dimensions and as animations. The objects lend themselves to holographic display and to experimentation with virtual reality and other scientific visualization techniques. Imagine donning goggles and wandering throughout these fractal landscapes. The method also can be used for the automatic generation of fractal music. Also, you probably can think of ways to improve the selection procedure. Additional human psychological studies would help to determine what numerical quantities best correlate with the aesthetics. There have been perceptual studies of fractal patterns, both visual[24] and auditory,[25] but relatively little work specific to aesthetics. Like the Rorschach ink-blot test, in which striking cultural differences have been reported, fractal patterns also might be useful in personality tests.[26] It would be interesting to see if the preferences are different for artists and scientists or for people of different ages, genders, or cultures. Some studies have suggested that musically sophisticated people prefer music with a higher fractal dimension.[27] Perhaps the same is true for fractal art.[28]

As a scientific tool, the technique provides an enormous assortment of strange attractors upon which statistical studies can be done. It would be interesting to see how the quantities that characterize the attractors, such as correlation dimension and Lyapunov exponent, scale with the parameters of the equations that generated them. There are probably many new and mathematically interesting examples of strange attractors waiting discovery. Certainly both mathematicians and artists will search for previously unknown examples of simple chaotic systems. They will explore the various routes by which systems evolve from periodic to chaotic by making small changes to parameters of the equations. They will see how the various measures of fractal dimension and other quantities compare. They will study the shape and dimension of the region of parameter space over which chaotic solutions occur. Mathematicians will become "biologists" confronted with an enormous variety of "species" as they catalog and find order and relations in these strange and beautiful patterns that so vividly illustrate the unexpected and provocative connections between art and science.

Notes

1. J. Gleick, *Chaos: Making a New Science* (New York: Viking, 1987).

2. B. B. Mandelbrot, *The Fractal Geometry of Nature* (New York: W. H. Freeman, 1982).

3. H. Peitgen and D. Saupe, *The Science of Fractal Images* (New York: Springer-Verlag, 1988).

4. A. K. Dewdney, "Probing the Strange Attractions of Chaos," *Scientific American* 257 (July 1987): 108.

5. H. G. Schuster, *Deterministic Chaos* (New York: Springer-Verlag, 1984).

6. A. Wolf, J. B. Swift, H. L. Swinney, and J. A. Vastano, "Determining Lyapunov Exponents for a Time Series," *Physica* 16D (1985): 285.

7. J. Kaplan and J. A. Yorke, "Functional Differential Equations and the Approximation of Fixed Points," *Springer Lecture Notes in Mathematics* 730 (1978): 228.

8. P. Grassberger and I. Procaccia, "Characterization of Strange Attractors" *Physical Review Letter* 50 (1983): 346.

9. W. Lauterborn and J. Holzfuss, "Evidence for a Low-Dimensional Strange Attractor in Acoustic Turbulence," *Physics Letters* A115 (1986): 369.

10. J. C. Sprott, "How Common is Chaos?" *Physics Letter* A173 (1993): 21.

11. J. C. Sprott, "Automatic Generation of Strange Attractors," *Computers & Graphics* 17 (1993): 325.

12. H.-O. Peitgen and P. H. Richter, *The Beauty of Fractals: Images of Complex Dynamical Systems* (New York: Springer-Verlag, 1986).

13. K. Falconer, *Fractal Geometry: Mathematical Foundations and Applications* (New York: Wiley, 1990).

14. J. C. Sprott and C. A. Pickover, "Automated Generation of Quadratic Map Basins," *Computers & Graphics* 19 (2) (1995): 309.

15. M. Barnsley and S. Demko, "Iterated Function Systems and the Global Construction of Fractals" *Proceeding of the Royal Society of London* A399 (1985): 243.

16. J. Hutchinson, "Fractals and Self Similarity," *Indiana University Journal of Mathematics* 30 (1981): 713. M. F. Barnsley, V. Ervin, D. Hardin, and J. Lancaster, "Solution of an Inverse Problem for Fractals and Other Sets," *Proceedings of the National Academy of Sciences* 83 (1986): 1975. M. F. Barnsley and L. P. Hurd, *Fractal Image Compression* (Wellesley, MA: A. K. Peters, 1993).

17. M. F. Barnsley, *Fractals Everywhere* (Boston: Academic Press, 1988).

18. J. D. Farmer, E. Ott, and J. A. Yorke, "The Dimension of Chaotic Attractors," *Physica* 7D (983): 143.

19. N. H. Packard, J. P. Crutchfield, J. D. Farmer, and R. S. Shaw, "Geometry from a Time Series," *Physical Review Letters* 45 (1980): 712. F. Takens, "Detecting Strange Attractors in Turbulence," *Lecture Notes in Mathematics* 898 (1981): 366.

20. A. Lempel and J. Ziv, "On the Complexity of Spatiotemporal Patterns," *IEEE Transactions on Information Theory* IT-22 (1976): 75. F. Kaspar and H. G. Schuster, "Easily Calculable Measure for the Complexity of Spatiotemporal Patterns," *Physics Review* A36 (1987): 842.

21. L. F. Anson, "Fractal Image Compression," *Byte* 18 (October 1993): 195.

22. J. C. Sprott, *Strange Attractors: Creating Patterns in Chaos* (New York: M&T Books, 1993).

23. J. C. Sprott, "Simple Programs Produce 3D Images," *Computers in Physics* 6 (1992): 132.

24. J. E. Cutting and J. J. Garvin, "Fractal Curves and Complexity," *Perception & Psychophysics* 42 (1987): 365. D. L. Gilden, M. A. Schmuckler, and K. Clayton, "The Perception of Natural Contour," *Psychological Review* 100 (1993): 460.

25. R. F. Voss and J. Clarke, "'1/f Noise in Music: Music from 1/f Noise," *Journal of the Acoustic Society of America* 63 (1978): 258. M. A. Schmukler and D. L. Gilden, "Auditory Perception of Fractal Contours," *Journal of Experimental Psychology: Human Perception & Performance* 19 (1993): 641.

26. M. Bleuler and R. Bleuler, "Rorschach's Ink-Blot Test and Racial Psychology," *Character and Personality* 4 (1935): 97. T. H. Cook, "The Application of the Rorschach Test to a Samoan Group," *Rorschach Research Exchange* 6 (1942): 51.

27. L. Short, "The Aesthetic Value of Fractal Images," *British Journal of Aesthetics* 31 (1991): 342.

28. D. J. Aks and J. C. Sprott, "Quantifying Aesthetic Preferences for Chaotic Patterns," *Emperical Studies of the Arts* 14 (1996): 1.

5

The Future of Fractals in Fashion

Danielle Gaines

―――――

I discuss the use of fractal patterns in clothing design and speculate about the future of mathematical art in the fashion world. I also include several of my own designs to give readers an indication of possible future trends.

―――――

My chosen field of business is fashion design, in particular, ladies' activewear. I also have an interest in mathematical patterns and metaphysics, so you can imagine my thrill upon discovering fractals. My first introduction to fractals occurred when I discovered John Briggs's book *Fractals: The Patterns of Chaos*.[1] As with most people first becoming familiar with fractals, I soon became aware of their vast diversity. They were everywhere I looked. Fractals were in me, on me, and around me. Whether it was the trees that inspired me, the computer art that "moved the earth under me," or the broccoli I ate, one thing was sure: My life took on new meaning because of my *gestalt fractalia*. As I have a creative mind, I instantly went to work in linking fractals to my chosen field of business—*fashion*. Since there is such a growing awareness and interest in fractals, and since fashion is my field of business, I thought I would incorporate some of Dr. Clifford Pickover's exciting fractal art into viable fashion products.[2] But first, for all of you math mavens, computer nerds, and science studs, I thought I'd give a brief description of what exactly fashion is and how fractals and fashion can have a beautiful union.

Clothing is wearing apparel or anything that covers the body. The basic functions of clothing include warmth, protection, and enhancement of beauty or sexual appeal. Clothing also gives a description of the wearer. A person can predict the social status, occupation, age, or sex of another person by his or her clothing. For example, the color Tyrian purple, derived from the abalone shell of ancient Phoenicia, was (and still is) considered a symbol of royalty. Due to the tedious nature of extracting the dye from the shell, purple was set aside to be worn only by royalty. Today, because of technology and mass production, purple is available to everyone. Various uniforms for professionals denote occupations; consider nurses, doctors, clergymen, and mechanics, to name a few. Clothing serves all of these functions, but what sets each culture, time period, and person apart is what we call fashion.[3]

According to the *American Heritage Dictionary*, fashion is defined as: "the current style or custom, as in dress or behavior; a piece of clothing that is in the current mode; to adapt, as to a purpose or occasion."

Fashion is what people refer to in order to describe a mode or style of behavior or, in our case, clothing representative of a particular time or culture. For instance, in the Western world, it wasn't fashionable for women to wear pants until the twentieth century; yet in many parts of the Eastern world, women began wearing pants centuries ago.

Throughout the ages, various factors have influenced and continue to influence fashion, including culture, tribal identification, religion, climate (for example, the grunge look that came out of Seattle relates to the city's weather), geography, material resources—natural and manmade (cotton, wool, hemp,

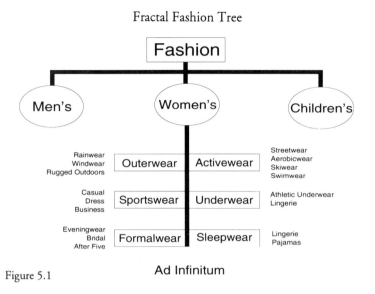

Figure 5.1

barkcloth, plastic, nylon, etc.), and even social occurrences. For example, higher castes had more expensive, less practical wardrobes. When men and women wore garments that were constricting and did not allow them to maneuver easily, it was understood that those people were wealthy and/or of noble birth. If one's clothing was constricting, the need to hire manual labor was apparent.

Art, another factor of fashion, stimulates design, whether the influence comes from ancient, classical, or contemporary art. Yves St. Laurent, inspired by the minimalist Piet Mondrian, designed his famous "Mondrian" dress using thick, black, perpendicular lines and primary color blocks strategically placed within the garment. Prints, incorporating influences from art such as Monet's *Water Lilies* and Escher's *Metamorphosis II*, are used on various fabrics.

War, another inevitable factor of fashion, has in the past enforced fabric rationing which limited the availability of certain fabrics while putting others in high demand. At the same time, however, fabric rationing forced designers to become more resourceful and cost effective. From chaos comes new fashion! War requires that clothing be useful as well. Body accessories and garments are designed for easy mobility and accessibility, not to mention protection and disguise, as in camouflage. Status and power, begotten of war or economic might, can influence fashion. Consider the effect of Westernization on the Eastern world. The United States has established its influence not only through wealth and political status but also through the media and entertainment. Japanese youth are heavily into American 1950s images such as Harley-Davidson motorcycles and James Dean.

Now, toward the end of the twentieth century, styles have changed several times within a decade in the United States alone. There are different categories of fashion (see figure 5.1)—activewear, sportswear, lingerie, underwear, formal wear,

starfish   jellyfish   wave

sea anemone   various flora

Figure 5.2 Fractal motifs on which my ladies' activewear is based.

evening wear, outerwear, swimwear, and so on. Each category influences the other. We now have subcategories that are melded together, as with active sportswear, sexy activewear, and athletic underwear. And within each of these subcategories are sub-subcategories leading to a dendritic (or fractal) world of clothing.

Fashion in the twentieth century also has been influenced by technology. Not only do we screenprint T-shirts with desired logos and graphics, but we denote our self-images through heat transfer and all sorts of intricate processes. Screenprinting is done not only by hand, but by machinery set up to print and embroider thousands of garments in a day. The images entail such subjects as people's names, characters from cartoons and movies, philosophical expressions, and scientific proofs. With the advent of fractals and the current growing interest in them, many people in various disciplines have incorporated fractals into their lines of work. Thus, it is no wonder that fractal fashion has emerged. Which brings me back to my primary subject: fractals in fashion.

There is a growing interest in computer-generated art based on mathematical equations, hence, fractals. Some leading researchers, scientists, and programmers have created diverse pieces of fractal art. Futurists have begun to make a business out of fractal products. People are wearing fractals on T-shirts and ties and drinking from fractal-printed coffee mugs. There's actually a store in London called Strange Attractions, which caters to the fractal intelligentsia looking for fractal paraphernalia—including posters, keychains, books, Christmas cards, and T-shirts. My company, AVES®, Inc., has added fractals to its latest collection.

I made use of fractal art from various articles and books by Clifford Pickover. (See figure 5.2.) What you see is an array of fractal art, which, to my eyes, resembles various forms of sea life. If you study the art, you'll see some resemblance to starfish, jellyfish, and waves.

So, what I decided to do was create two small collections of ladies' activewear with a "beach" theme for summer (see figures 5.3a-c) and a "winter solstice" theme for fall (see figures 5.4a-b). As you can see, the fractal theme is not only incorporated through the computer graphic art, but also through construction and composition in some of the pieces.

FUTURE TRIPPING

In the future, designers will make increasing use of computer graphics on clothing, incorporating fractals and other graphics within the theme of their collections. Because we're living in a rapidly advancing, technological society, designers can't resist incorporating scientific influences into clothing. The various prints and motifs we use on our fashions represent, and are influenced by, the current issues in our lives. As in the past, we still wear symbols denoting our religious preferences, cultural originations, hobbies, occupations—in short, our identities.

Fractals in fashions of the future could be very interesting from technological, functional, and artistic standpoints. Fractals and fashion, hand in hand, could help pave strange, new roads in the medical industry. For example, I can envision a one-piece "skin" that will cover the entire body like a unitard. Then, on that skin would be a whole network of life-supporting dendrites that could help sustain the life of burn victims by protecting the damaged skin while allowing oxygen and nutrients to be distributed efficiently along the outer dermis, thus enabling the skin to heal properly. Real skin could be used. We could actually produce organic "Band-Aids" for small wounds that need strictly regulated amounts of nutrients and oxygen. Future laws could permit the inclusion of skin on the organ donor cards. After all, skin *is* the largest organ.

Figure 5.3a

Figure 5.3 (a, b, c) Ladies' summer activewear collection using fractal patterns.

Figure 5.3b

Figure 5.3c

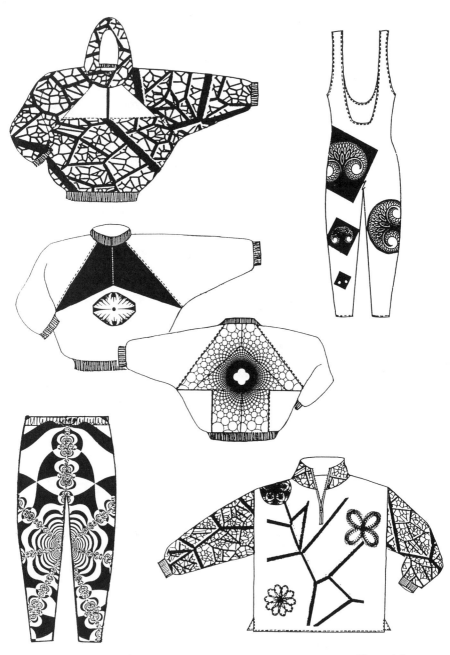

Figure 5.4a

Figure 5.4 (a, b) Ladies fall activewear collection using fractal patterns.

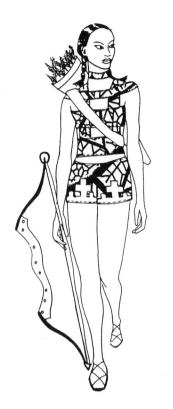

Figure 5.4b

Fractal fashion will certainly make good camouflage uniforms for soldiers in the twenty-first century. Perhaps, when humankind can render holographic images onto fabric efficiently, soldiers will be able to camouflage themselves with outfits that display different (two or more) environments depending on from which perspective the fabric is viewed. And for a quick change of scenery, they could have reversible fractal jumpsuits camouflaged in their designated environment. The fractal patterns could be desert or jungle patterns to be interchangeable in time of war or used only in an advanced game of hide and seek. We also could create an environment-reflective fabric that would imitate whatever environment the wearer was in: chameleon wear!

There are advantages for other occupations as well. Costume design would take on, literally, new dimensions. But teachers also could use fractal fashion to illustrate points. For instance, biology teachers could wear a fractal jumpsuit with a "living" cardiovascular or nervous system to better demonstrate those systems. Fractals are already used for special effects in movies and to decorate cups, T-shirts, jigsaw puzzles, and ties. No doubt they will find increasing use in virtual reality fashion as we move into the twenty-first century.

It will be interesting to see how copyright laws might change as a result of future fashion designs created not by the human hand but by mathematical formulas processed on a computer.

Notes

1. J. Briggs, *Fractals: The Patterns of Chaos* (New York: Simon & Schuster, 1992).
2. C. Pickover, *Computers, Pattern, Chaos and Beauty* (New York: St. Martin's Press, 1990). C. Pickover, *Computers and the Imagination* (New York: St. Martin's Press, 1991). C. Pickover, *Mazes for the Mind: Computers and the Unexpected* (New York: St. Martin's Press, 1992). C. Pickover, *Chaos in Wonderland: Visual Adventures in a Fractal World* (New York: St. Martin's Press, 1994).
3. *The New Encyclopaedia Britannica* (Micropaedia Ready Reference), vol. 4, 15th ed. (Chicago: Encyclopaedia Britannica, 1993).

6

Knight Life

Ronald R. Brown

———

I describe how the movement of a knight on a chess board can be used to produce visually appealing fractal designs using a commercially available computer program. In the future, designs such as these will be exhibited in virtual reality art galleries as well as in traditional galleries.

———

Many have become chess masters—no one has become the master of chess.

—Tarrasch

Chess is a game for two players each of whom move pieces across a checkerboard to capture the opponent's king. It's an ancient game whose origins can be traced back to the sixth century A.D. in India. Recently I attempted to bridge my interest in chess and in fractals using well-known software called FractalVision.[1] The software has since provided me an ideal tool to explore and expand my algorithmic approach to art. It allows me (and possibly, you) to see and contemplate structures never considered before.

Simply stated, all my art is based on the way a knight moves on a chess board. Each of my art pieces originates from a solution to what is called the "knight's tour" problem: to move a knight on a chess board so that all 64 squares are jumped on only once. I have described my approach in various articles.[2] The knight's tour used in those articles is also used in this chapter. A schematic illustration is shown in figure 6.1. I have also used a similar approach to generate musical compositions.[3]

You will observe that a knight, starting at position 1, can move on all 64 squares by following the indicated numbers in consecutive order. One reason I am attracted to this particular knight's tour (as opposed to any of the other millions that exist) is that the sum of each row and each column is the same, namely, 260. My articles provide other interesting patterns I have discovered in this knight's tour.

50	11	24	63	14	37	26	35
23	62	51	12	25	34	15	38
10	49	64	21	40	13	36	27
61	22	9	52	33	28	39	16
48	7	60	1	20	41	54	29
59	4	45	8	53	32	17	42
6	47	2	57	44	19	30	55
3	58	5	46	31	56	43	18

Figure 6.1 The knight's tour used in this discussion. Of particular importance are the cells numbered 1-4.

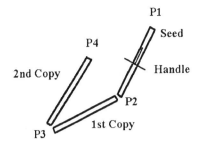

Figure 6.2 Cells 1-4 of the knight's tour connected with three rectangles: a seed and two copies of it, which make an initial template. All the fractals shown were derived from such a template.

Figure 6.2 shows the first four positions of the knight's tour, designated P1, P2, P3, and P4, connected with three rectangles. These positions are significant because all the fractals discussed in this chapter were derived from the three components connecting them. Before I describe my method for creating the fractals, I must tell something about the software.

The software I use allows the user to define a seed shape and copies of it, which together constitute a template. The software uses the template to produce an image based on the number of levels of detail that the user has set and the color associated with each level. Figures 6.3 to 6.9 were created using seven levels of detail with the color for each level being black. The background color was white. Selecting different colors for each level and selecting a black background will produce even more beautiful fractals.

In figure 6.2, the seed is the rectangle connecting P1 with P2. Two copies of the seed have been created and placed so P2 connects with P3 (1st Copy) and P3 connects with P4 (2nd Copy). The "T" superimposed on the seed represents its "handle." Each component has a handle associated with it, which is used for manipulation purposes. Once the handle is accessed, the component can be modified by selecting an option from a menu. Two options, Shrinking and Rotating (the software uses the term "Spin"), along with information on how the handles of the components are initially oriented, are more than adequate to create a large variety of fractals.

I used three variables, H, S, and R, to create these fractals. The variables refer to the Handle orientation, Size reduction, and amount of Rotation of each component from the initial template where all the components are the same size and positioned properly.

These variables can have the following values:

For H =
> : The Handle is positioned so that the leg of the T points toward the next position.

< : The Handle is positioned so that the leg of the T points away from the next position.

(In figure 6.2, the handle of the seed is oriented with a < value.)

For S =
0 : No change in size.
- : Perform the Shrink selection twice. (This will reduce the component's size to about three-fourths of its initial size.)
= : Perform the Shrink selection four times. (This will reduce the component's size to slightly more than half of its initial size.)

For R =
0 : No rotation.
+ : Perform a clockwise Rotation twice. (The angle of rotation will be 22 ½ degrees.)
# : Perform a clockwise Rotation four times. (The angle of rotation will be 45 degrees.)
- : Perform a counterclockwise Rotation twice.
= : Perform a counterclockwise Rotation four times.

I use the notation [Seed HSR, 1st Copy HSR, 2nd Copy HSR] to describe the final template used to create the fractal. The legends accompanying figures 6.3 to 6.9 define the templates used to create them. The seed's size and rotation is always 00 because the values for the copies' sizes are relative to it and the seed is never rotated from its initial position.

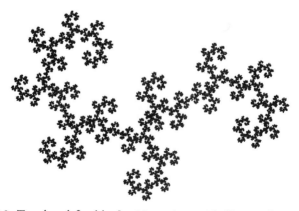

Figure 6.3 Template defined by [< 00, < - 0, < = 0]. (See text for template description.)

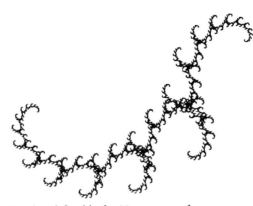

Figure 6.4 Template defined by [< 00, < - -, < = =]

Figure 6.5 Template defined by [< 00, < - 0, > = 0]

Figure 6.6 Template defined by [< 00, < - -, > = =]

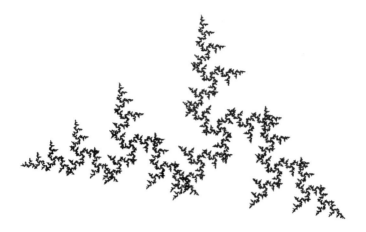

Figure 6.7 Template defined by [< 00, > - 0, > = 0]

Figure 6.8 Template defined by [< 00, > = 0, > - 0]

Figure 6.9 Template defined by [< 00, > = #, > - +]

Let's examine figure 6.7 as an example. The creation process can be described by positioning the Seed handle away from P2 (<), the 1st Copy handle toward P3 (>), and the 2nd Copy handle toward P4 (>), shrinking the 1st Copy two times and no rotation (- 0); and shrinking the 2nd Copy four times and no rotation (= 0).

In the future, I envision a virtual reality art gallery where two- and three-dimensional fractals will be generated in real time while the observer wanders through the virtual gallery and each of the three-dimensional fractals. In a real art gallery setting or a private/corporate setting, I envision flat-panel displays where a variety of fractals will be generated in real time. Perhaps fractal music will play in the background. No doubt, using more moves of the chess knight will provide for an endless collection of complex designs.

FOR THE READER

You may want to conduct the following experiments with the same or similar fractal design software.

1. All the fractals in this discussion have the seed's handle being oriented in the < direction. Create fractals with the seed's handle being oriented in the > direction.
2. Start at a different location, such as Position 21.
3. Use four rectangles instead of three. Shrink the middle two rectangles so they are just visible.

Notes

1. D. Oliver, *FractalVision: Put Fractals to Work for You* (Indianapolis: SAMS Publishing, 1992).
2. R. Brown, "The Use of the Knight's Tour to Create Abstract Art," *Leonardo* 25 (1) (1992): 55-58. R. Brown, "Using the Knight's Tour to Create Abstract Art," *The Chess Collector* 1 (4, 5) (1990): 17-21 (abridged version of the *Leonardo* article). R. Brown, "Explorations Using the Knight's Tour," *Proceedings of the Fourth Biennial Symposium on Arts and Technology* (New London, CT: Connecticut College, 1993), 106-113. R. Brown, "Explorations Using the Knight's Tour," *Proceedings of the Twelfth Annual Symposium on Small Computers in the Arts* (Philadelphia, PA: Franklin Institute, 1992), 6-11.
3. C. Pickover, *Mazes for the Mind: Computers and the Unexpected* (New York: St. Martin's Press, 1992): 204-205.

III

Fractal Models and Metaphors

7

One Metaphor Fits All: A Fractal Voyage with Conway's Audioactive Decay

Mario Hilgemeier

––––––

Fractals seem to provide a unifying metaphor for many phenomena in science, art, management, and other disciplines. In this chapter, an informal collection of essays, I will show some structural similarities among DNA, the weather, robots, and plants using "audioactive decay" as an example. The topic of aesthetics is also discussed. Finally, fractals are presented as a stage in the evolution of mathematical models of reality.

––––––

We shall not cease from exploration
And the end of all our exploring
Will be to arrive where we started
And know the place for the first time.

—T. S. Eliot[1]

THE VOYAGE

DNA, the weather, robots, and plants—what have they in common? These and other phenomena, and their structure or behavior, can be described by *iterated systems,* a mathematical metaphor. In fact, I think the general intellectual and cultural climate of our era is at least partly expressed by the mathematical metaphor of iterated systems of which fractals and deterministic chaos are special cases. There is nothing new in the fact that a mathematical metaphor expresses the zeitgeist. (In this context use *zeitgeist* to mean include not only epoch-specific art forms, fashions, lifestyles, and values, but also the scientific perception of reality.)

The zeitgeist of the baroque was expressed in a mathematical curve, the ellipse. The ellipse became popular and was used in physics, astronomy, engineering, and art. This was a time characterized by Leibniz's idea of the prestabilized harmony (and the "best of all possible worlds"). In the mind of a cultivated person of that time, the planets traveled along perfect ellipses, and people were certain about the stability of solar system. (Today we are not so certain, a fact I discuss later.) The ellipse permeated the understanding of the structure of reality as well as the architect's construction of churches and palaces.

Today the pioneering work of Mandelbrot is echoed in many fields: biology,[2] economics, art and design, management,[3] marketing, engineering,[4] government and international relations,[5] and others.[6] Although the beginning of fractals can be traced back to ideas of Leibniz, Laplace, Delboeuf and others,[7] these great thinkers did not perceive the range of fractal applications and imaging as explored by many people today. The advent of the computer opened up the area of fractals and brought exciting new visualization tools for functions and dynamic systems.[8]

We have just started our exploration by looking back to where we came from. Now we will embark on our ship, called *Audioactive Decay,* and begin our metaphorical voyage, visiting many places old and new. (See figure 7.1.) In the end, we will try to look over the event horizon and speculate what may lie beyond.

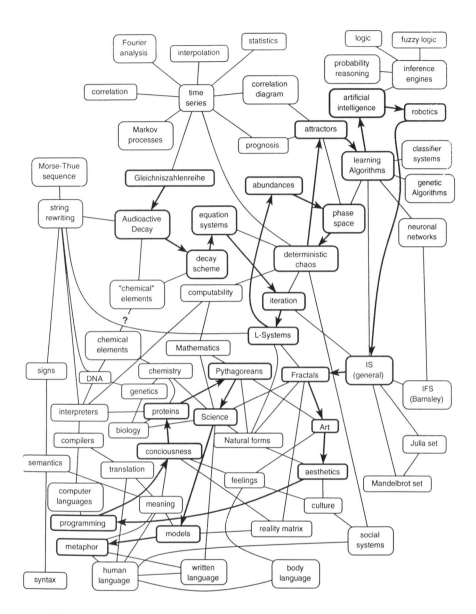

Figure 7.1 Partial map of fractal land and adjacent countries. Nodes with fat border are visited in this article, starting at "Gleichniszahlenreihe" (left, center). Not all interconnections are shown.

Please forgive me if the trip is too fast or if some places are sketched insufficiently or even inaccurately. But I think it is better to show roughly where something might be than be silent and give no hint at all.

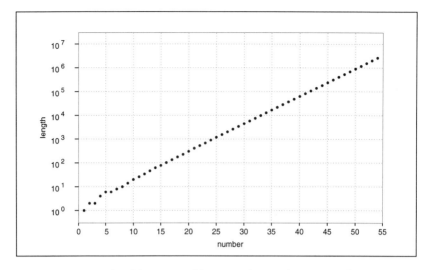

Figure 7.2 Lengths of the iterates of the GZR. The straight line in this logarithmic plot indicates the exponential growth of the strings.

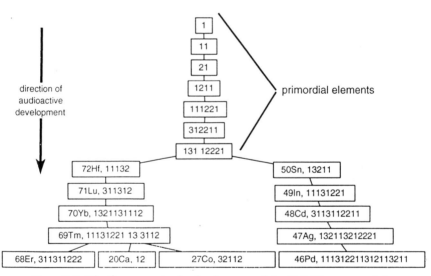

Figure 7.3 The first stages in the development of the "1." The names of the elements are given, together with the strings of numbers.

AUDIOACTIVE

Consider these strings: 1, 11, 21, 1211, 111221, 312211, . . . What is the next item? If you think it's 13112221, you're right. Why? Speak aloud, "one" 1, "this is one one" 11, "now we have two ones" 21, "that's one two and one one" 1211—you get the idea. Because of this Conway has called the iteration rule *audioactive*. I've called it *Gleichniszahlenreihe* (GZR) because of its analogy with *meaning* (what does 11 mean? 1), *description* (what is 11? 21), and *growth* because the string length (i.e. the number of characters) grows regularly (exponentially) with a factor of about 1.303577269 (in the limit), according to Conway.[9] (See figure 7.2.)

DECAY

Conway was the first to discover that these strings, growing exponentially in length, do split "naturally" at certain points. After such a split, each of the fractions develops without influencing the other any more; then it splits again. Such a fraction is called an element. An example of this split is shown in figure 7.3. The last string of the primordial elements, 13112221, splits into 131 and 12221; this is indicated by a space. In the following audioactive decay the pair 11 (where the split occurs) is changed to 21. But the resulting two parts 11132 and 13211 can be considered as independent in the next stages. For instance, the string called 71Lu in figure 7.3 develops from 72Hf; we do not need to know that the whole string is longer.

By analogy to the decay of radioactive elements, Conway called the splitting process decay. If you start with the number 1 (or any other character) and let it grow for some hundred iterations, you get only 92 (mostly instable) elements of the audioactive decay. Because there are also 92 known (stable) chemical elements in nature, and some of them change by radioactive decay, Conway named them uranium (U), protactinium (Pa) down to helium (He) and hydrogen (H). Each of these elements (except hydrogen) transmutes into another or splits into two or more elements. In this "nuclear-chemical" view, the Gleichniszahlenreihe starts with a primordial element 1, which rapidly develops into the elements Hafnium (Hf,11132) and Tin (Sn, 13211). A "primordial element" (see figure 7.3) is an element that never recurs in audioactive decay. Figure 7.4 shows the complete development scheme for all 92 elements.

Table 7.1 (a table of elements) contains a column called "element abundance." The abundance of an element is defined as the number of atoms of each element per million atoms. The appropriate measuring unit is ppm, parts per million. How can we compute the abundance of an element? Why do elements have particular abundances anyhow? We consider these questions later.

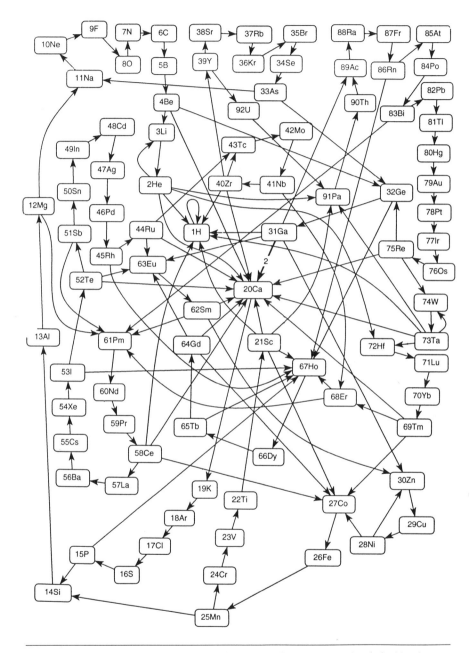

Figure 7.4 "Development scheme" of Conway's elements. For a detailed table of all elements and their abundances, see text. The arrow marked "2" from 31Ga to 20Ca means: 31Ga "decays" into two atoms 20Ca (among others).

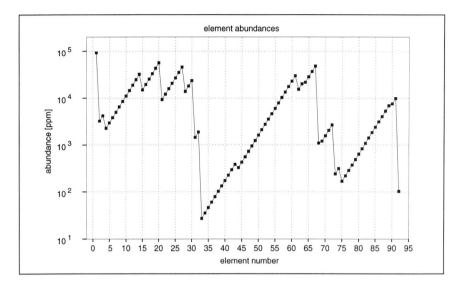

Figure 7.5 Abundances of Conway's elements, ordered by element number. Note the logarithmic scale. The sawtooth pattern is explained in the text.

The sawtoothlike form of the element abundances when drawn as a function of element number (figure 7.5) is caused by the decay process. There are certain development chains in which one element decays into exactly one other element. One instance of such a loop is the chain starting at 61Pm (middle left in figure 7.4). Because generally the development is from higher to lower element numbers, and because the total string length grows with each decay step, the higher element numbers in a decay chain are "better fed." Because of the many decay chains, the sawtooths in figure 7.5 occur. You can see this also in table 7.1: Abundances of elements with only one successor are exactly 1.303577269 (the growth factor) times greater than their successor (for example, 60Nd and 59Pr). When the abundances of the elements are ordered by rank (starting with the highest abundance), nothing special can be seen. (See figure 7.6.)

A context-sensitive string-rewriting system can be defined in this way: Every character of the previous string is transformed according to its context (its neighboring ciphers or characters). This is exemplified in the GZR by the fact that a 1 doesn't always develop in the same way, depending on the ciphers that are before and after it in the string. In contrast, in a context-free grammar each character develops in a certain way, independent of its neighbors. (See "Element Development as Equation System" later in this chapter) By defining his elements, Conway achieves something remarkable: A context-sensitive grammar can be seen as a context-free grammar by going to the higher level of the "elements." *Both grammar types are present in the same iterated system.* At the start, when primordial elements develop, everything is context-sensitive. But as soon as the first elements appear, the context-free view becomes possible.

Table 7.1
Abundances, Lengths, and Strings of Conway's Elements

Element	abundance	length	string
92 U	102.56285249	1	3
91 Pa	9883.5986391	2	13
90 Th	7581.9047124	4	1113
89 Ac	6926.9352045	4	3113
88 Ra	5313.7894999	6	132113
87 Fr	4076.3134078	10	1113122113
86 Rn	3127.0209328	12	311311222113
85 At	2398.7998311	7	1322113
84 Po	1840.1669683	10	1113222113
83 Bi	1411.6286100	10	3113322113
82 Pb	1082.8883286	9	123222113
81 Tl	830.70513293	12	111213322113
80 Hg	637.25039755	14	31121123222113
79 Au	488.84742983	18	132112211213322113
78 Pt	375.00456739	24	111312212221121123222113
77 Ir	287.67344775	28	1131122113222112211213322113
76 Os	220.68001229	34	1321132122211322212221121123222113
75 Re	169.28801808	42	111312211312113221133211322112211213322113
74 W	315.56655252	27	312211322212221121123222113
73 Ta	242.07736666	32	13112221133211322112211213322113
72 Hf	2669.0970363	5	11132
71 Lu	2047.5173200	6	311312
70 Yb	1570.6911808	10	1321131112
69 Tm	1204.9083841	14	11131221133112
68 Er	1098.5955997	9	311311222
67 Ho	47987.529438	7	1321132
66 Dy	36812.186418	12	111312211312
65 Tb	28239.358949	16	3113112221131112
64 Gd	21662.972821	11	13221133112
63 Eu	20085.668709	7	1113222
62 Sm	15408.115182	6	311332
61 Pm	29820.456167	3	132
60 Nd	22875.863883	6	111312
59 Pr	17548.529287	8	31131112
58 Ce	13461.825166	10	1321133112
57 La	10326.833312	5	11131
56 Ba	7921.9188284	6	311311
55 Cs	6077.0611889	8	13211321
54 Xe	4661.8342719	14	11131221131211
53 I	3576.1856107	18	311311222113111221
52 Te	2743.3629717	13	1322113312211
51 Sb	2104.4881933	7	3112221
50 Sn	1614.3946687	5	13211
49 In	1238.4341972	8	11131221
48 Cd	950.02745645	10	3113112211
47 Ag	728.78492056	12	132113212221

Element	abundance	length	string
46 Pd	559.06537945	18	111312211312113211
45 Rh	428.87015042	24	311311222113111221131221
44 Ru	328.99480576	21	132211331222113112211
43 Tc	386.07704943	15	311322113212221
42 Mo	296.16736852	20	13211322211312113211
41 Nb	227.19586752	29	1113122113322113111221131221
40 Zr	174.28645997	23	12322211331222113112211
39 Y	133.69860315	7	1112133
38 Sr	102.56285249	7	3112112
37 Rb	78.678000089	10	1321122112
36 Kr	60.355455682	14	11131221222112
35 Br	46.299868152	16	3113112211322112
34 Se	35.517547944	20	13211321222113222112
33 As	27.246216076	26	11131221131211322113322112
32 Ge	1887.4372276	23	31131122211311122113222
31 Ga	1447.8905642	17	13221133122211332
30 Zn	23571.391336	3	312
29 Cu	18082.082203	6	131112
28 Ni	13871.124200	8	11133112
27 Co	45645.877256	5	32112
26 Fe	35015.858546	8	13122112
25 Mn	26861.360180	12	111311222112
24 Cr	20605.882611	5	31132
23 V	15807.181592	8	13211312
22 Ti	12126.002783	14	11131221131112
21 Sc	9302.0974443	16	3113112221133112
20 Ca	56072.543129	2	12
19 K	43014.360913	4	1112
18 Ar	32997.170122	4	3112
17 Cl	25312.784217	6	132112
16 S	19417.939250	10	1113122112
15 P	14895.886658	12	311311222112
14 Si	32032.812960	7	1322112
13 Al	24573.006695	10	1113222112
12 Mg	18850.441227	10	3113322112
11 Na	14481.448773	9	123222112
10 Ne	11109.006821	12	111213322112
9 F	8521.9396539	14	31121123222112
8 O	6537.3490750	18	132112211213322112
7 N	5014.9302464	24	111312212221121123222112
6 C	3847.0525419	28	3113112211322112211213322112
5 B	2951.1503716	34	1321132122211322212221121123222112
4 Be	2263.8860324	42	111312211312113221133211322112211213322112
3 Li	4220.0665982	27	312211322212221121123222112
2 He	3237.2968587	32	13112221133211322112211213322112
1 H	91790.383216	2	22

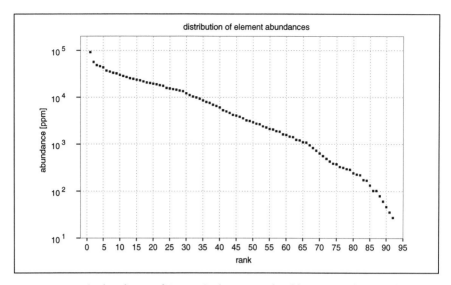

Figure 7.6 Abundances of Conway's elements, ordered by magnitude. Note the logarithmic scale.

Table 7.2
Numerical Interpretation of the GZR

string	number	factorized
1	1	1
11	4	2^2
21	7	7
1211	49	7^2
111221	376	2^3 * 47
312211	886	2 * 443
13112221	4777	17 * 281

STOP AND THINK

Is the ordering of the elements in table 7.1 really forced? That is, Is it the only one possible? When you look at figure 7.4, you will find elements that have two or more outgoing arrows (2He, 4Be, 31Ga, 32Ge). These elements develop into two or more elements. At these junctions Conway's ordering could be changed. This is the "labyrinthine" aspect of the GZR. I would be interested to hear from readers who find an alternative ordering.

The generated strings 1, 11, 21, 1211, and so on can be interpreted as a number system with the base 3 (but without the zero). When position is counted from right to left, the first position has the multiplier $1(= 3^0)$, the second $3(= 3^1)$, the third $9(= 3^2)$, the fourth $27(= 3^3)$, and so on. The number 10 in this system is written as $31(= 3*3 + 1*1)$. The decimal number 23 can be expressed as $212 (= 2*9 + 1*3 + 2*1)$. A numerical interpretation of the GZR can be found in table 7.2. Can you find any interesting number interpretations of the GZR-strings? Are there significant number-theoretic aspects?

ELEMENT DEVELOPMENT AS EQUATION SYSTEM—AN EXEMPLARY L-SYSTEM

Now let's determine where the abundances come from. First, we have to see that Conway's audioactive decay can be understood as a system of equations. The solution of this system of equations (by iteration) gives the abundances.

Let me clarify this by a simpler string-rewriting system (such as GZR). The GZR operates on ciphers, while this example uses characters. There are five somewhat unusual characters to be iterated: "F", "[", "]", "+" and "-". The iteration rule is: "F" is expanded to "F[+F][-F]". All other characters remain as they are.

Consider the program code of the following system of equations. It describes the growth of the number (abundance) for each of the five characters from one development stage to the next. One "F" generates three new "F"s. The number of left brackets "[" is the sum of the number of existing (old) left brackets plus those (new) generated by the expansion of the "F" characters.

```
Fnew = 3 * Fold
[new = [old + 2*Fold
]new = ]old + 2*Fold
+new =+old + Fold
-new = -old + Fold
```

This system is visualized in figure 7.7 (in the style of figure 7.4). If you iterate this system, what will the abundances of the five characters be? If you don't see it at once, you could do the computation on a computer and see that the abundances

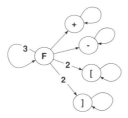

Figure 7.7 The tree-generating L-system described in the text. The equation system is visualized in the style of figure 7.4.

Program code 7.1   A tree- or bush-generating L-system.

```
set
L   1/10 of your graphical display screen
r   1.3
alpha   30 degrees
beta    20 degrees
declare a string structure of adequate length (remember the rapid growth)
```

There are five signs to be iterated: "F", "[", "]", "+", and "-". Start with the string "F" (the so-called "axiom" of the L-system). The iteration rule is: "F" is expanded to "F[+F][-F]". All other signs remain as they are.

You can interpret this string graphically in the following way (imagine a drawing pen moving):

```
F   draw a line (forward) of length L
[   store this position and divide L by r (r1)
+   turn right by an angle alpha
-   turn left by an angle beta
]   return to the position previously stored and multiply L by r.
```

The factor r causes shortening of subsequent branches.

---

of the characters eventually settle to $1/4$ for F and each of the brackets and $1/8$ for the + and characters. Of course, the growth factor is 3 because only F is responsible for the development of the system. This string-rewriting system is one of the simpler L-systems.[10] When you experiment with program code 7.1, you will generate fractal images of trees and bushes. Can you find similar interpretations for the strings of the GZR?

Equation systems (almost always) have characteristic limit vectors and growth factors. This is exemplified by the above system and also by the GZR. The abundances of the GZR can be computed by program code 7.2.

Program code 7.2

To compute the abundances of the elements (in table 7.1) the following code has been used (in C-language). The code lines like "new = old . . ." can be taken directly from figure 7.4. There is a vector abd that stores the old and new abundances. All initial abundances are set to 1.0. On my machine, I needed 330 iterations before the growth factor gf converged. Here gf is computed from the rarest element (33As) but you could use any element for the computation. To avoid overflow, the abundances are normalized to 1.0. When you print the final abundances multiply by 1,000,000 to get ppm (parts per million).

```
for (i=0; i < iterations; i++)
{
    abd[1].new = abd[2].old + abd[58].old
        + abd[21].old + abd[73].old + abd[31].old + abd[40].old + abd[1].old;
    abd[2].new = abd[3].old;
    abd[3].new = abd[2].old + abd[4].old;
    abd[4].new = abd[5].old;
    abd[5].new = abd[6].old;
    abd[6].new = abd[7].old;
```

```
abd[7].new = abd[8].old;
abd[8].new = abd[9].old;
abd[9].new = abd[10].old;
abd[10].new = abd[11].old;
abd[11].new = abd[12].old + abd[33].old;
abd[12].new = abd[13].old;
abd[13].new = abd[14].old;
abd[14].new = abd[15].old + abd[25].old;
abd[15].new = abd[16].old;
abd[16].new = abd[17].old;
abd[17].new = abd[18].old;
abd[18].new = abd[19].old;
abd[19].new = abd[20].old;
abd[20].new = abd[21].old + abd[69].old + abd[73].old + abd[75].old
   + 2*abd[31].old + abd[40].old + abd[4].old + abd[2].old + abd[44].old
   + abd[52].old + abd[62].old + abd[64].old + abd[58].old;
abd[21].new = abd[22].old;
abd[22].new = abd[23].old;
abd[23].new = abd[24].old;
abd[24].new = abd[25].old;
abd[25].new = abd[26].old;
abd[26].new = abd[27].old;
abd[27].new = abd[28].old + abd[69].old + abd[21].old + abd[64].old +
   abd[58].old;
abd[28].new = abd[29].old;
abd[29].new = abd[30].old;
abd[30].new = abd[31].old + abd[62].old + abd[28].old;
abd[31].new = abd[32].old;
abd[32].new = abd[33].old + abd[4].old + abd[75].old;
abd[33].new = abd[34].old;
abd[34].new = abd[35].old;
abd[35].new = abd[36].old;
abd[36].new = abd[37].old;
abd[37].new = abd[38].old;
abd[38].new = abd[39].old;
abd[39].new = abd[40].old;
abd[40].new = abd[41].old;
abd[41].new = abd[42].old;
abd[42].new = abd[43].old;
abd[43].new = abd[44].old + abd[40].old;
abd[44].new = abd[45].old;
abd[45].new = abd[46].old;
abd[46].new = abd[47].old;
abd[47].new = abd[48].old;
abd[48].new = abd[49].old;
abd[49].new = abd[50].old;
abd[50].new = abd[51].old;
abd[51].new = abd[52].old;
abd[52].new = abd[53].old;
abd[53].new = abd[54].old;
abd[54].new = abd[55].old;
abd[55].new = abd[56].old;
abd[56].new = abd[57].old;
abd[57].new = abd[58].old;
abd[58].new = abd[59].old;
abd[59].new = abd[60].old;
abd[60].new = abd[61].old;
abd[61].new = abd[62].old + abd[12].old + abd[51].old + abd[83].old +
   abd[68].old;
abd[62].new = abd[63].old;
abd[63].new = abd[64].old + abd[52].old + abd[44].old + abd[31].old;
abd[64].new = abd[65].old;
abd[65].new = abd[66].old;
abd[66].new = abd[67].old;
```

```
abd[67].new = abd[68].old + abd[15].old + abd[65].old + abd[45].old +
   abd[53].old + abd[21].old + abd[86].old + abd[32].old;
abd[68].new = abd[69].old + abd[41].old;
abd[69].new = abd[70].old;
abd[70].new = abd[71].old;
abd[71].new = abd[72].old;
abd[72].new = abd[73].old + abd[2].old;
abd[73].new = abd[74].old;
abd[74].new = abd[75].old + abd[73].old;
abd[75].new = abd[76].old;
abd[76].new = abd[77].old;
abd[77].new = abd[78].old;
abd[78].new = abd[79].old;
abd[79].new = abd[80].old;
abd[80].new = abd[81].old;
abd[81].new = abd[82].old;
abd[82].new = abd[83].old;
abd[83].new = abd[84].old;
abd[84].new = abd[85].old;
abd[85].new = abd[86].old;
abd[86].new = abd[87].old;
abd[87].new = abd[88].old;
abd[88].new = abd[89].old;
abd[89].new = abd[90].old + abd[31].old;
abd[90].new = abd[91].old;
abd[91].new = abd[92].old + abd[2].old + abd[21].old + abd[73].old;
abd[92].new = abd[39].old;

sum = 0.0; /* sum abundances for normalization to 1 */
for (j=1; j<=92; j++)
    sum += abd[j].new;

if (((i+1)%10)==0) /* show every 10th growth factor */
{
    gf = abd[33].new / abd[33].old; /* growth factor */
    fprintf(stderr,"%d gf = %.16lf\n", i+1, gf);
}

for (j=1; j<=92; j++)
    abd[j].old = abd[j].new / sum; /* normalization to 1 */
}
```

Now you can see the relation of the exemplary L-system to the audioactive development. For instance, interpret pairs of ciphers of the GZR as drawing instructions. There are eight different pairs: 11, 12, 13, 21, 22, 23, 31, and 32. (To see the proof of why 33 never occurs, see Conway.[11]) Now interpret these eight pairs as letters of a graphical alphabet analogous to program code 7.1, possibly with some extensions or redundancies. I've tried some simple interpretations, but they didn't show the (fractal) self-similarity that should be there because the iteration rule of the GZR is pure self-description. Can you find a beautiful interpretation of the GZR?

L-systems are used in biology, to model cell division patterns and morphogenesis research; robotics, for structural pattern recognition and speech recognition; and computer science, in the semantics of programming languages.[12]

Beautiful pictures of plants can be found in the book by Lindenmayer and Prusinkiewicz.[13] The complex L-systems described there model the development patterns of plants, their growth and flowering patterns.

FROM CHAOS TO ROBOTS

We see that a string can be the description of the state of a system. If the system is a plant, then L-systems may be the appropriate string-rewriting systems to model development patterns. Other dynamical systems, in celestial mechanics, for instance, are described by a set of numbers (three-dimensional coordinates and impulses) and the equations governing the relations of the bodies (gravitation). The next set of numbers is derived from the previous by iteration because gravitational systems of more than two celestial bodies are not exactly solvable by symbolic differential equations.

If a state is characterized by n numbers, it can be represented as a point in n-dimensional space, the so-called phase space. As the system evolves, the point hops through phase space, not always very smoothly. (Large jumps happen.) Although the description of the dynamical system by differential equations yields continuous movement, for reasons of computability the representation is transformed into a discrete dynamical system where iteration is used. But the introduction of iteration leads to discrete "jumps" of the system state (quantization).

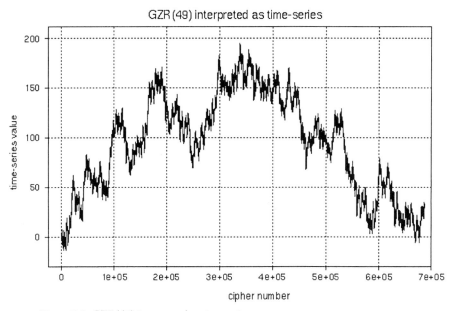

Figure 7.8 GZR(49) interpreted as time series.

Watching only one coordinate of the phase space is like measuring any parameter (for example, the rainfall per day) of a dynamical system (say, the weather). What you get is a time series, sometimes a fractal curve. (This depends on the type of dynamical system.) If you enlarge portions of a fractal curve, they look like the whole curve; you find the same patterns after each renewed magnification.

Any iteration of the GZR can be seen as a time series. Figure 7.8 shows an example for the forty-ninth iteration. How I arrived at this curve is shown in program code 7.3. This curve approaches self-similarity because of the iteration rule of the GZR. But the self-similarity is impure, because the line segments are finite.

Program code 7.3

Conversion of an iterate of the GZR into a time-series-like fractal graph. The abundances of ciphers in large iterates of the GZR are known to be about 49.5101% "1", 32.0352% "2" and 18.4547% "3". In the following piece of code, ciphers are interpreted as "1" = up, "2" = down, "3" horizontal (no change). The amounts of increase or decrease are adjusted to give an approximately horizontal line—at least to get an end value of the time series near the start value (0.0).
   Read the next cipher from an iterate of the GZR
if the cipher is

```
   {
   1:value = value    +    0.320352 / (0.320352 + 0.495101); /* up */
   2: value = value   -    0.495101 / (0.320352 + 0.495101); /* down */
   3: /* 0.184547 do nothing */
}
```

Note: The cipher abundances of the GZR yield (0.495101 * 0.320352) / (0.184547 * 0.184547) = 4.65701, only 0.26% smaller than Feigenbaum's constant (about 4.66920) [47]. Can you explain this, or is it just chance?

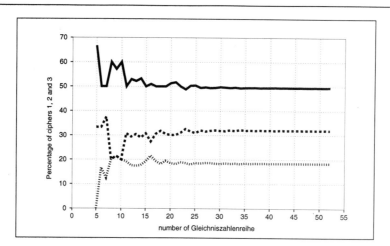

Figure 7.9 Cipher abundances in iterates of the GZR. The approximate limit abundances are "1": 49.51 percent, "2": 32.04 percent and "3": 18.45 percent.

When you examine the phase space for a given dynamical system, you find that certain regions will never be visited by the moving system state. In other regions, however, the point density is very high—these are preferred states, visited often by the system. Phase spaces with two or three dimensions can be represented graphically very easily. Plot a point for each iteration of a nonlinear dynamical system. The geometrical shapes that evolve may be peculiarly beautiful or fractal.[14] If the shapes are fractal, they are called *strange attractors*. Many readers may be familiar with the beautiful fractal images from the theory of dynamic systems.[15]

The length of time for which predictions are valid for nonlinear dynamical systems (weather, the solar system) depends on our (limited) knowledge of the initial conditions. Because system states that are very close in phase space can get farther and farther apart in the future, our prediction ability is as limited as our knowledge of the initial conditions. Systems of orbiting bodies in celestial mechanics that have rational proportions of orbital periods become unstable sooner or later. Therefore, it has been said that the stability of the solar system crucially depends on the irrationality of the proportions of the orbital periods. Well, for the foreseeable future, no disaster looms, as far as computer models go today. Computers often model states of the real world. By computation, limited prediction and explanation of the phenomena of the world become possible.

ROBOTICS

A robot must have an internal model of the world around it—at least of those qualities of the environment that it needs for its function. For example, a washing machine is concerned with laundry type, water, temperature, detergent, time, and the like. The model of the robot's environment is represented in the tiny computer by a set of numbers and rules; hence this is another kind of phase space. The robot's state has a position in this many-dimensional phase space. Its internal representation of its "position in the world" moves in phase space. If it is to be a successful robot, it will learn what the best actions are in certain regions of the phase space (for example, "if the is water too cold for linen, start heater"). Of course, if the robot is to learn by trial-and-error, it must somehow evaluate which actions were successful.

We've entered the world of artificial intelligence (AI). AI is concerned with reasoning, learning, and making analogies in new situations. The kinds of logic used are not always of the simple yes-and-no type, and learning paradigms abound. Sadly, I can't go further into this now. If you want to read more, see Hofstadter and Mitchell.[16] The point I wanted to make here is that fractals may have something to do with (machine) intelligence. Iterated systems (IS), such as computers, can show chaotic (fractal) behavior.

Many questions arise. Here are a few: Does the robot's state in phase space move toward a strange attractor? What are the consequences? For what purpose can we use the dynamics of learning systems if they are chaotic? Is there a danger in the low predictability of chaotic systems? Engineers have just begun to exploit nonlinear chaotic systems. Some chaotic systems can be controlled by a chaotic signal and forced to express the desired behavior.[17] Can chaotic systems be used to generate "creativity" of robot artists? Or even for the automatic generation of unique stories, films, architectural designs, virtual reality scenarios or other cultural artifacts? In Birkhoff's aesthetic theory, art is defined as something not boring, but at the same time not too surprising.[18] Fractals surely fit this simple criterion. So they may be useful for everyday design (and—in the hand of humans—for works of art).

THE TWO FACES OF INTELLIGENCE

"Using *metaphor,* we say that computers have senses and a memory" said William Jovanovic.[19] In the dictionary,[20] intelligence is defined in two principal ways: (1) as consciousness, an intelligent entity (as in humans), intelligent minds or mind (as in cosmic intelligence), the act of understanding, comprehension; and (2) as mechanical thinking, which includes the ability to learn or to deal with new situations, the ability to apply knowledge to manipulate one's environment or to think abstractly (*as measured by tests*).

I believe that intelligence in the sense of the first meaning creates mathematical ideas. It is a mind that looks at itself and says, "Well, I am one 1; there is one of me 11, and this description has two parts 21. . . ." However, prevalent theories insist that even human intelligence developed out of simple information-carrying molecules (the DNA) via evolution of the genes in the contexts of cell, organism, and ecosystem. We do not know enough yet to conclude with absolute certainty that the intelligence of humans is only that of a biological (molecular) machine. Molecules are subject to quantum processes. And we do not know whether consciousness is controlled or influenced by quantum processes.[21]

DNA AND PROTEIN FOLDING

When DNA is decoded, the chain of its four bases is interpreted to produce proteins that can influence the environment. These proteins may be considered as a "meaning" of the DNA. For instance, enzymes, hair, and nails are made of proteins. Seen chemically, proteins are strings of amino acids that are folded in three-dimensional space to give rise to the most complex forms. The folding is governed by the attraction between certain parts of the amino acids and the environment of the protein molecule.

Ribosomes interpret DNA. Ribosomes are cell particles that work with partial copies of the DNA to produce proteins. Each triplet of DNA bases can be the code for an amino acid or a control sign, such as "stop here." Again, an iterated system (the cell with its DNA in a biological context) produces beautiful fractals.[22]

STOP AND THINK

Molecular scientists often use "folding" programs to join particular regions of a linear sequence into a two-dimensional pattern. Can you fold any of the longer strings of the Gleichniszahlenreihe, let's say the twenty-third iteration or so? Are there beautiful two-dimensional (or even three-dimensional) foldings of the GZR? You can use the following laws (or a variation of them) as attracting force:

- Equal ciphers (or elements) attract with a strength of 1.
- A pair of mutually attractive ciphers attracts more strongly than single ciphers, a trio more strongly than a pair, and so on.
- The "bonds" between the ciphers cannot be overstretched (or else they would break up the "chemical bond").
- Different ciphers (or elements) repel each other.

I've tried to fold the element 4Be, which is one of the two longest elements. The result is shown in figure 7.10. Thin dashed lines represent weak bonds between the ciphers. The fat lines denote the order of the string (strong bonds). Note the shorter strong bonds between equal ciphers (because of their attraction). The folding is essentially governed by the weak bonds.

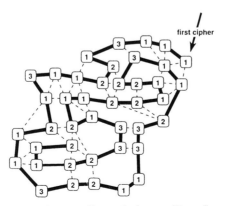

Figure 7.10 Attempt to interpret Conway's element 4Be as "protein" and to fold it.

Can you find similar patterns for other cipher (or element) strings? I've shown a folding in the plane (two dimensions). What happens if three or more dimensions are allowed?

THE (NEARLY) INFINITE
PARADISE OF FRACTAL IMAGES

We wish to search for beautiful images by computer. Probably you are using a computer monitor to watch the program-generated images. Let's suppose the monitor has a resolution of 1,000 pixels (picture elements) horizontally and vertically, with 256 possible colors for each pixel. How big is the picture space to be traveled by the beauty-seeking program? That's easy to compute from combinatorics: 256 to the 1,000,000th power, which is roughly equal to $10^{2,408,234}$. This is more than the square of a gigaplex, which Rudy Rucker defines as the sum of all possible thoughts of a man.[23] Even if we settle for a low-resolution black-and-white monitor with 200*200 pixels, we still have to cope with a picture space of $2^{40,000}$ or about $10^{12,041}$. If 10 billion people watched one picture each second for 80 years (without sleeping), this enormous effort could scan only $2.5*10^{19}$ pictures.

Let's suppose, a "picture space of all fractal images" can be defined in some way. Because the number of fractal images may be large, the time for "just looking at all the beautiful fractal images" is a lot longer than many lifetimes. This is true even if you had a filter program to select for you "only the beautiful ones, and only one of each group of similar-looking ones." If you get bored, you could modify your filter program to exclude "the boring ones" (and wait a little longer) or even trim your search algorithm to seek out realms in image space where "things are most unsimilar to things already discovered." Provided you have a "dissimilarity measure," you may find something new. The ultimate limits, of course, are (1) your picture-generating program, (2) time and (3) the number of pictures that Homo sapiens can distinguish. (As far as I know, this is an unknown number but probably very large, depending on the number of rods and cones on the retina and the number of neurons in the visual cortex.)

TRAVELING THE OCEAN

What holds true for fractal images is also true for sounds, melodies, and musical compositions, as other chapters in this book illustrate. In your quest for pattern and beauty, how do you travel this endless fractal ocean? You realize that you cannot see everything in your lifetime—you cannot "have it all." You must omit an almost infinite number of things. This reminds me of Stanley Clarke's song comparing life to a game with many different rules to choose from. [24]

ART UNLIMITED
AND THE RETURN OF THE PYTHAGOREANS

We can see now that the possibilities for art are endless, even in its classical dimensions (design, painting, sculpture, architecture, poetry, literature, history, music, dance, drama, and science). Any successful artist today can hardly avoid the digitizing of his or her art in video images, CD recordings, or files of design programs. In modern media, all is converted to signs or symbols (numbers in most cases). Pythagoreans saw numbers and their proportions as the nature and essence of all things.[25] Some of their discoveries were the harmonical laws in music and geometry (the Pythagorean theorem). Today, numbers form the basis for the recording, storing, and retrieving of information (and a lot of beautiful artistic compositions), regardless of the notion that art has a nonrational component that may not be computable at all. All of modern physics, genetics, "artificial life,"[26] and automata psychology (another term for artificial intelligence) can be seen as a quest for finding the ordering principles in nature by using mathematics as a tool to operate on models of nature or even generate models.

MODELS AND METAPHORS

What is the difference between a model and a metaphor? A metaphor carries meaning beyond the obvious; therefore a metaphor is more than a theory. A metaphor can be a kind of guiding principle for thought. A metaphor could be explained as a kind of high-level analogy. For instance, the Gleichniszahlenreihe can be seen as a curious mathematical object—as just another string-rewriting system. However, we have already seen some of its metaphorical qualities.

Here are some exemplary metaphorical meanings of the GZR:

- A metaphor for an iterated system producing a chaos of fractals.
- A mind looking at itself and bringing a "world with elements" into being.
- A meta-metaphor that shows what a metaphor is, a metaphor for a metaphor, for "jumping out of the system" and looking at it from the outside.

The original title of this article "One Metaphor Fits All" was in quotes, which the editor removed. The title was intentionally set in quotes to be distinguished as an example (metaphor) for the view that there is something like a "world formula." Is it really possible to capture the phenomena of the physical and mental world in a finite set of mathematical models? The evidence of the last centuries looks promising. Scientific fields such as celestial mechanics, quantum theory, and genetics are good models in their domains. And I feel iterated systems (fractals) to be a most promising candidate for what the Pythagoreans of today (physicists, geneticists, and artificial intelligence researchers) are after, with some caveats described in the next sections.

STOP AND THINK

Look again at figure 7.1. Is it possible to iterate the matrix in ways similar to figures 7.4 and 7.7? Which conceptual slippage[27] is needed for such an operation? Can semantic nets[28] be used for prediction or scenarios?

A LOOK BEYOND THE EVENT HORIZON

Today the ellipse is the tool of the engineer and scientist, and fractal mathematics are becoming more widely used in solid state physics (quasicrystals, spin glasses) and in control theory (chaotic). French researchers have produced aerogels, which they claim the first totally fractal material of the world (down to the molecular level). Aerogels are known for their diverse fields of application from detectors to solar architecture.[29]

Let me speculate about developments in the next century. The fractal metaphor has now grown into a mental tool like the ellipse but is still in its youth (or infancy). This mental tool is already used not only by engineers[30] but by artists as well.[31] Consumer products with more "artificial intelligence" than exist today will surely exhibit fractals in a variety of ways in their programs and designs. Maybe there will not even be such a thing as a consumer product, as people order directly from a factory that can produce millions of variations.

The zeitgeist of the next century might be strongly influenced by the fractal metaphor that is already so applicable and useful. Many natural-scientific and artistic disciplines will rediscover what they have in common. We may be tempted to believe that a unifying theory, based on iterated systems, will reunite different sciences. Science will not only mean natural science, but also the science of art, social science, mental science, and others. Science may even mean something like religion in the end, with some caveats to be described.

As the classical architecture of the Romans and Greeks liked circles and golden rectangles, as the baroque liked the ellipse, so the present likes the fractal metaphor. What will be the next metaphor? Nonlinear dynamical systems were the "blind spot" (or taboo theme) that was touched on only slightly by the grand mathematicians of previous centuries. The metaphors of fractals and chaos have been increasingly applied to diverse fields of human knowledge. However, I believe that in the future we will need to refine our metaphors and develop new ones that permit a greater range of predictions and produce even more realistic models of nature. Hopefully, we will find new metaphors that are more easily applied than the current number-crunching computer methods and that further enhance our understanding of the universe.[32]

Remember, we're past the event horizon—a place in space time about which nothing can be known now; this is just my hunch. Maybe we will see the end of the kind of natural science that is looking for natural laws and the finite "model of the world." Fractals promise that such a model may exist, if we lower our expectations for predictability. But the metaphor could carry us beyond itself to the realization that "one metaphor fits all" is just a metaphor for the human quest for understanding, trying to fit infinity into a nutshell.

The voyage does not really end here. Rather, it has just begun.

> To see a World in a Grain of Sand,
> And a Heaven in a Wild Flower,
> Hold Infinity in the palm of your hand,
> And Eternity in an hour.
>
> —William Blake[33]

Further Reading

J. L. Casti, *Searching for Certainty—What Science Can Know About the Future* (London: Abacus, 1993).

B. Chatwin, *Traumpfade* (Frankfurt am Main: Fischer, 1992). German translation of *The Songlines* (London: Cape, 1987).

F. Cramer, *Der Zeitbaum—Grundlegung einer allgemeinen Zeittheorie* (Frankfurt am Main: Insel, 1993).

M. Hilgemeier, "Die Gleichniszahlen-Reihe," *Bild der Wissenschaft* (12) (December 1986): 194.

O. E. Rössler, cited by P. Brügge, "Der Kult um das Chaos," *Der Spiegel* (41) (1993): 252.

Notes

The following persons helped by discussions or comments during the writing of the chapter: Professor Dr. Michael Hortmann, Professor Dr.-Ing. Hans-Joerg Kreowski, Dr. Karin Meißenburg, Professor Dr. Manfred Schroeder, Chris Stevens, and my dear wife, Eva Schibel-Hilgemeier. Last but not least, I want to thank *megatel* Informations-und Kommunikationssysteme GmbH, Wiener Straße 3, D-28359 Bremen, Germany, for letting me use their computers on the weekends.

1. K. Smith, *Digital Quotations,* Release (v28) (NeXT Computer, Inc., 1988, 1990).

2. H. Ehrig, H.-J. Kreowski, and G. Rozenberg, eds., *Graph Grammars and Their Application to Computer Science—Proceedings of the 4th International Workshop, Bremen, Germany, March 1990. Lecture Notes in Computer Science* 532 (Berlin: Springer, 1991). A. Lindenmayer and P. Prusinkiewicz, *The Algorithmic Beauty of Plants* (New York: Springer, 1990).

3. G. Binnig, *Aus dem Nichts—über die Kreativität von Natur und Mensch* (München: R. Piper GmbH & Co KG, 1989).

4. W. L. Ditto and L. M. Pecora, "Mastering Chaos," *Scientific American* (8) (August 1993): 62. J. Fricke, "Aerogele—die leichte Alternative," *Bild der Wissenschaft* 9 (September 1993): 78.

5. G. Mayer-Kress, "Chaos and Crises in International Systems." Paper presented at SHAPE technology symposium on crisis management, Mons, Belgium (1993). Available via anonymous ftp from ftp.ncsa.uiuc.edu in directory GlobalModels/Papers.

6. C. Pickover, *Mit den Augen des Computers* (München: Markt & Technik, 1992). Also available as *Computers and the Imagination* (New York: St. Martin's Press, 1991). C. Pickover, *Mazes for the Mind: Computers and the Unexpected* (New York: St. Martin's Press, 1992).

7. B. B. Mandelbrot, *The Fractal Geometry of Nature* (San Francisco: W. H. Freeman and Company, 1982).

8. H.-O. Peitgen, H. Jürgens, and D. Saupe, *Chaos and Fractals—New Frontiers of Science* (New York, Berlin: Springer, 1992). M. Schroeder, *Fractals, Chaos, Power Laws—Minutes from an Infinite Paradise* (New York: W. H. Freeman and Company, 1991).

9. J. Conway, "Weird and Wonderful Chemistry of Audioactive Decay," in T. M. Cover and B. Gopinath, *Open Problems in Communications and Computation* (New York: Springer, 1987). A. Lindenmayer, and P. Prusinkiewicz, *Algorithmic Beauty of Plants.*

10. Ehrig, Kreowski, and Rozenberg, eds., *Graph Grammars.* Lindenmayer and Prusinkiewicz, *Algorithmic Beauty of Plants.* Peitgen, Jürgens, and Saupe, *Chaos and Fractals.*

11. Conway, "Weird and Wonderful Chemistry of Audioactive Decay."

12. Ehrig, Kreowski, and Rozenberg, eds., *Graph Grammars.*

13. Lindenmayer and Prusinkiewicz, *Algorithmic Beauty of Plants.*

14. C. Pickover, *Computers and the Imagination.* C. Pickover, *Mazes for the Mind.*

15. Peitgen, Jürgens, and Saupe, *Chaos and Fractals.* M. Barnsley, *Fractals Everywhere (Boston: Academic Press, 1988).*

16. D. R. Hofstadter, *Gödel, Escher, Bach: an Eternal Golden Braid* (New York: Random House/Vintage, 1980). M. Mitchell, *Analogy-Making as Perception—A Computer Model* (Cambridge, MA: MIT Press, 1993).

17. Ditto and Pecora, "Mastering Chaos," 62.

18. Schroeder, *Fractals, Chaos, Power Laws.*

19. Smith, *Digital Quotations.*

20. NeXT Computer, Inc., and Merriam-Webster Inc., *Webster's Ninth New Collegiate Dictionary and Webster's Collegiate Thesaurus,* 1st digital ed. Version 3.0 (Springfield, MA: NeXT Computer, Inc., and Merriam-Webster Inc., 1988, 1992).

21. R. Penrose, *The Emperor's New Mind—Concerning Computers, Minds and the Laws of Physics* (New York: Oxford University Press, 1989). J. Eccles, "Brain and Mind, Two or One?" in C. Blakemore, and S. Greenfield, *Mindwaves* (London: Basil Blackwell, 1989).

22. L. Rensing and A. Deutsch, *Natur und Form—Schönheit und Gesetzmäßigkeiten rhythmischer Strukturen* (Bremen: Zentraldruckerei der Universität, ca. 1990).

23. R. Rucker, *Der Ozean der Wahrheit—über die logische Tiefe der Welt* (Frankfurt am Main: Fischer Logo, 1990).

24. S. Clarke, *School Days* (New York: Nemperor Records/Warner, 1976).

25. K. Simonyi, *Kulturgeschichte der Physik* (Thun: Frankfurt am Main: Verlag Harri Deutsch, 1990).

26. J. Horgan, "The Death of Proof," *Scientific American* (10) (October 1993): 72. C. G. Langton, ed., *Artificial Life—Proceedings* (Redwood City, CA: Addison-Wesley, 1989). C. G. Langton, C. Taylor, J. D. Farmer, and S. Rasmussen, eds., *Artificial Life II—Proceedings* (Redwood City, CA: Addison-Wesley, 1992).

27. Mitchell, *Analogy-Making as Perception.*

28. Hofstadter, *Gödel, Escher, Bach.*

29. J. Fricke, "Aerogele—die leichte Alternative," *Bild der Wissenschaft* (9) (September 1993): 78.

30. Ditto and Pecora, "Mastering Chaos," 62.

31. Pickover, *Computers and the Imagination.* Pickover, *Mazes for the Mind.*

32. Horgan, "Death of Proof," 72.

33. Smith, *Digital Quotations.*

8

Sponges, Cities, Anthills, and Economics: Modeling with Fractals in the Future

Timothy D. Greer

Fractals provide new tools for builders of mathematical models. The fractal nature of many physical objects has been recognized and exploited to create realistic computer graphics images. However, in addition to making pictures, it is possible to learn about the underlying physical processes by using fractals to represent specific aspects of the system. It is also possible to create simple models that, without fractals, would be complex or difficult to define. For those systems that can be defined succinctly using fractals, there is hope that we can learn to manipulate the fractal definitions directly. If we can, doing so will open possibilities for investigation that are currently computationally expensive or impractical.

In this informal chapter, fractal modeling is demonstrated with an iterated function system showing the market share of two competing products. Growth models of sponges, cities, and anthills provide additional examples.

FRACTAL MODELS

Fractals are wonderful for creating realistic-looking images of natural objects. Coastlines, craters,[1] and cornstalks[2] all have been beautifully rendered. We all have admired and are astounded by the use of fractals in art. But fractals can be a useful tool for the "greasy" engineer as well as the eloquent artist. In this chapter I give a few examples of successful fractal models for physical systems and discuss some of my own attempts at using fractals as a modeling tool.

Fractal models are most useful when models using Euclidean constructs such as lines and spheres begin to break down or become unwieldy. As the mathematics of fractals is further developed, model builders will have an easier time employing fractals. In the future fractal modeling tools will become more common. I like to use the *finite element method* as a metaphor: Finite element modeling is used often to study mechanical characteristics of proposed designs, usually by people who haven't had to learn all the math and programming tricks behind the tools used to implement the model. I expect fractal-modeling toolkits to appear during the next few decades, toolkits that will give users a similar boost in modeling capability and efficiency.

Today, fractal models sometimes help us understand something about a natural object. For instance, the roughness of a simulated mountain range can be varied by altering its fractal dimension. Viewing computer-generated mountain ranges of various fractal dimensions and comparing these with pictures of real mountains gives a person a feeling for the range of roughness to be found naturally. Numbers can be assigned to that "feeling," making comparisons more objective and more easily communicated to others. But Kaandorp, referring to fractals that look like plants, points out a problem: "Although the objects shown . . . sometimes have an amazing resemblance to real objects, the procedure for their generation is not based on a model of a biological process. The plant images are not generated, for example, by modeling the growth of meristems but by artificial constructions."[3] Likewise, there are no earthquakes or plate tectonics creating our fractal mountains. Kaandorp goes on to describe his own modeling of sponges, models that *are* based on biological processes. His sponge models are fractal because they include the iterative process of growing and then branching, growing and branching. . . .

Results similar to Kaandorp's have been attained using *Lindenmayer systems* (L-systems). Knox uses L-system techniques to produce cornstalks.[4] He describes creating a video of cornstalks growing from seedlings to mature plants and finally dying. As with Kaandorp, the growth was included as part of the model, so that Knox was able to modify some parameters and obtain a flowering plant rather than corn.

A different sort of fractal modeling technique used today is known as diffusion limited aggregation (DLA),[5] where the iteration is on time itself. For example, model snowflakes and such crystals are grown by starting with a single point in a lattice as a seed and then sending additional particles randomly through the lattice. If a new particle hits an existing one, it sticks, and thus the crystal grows. Changing the rules determining how a particle floats through the lattice and whether it sticks causes different crystal patterns to appear. Batty applies this and related techniques to modeling the growth of cities.[6]

Batty's cities bear a striking resemblance to images I produced when attempting to model anthills.[7] By "anthill" I am referring to the underground ant nest, not the conical structure that some species build aboveground. It may be stretching definitions a bit to call my anthills fractals, but the resemblance to Batty's cities is not happenstance; our generation techniques were similar. For the anthills, I first generated a two- or three-dimensional array of cells, each with some random "digging difficulty." The distribution of this digging difficulty variable was based inversely on soil particle size: Sand is easier to dig in than clay because the particle sizes are bigger. This may seem counterintuitive because we are accustomed to thinking on a macroscopic scale, but any backyard gardener will attest to the fact that clay sticks together better than silt, and silt better than sand, so clay is the hardest to dig in. The reason for this is that the ratio of surface area to volume increases as particle size gets smaller, and cohesion depends on surface area. Adding vermiculite to clayey soil will make it easier to work with and let plants grow easier because the roots can push their way more easily through the soil. And so it is with ants—the smaller the particle sizes of the soil, the harder it is for an ant to tear out a chunk to haul to the surface.

After defining this two- or three-dimensional array to simulate dirt for the ants to dig in, I set my ants to work. Starting at the top of the array and in the middle, so that they won't bump into an edge too soon, they dig the first cell. From then on, the ants can choose to dig next any cell that is adjacent to one already dug. While they try to minimize the excavation effort, real ants' digging is directed toward the goal of creating "useful space." Useful space for ants appears to be the center of rooms (ants pile stuff in the center of rooms, not on bookshelves around the edges), so a cell becomes "useful" only when it and all four adjacent cells on the same level have been dug. The effort required to excavate a particular cell is the sum of its "digging difficulty" as defined plus some constant times the distance in cells back to the original excavated cell. This distance is measured along the tunnels that the ant must walk; if a tunnel meanders, the excavation effort for cells at its end will be greater. First I let my simulated ants always dig the next cell that was easiest, based on digging difficulty and how far they would have to drag the dirt back to the opening of the nest. Always digging the easiest cell next created a mess (figure 8.1) not much resembling a real anthill. Then I tried various weightings of the excavation effort to encourage digging of cells adjacent to

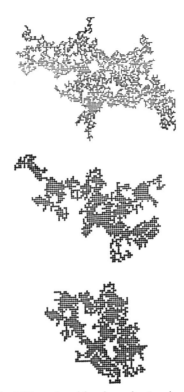

Figure 8.1 (top) "Anthill" produced by always having the ants dig wherever it's easiest, in clayey soil. View is from above, and the whole ant nest is contained on one level, as one might find upon turning over a rock. Each of figures 8.1-8.3 contain 500 "useful cells." A cell will be useful (for storage) for the ants if each of the four adjacent cells are also dug.

Figure 8.2 (middle) Anthill produced by digging in clayey soil, with weightings to encourage digging of adjacent cells. Notice how much more compact this anthill is, although it has the same number of useful cells as the anthill of figure 8.1.

Figure 8.3 (bottom) Anthill produced by digging in silty soil, with weightings to encourage digging of adjacent cells. Like figure 8.2, this anthill is much more compact—more like real anthills—than is the anthill in figure 8.1.

multiple cells already dug. Thus the ants would dig around a rock but would create one huge room if the soil was uniform. This improved results (figures 8.2 and 8.3), although it's clear that there is opportunity for further research in this field.

The anthill figures are views from above, as if the top of the ant nest has been mowed off with a lawn mower, or as if you lifted a large rock and found an ant nest tunneled underneath. To create similar patterns, you should first generate the array of "digging difficulties" for the cells. Soil particle size distributions can be found in D. Hillel's *Soil and Water*.[8] I recommend working only in two dimensions to begin with. Around the initial cell dug, the four adjacent cells can each be assigned a total excavation effort of their digging difficulty plus whatever cost you choose for carrying a cell of dirt one unit. Keep track of these cells in a stack, sorted by total excavation effort. Pull from the stack the cell of least excavation effort, mark it "dug," and update the stack by adding the three newly accessible cells. Calculate the total excavation effort of each of the newly accessible cells by adding one to the just-dug cell's distance from the original cell to get the new distance to the original cell, and then add the carrying effort based on this number to the digging difficulty. That is the technique used to generate figure 8.1.

For figures 8.2 and 8.3, I kept four stacks, one each for cells with one, two, three, and four adjacent cells already dug. Then I weighted the choice of which of the four stacks to pick my next cell from, according to the likelihood of that cell eventually becoming a "useful cell." The total excavation cost of digging a cell surrounded by four already-dug cells was weighted by one, since digging it would immediately produce a useful cell. The total excavation cost of a cell with three dug neighbors gets weighted by two, and so on. The total costs are multiplied by the weightings and then the comparison is made to determine which cell to dig. This will produce more compact structures similar to figures 8.2 and 8.3. I also experimented some with other weightings, with limiting the sizes of the stack (in effect causing the ants to forget about cells that have been on the stack a long time), and of course with three dimensions. In three dimensions, the primary additional consideration is that if the ants dig the cell immediately below a "useful cell," that cell can no longer be useful because, obviously, it no longer can be used as storage space.

From the examples I've given so far, you might get the idea that modeling with fractals consists of finding a way to make a fractal look like a real-world object. But what about situations where there is no real-world object? Now I'd like to present such an example, where the fractal appears solely as a way of defining the model. We don't justify our use of fractals this time by pointing out how good the results look. Instead, we use the fractal as the most convenient way to describe what is going on, and never mind how it looks (rather ugly, it turns out). However, first a little digression to explain iterated function systems.

An *iterated function system* (IFS) is defined by M. Barnsley.[9] I usually think of an IFS as simply a collection of functions that each has its domain and range in some space. I also usually think in Euclidean terms: a line, plane, or three-dimensional space. The functions are also known as mappings. If you start with a point in the space, you can apply the functions successively, in any order, and make that point "hop" around. You also can start with a collection of points and watch them all hop together. It turns out, though, that although an initial point may hop to many different positions, it won't be able to go to an arbitrary point in space.

Consider the case where the space is a number line and the IFS consists of the two functions $f(x) = x/3$ and $f(x) = x/3 + 2/3$. If you start at the point 1 and apply these two mappings successively, in whatever order you choose, the resultant points bounce around between zero and one, and can hit lots of numbers, such as $1/3$, $7/9$, $25/27$, and so on. But the points will never get too close to $1/2$ (no closer than $1/6$ of a unit, in fact). Since you are reading a book about fractals, you may have already recognized this as the *Cantor set.* It turns out that the points you can reach are those making up the Cantor set. Of course, you can start with some point outside the set—714 for Babe Ruth fans, for example. But any point you choose will get pulled toward the Cantor set as you apply the mappings over and over, and soon you will be so close that if you plot the points on your computer, you won't be able to tell them from the Cantor set you would have plotted by starting at 1. Barnsley proves a theorem about this; it's no accident.

We can do the same thing in higher dimensions, but to keep things simple we will stick with two dimensions. We now introduce an economic model as an example, initially quoting from *Envoy* magazine.

> Computer hardware maker XYZ makes both Personal Computers (PCs) and intelligent terminals. From time to time, XYZ comes out with new models of each product, affecting its market share of both markets. The markets for PCs and intelligent terminals are interrelated; PCs are frequently used as terminal emulators, and intelligent terminals often conveniently embody functions that are not conveniently available on the PC used as an emulator. So sometimes a new PC product will lure customers who would have otherwise bought intelligent terminals, and sometimes customers would buy a new model of intelligent terminal rather than an older model PC.
>
> The market share of XYZ in these two markets at a given time can be represented as a point in 2-dimensional space; let the X coordinate represent XYZ's market share of PCs, and the Y coordinate represent XYZ's market share of intelligent terminals. When a new product of either type is introduced, XYZ's market share should change. The introduction of a new product will then be a function mapping 2-D space to itself—introducing a new product changes the XYZ's point in the coordinate system. It is reasonable to expect that the new market shares will depend partly on current market shares and partly on the new product.[10]

If we plot the market shares at sequential times, we get a series of points $(x, y)_1$, $(x, y)_2$, $(x, y)_3$, and so on. The model is an attempt to relate one point with the next—$(x, y)_{June}$ becomes $(x, y)_{July}$ and so on. Or more generally, $(x, y)_{new} = f(x, y)$. To begin with, we don't know much about the function f, but we can learn a lot about it by noticing what restrictions must apply. For instance, over time market share would be expected to decrease if XYZ does nothing, since competitors would be expected to improve their product lines and otherwise woo customers. So we might write $x_{new} = ax$, where a is greater than or equal to zero and less than one. But of course XYZ doesn't sit idle. They might introduce a new version of X, which should boost sales. Perhaps the new X version grabs e additional market share, so that $x_{new} = ax + e$. What about the effect of the new product on sales of Y? Probably some would-be Y customers will switch to X instead. So perhaps $y_{new} = dy + f$, where f is less than or equal to zero, and d is smaller than it would have been if XYZ had simply done nothing.

To keep things fairly simple we restrict our consideration to versions of $f(x, y)$ which are *affine transformations*. In two dimensions this looks like

$$\begin{pmatrix} x \\ y \end{pmatrix}_{new} = \begin{bmatrix} a & b \\ c & d \end{bmatrix} \begin{pmatrix} x \\ y \end{pmatrix} + \begin{pmatrix} e \\ f \end{pmatrix}$$

If XYZ only comes out with products of type X, the market share for products of type Y should go to zero, and vice versa. To satisfy this, the matrix entries c and f must be zero for mappings representing new releases of product X, and the entries b and e must be zero for new releases of product Y. (For a more complex model we might argue that this isn't quite true—the prestige XYZ obtains from good sales of X might sustain some market for Y indefinitely. You can use the display algorithm, program code 8.1, to explore the effects of this.) We also know that, no matter what the values of x and y are, the value x_{new} cannot be less than zero nor greater than one, since market share must be between 0 and 100 percent. That is, $0 \leq ax + by + e \leq 1$. Using the boundary values $(x, y) = (0,0)$ and $(x, y) = (1, 1)$ we get $0 \leq e \leq 1$ and $0 \leq a + b + e \leq 1$. Similarly, we have $0 \leq y_{new} \leq 1$, so $0 \leq f \leq 1$ and $0 \leq c + d + f \leq 1$. Using $(x, y) = (0, 1)$ we get $0 \leq b + e \leq 1$ and $0 \leq d + f \leq 1$. Likewise $(x, y) = (1, 0)$ gives us $0 \leq a + e \leq 1$ and $0 \leq c + f \leq 1$. Notice that while e and f must be positive or zero, a, b, c, and d are not restricted from being negative. In fact, it might make sense for b or c to be negative. When XYZ introduces a new version of X, it might, in order to protect its market share of Y, take certain actions such as modified advertising or suppression of certain desirable features of the new X. So the larger the value of y, the smaller the new value of x should be, which would be accomplished by using $b < 0$.

Tables 8.1 and 8.2 (at the end of this chapter) show some reasonable possible values of the parameters based on various actions of the company. The possible market shares in this model make up a fractal that we can plot, as in figures 8.4 and 8.5.

Plotting iterated function systems is easy—if you can find one point on the fractal, applying any function in the system to that point gives another point on the fractal. So you can simply start applying the functions at random, and plot each new point. Barnsley calls this the *Chaos game* and also the *random iteration*

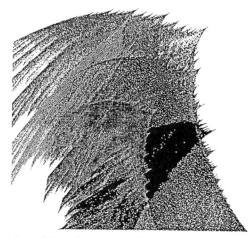

Figure 8.4 All possible market share combinations for the model defined by table 8.1. The straight edges at the bottom and the left side are the X and Y axes, respectively. Market share can't drop below zero, so the coordinate axes must bound the fractal. The fractal really does have straight edges on these two sides. This figure allows us to visualize the trade-offs possible between market share of the two products. It is clear, for instance, that market share for both products cannot be maximized simultaneously and that the point at the "upper right-hand corner" is best if the value of market share for each product is about equal. It is also apparent that the maximum market share possible for product Y occurs at a nonzero market share for product X and that, by giving up just a little of the Y-share at that point, significant increases in X-share are possible.

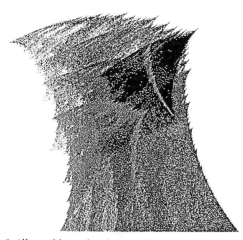

Figure 8.5 All possible market share combinations for the model defined by table 8.2. Again, the flat bottom is really part of the fractal. It's also the X axis, reflecting the fact that market share of Y can't drop below zero. From the appearance of this fractal, it's apparent that there is only one point at which market share for product X can go to zero.

algorithm. By sticking to affine transformations, as we did with the last model, a fairly general program can be written to read in the coefficients *a* through *f* for all the functions and then display the fractal. Usually probabilities also are included with the data, so that you can adjust the relative likelihood that each function is called. Since this method is available in many books, including Barnsley's, I won't include it here. Instead, I offer program code for what I jokingly call Senile Depth First Search.[11] (See program code 8.1.) This algorithm avoids some of the frustrations I encountered using the random iteration algorithm. It is also an example of a new fractal-related tool, the sort of development that will make fractals more useful and easier to use in the future.

Program code 8.1

```
{

  /* Array size declarations */
  float A[10][6];          /* Parameters for the IFS mapping.  The 10  */
                           /*  is arbitrary -- allows up to 10 Mappings */
  double DFSX[Senility];   /* (x,y) coordinates after DFS to Ith level  */
  double DFSY[Senility];        /* Senility=300 works ok                */
  int DFSmap[Senility];         /* DFS map applied at Ith level of DFS  */

  /* Declarations for other variables */
  int TOP=0, BOT=0;             /* Top and Bottom of stack DFSmap */
  int U, V;
  float InitialX, InitialY;

  /* Function declarations */
  void DFSXYset(int,int);
  int Increment(int);
  int Decrement(int);

  /* Initializations (Read in A[.][.] seperately) */
  DFSmap[0]=0;
  InitialY = (A[0][0]-1.)*(A[0][3]-1.)-A[0][1]*A[0][2];
  InitialX = (A[0][1]*A[0][5]-(A[0][3]-1.)*A[0][4])/InitialY;
  InitialY = (A[0][2]*A[0][4]-(A[0][0]-1.)*A[0][5])/InitialY;
  DFSX[0] = InitialX;
  DFSY[0] = InitialY;

  /* Main algorithm begins here */
  do {
      do {
          DFSXYset(TOP,DFSmap[TOP]);
          TOP=Increment(TOP);
          if (TOP==BOT) BOT=Increment(BOT);
          DFSmap[TOP]=0;
          U = DFSX[TOP];
          V = DFSY[TOP];
          U = _setpixel(U,V);
      } while ( U==0 ); /* enddo */
      do {
          TOP=Decrement(TOP);
      } while ( (TOP!=BOT) && (DFSmap[TOP]==NMAPS-1) ); /* enddo */
      DFSmap[TOP]=DFSmap[TOP]+1;
  } while ( (TOP!=BOT) || (DFSmap[TOP]<NMAPS-1)); /* enddo */

}
```

```
                    Program code 8.1 (continued)


/* Subroutines */

void DFSXYset(IX,Mapping)    /* Apply "Mapping"th transformation */
{
    int IX2;

    IX2 = Increment(IX);
    DFSX[IX2] = A[Mapping][0]*DFSX[IX]
              + A[Mapping][1]*DFSY[IX]
              + A[Mapping][4];
    DFSY[IX2] = A[Mapping][2]*DFSX[IX]
              + A[Mapping][3]*DFSY[IX]
              + A[Mapping][5];
    _setcolor(15-Mapping);
}

int Increment(IX)    /* Circular Addition */
{
    int IX2;
    IX2 = IX+1;
    if (IX2>=Senility) IX2=0;
    return(IX2);
}

int Decrement(IX)    /* Circular Subtraction */
{
    int IX2;
    IX2 = IX-1;
    if (IX2<0) IX2=Senility-1;
    return(IX2);
}
```

In addition to needing to specify a set of probabilities—one for each function—in an iterated function system, an additional annoyance with using the random iteration algorithm is that you never know when it is finished. How do you know when all of the points have been plotted? Of course, you can't plot all the points on a fractal, but you can plot to the limit of the resolution of your display! To ensure that you have all the points plotted, you need to make sure you have plotted all the combinations of sequences of the functions applied to the original point.

Senile Depth First Search uses an ordinary *depth first search* algorithm to plot points, backing up when it tries to plot a display point that has already been plotted. Ordinary depth first search used in this way on fractals requires an excessive amount of storage, though, so I added a provision to start overwriting storage after a while. Thus the term "senile": The algorithm forgets some of the data. But

Notes on the Program Code for Senile Depth First Search

I've used C notation for the program code. Five subroutines are used. `Incre-ment()` and `Decrement()` are very simple, merely providing circular addition and subtraction. `DFSXYset()` is matrix multiplication—the application of one of the mappings of the IFS to the point (DFSX[IX], DFSY[IX]). `_setpixel()` and `_setcolor()` are system-provided functions for plotting a point and choosing its color, respectively. I used "15-Mapping" instead of "Mapping" for the color because the color zero usually defaults to black, and I find black-on-black graphics rather uninteresting. The dynamics of drawing a fractal like the galaxy in figure 8.6 in color (using a different color for each mapping, as I have done in the program code) are pretty to watch. "Senility" is a constant you can choose to fit your available storage size; I've found 300 works fine; smaller would probably be okay. Arrays are subscripted with brackets []. The two-dimensional array A[][] must be initialized with the functions of the IFS. Counting starts at zero, so A[1][4] would be the fifth parameter of the second mapping. In the tables, the fifth parameter is e, so for table 8.1, A[1][4] = 0.25. I've allowed for up to ten functions; change the declaration of A[][] for more. `InitialX` and `InitialY` are calculated by finding the fixed point of the first mapping, that is, the point (x, y) where

$$\begin{pmatrix} x \\ y \end{pmatrix}_{new} = \begin{bmatrix} a & b \\ c & d \end{bmatrix} \begin{pmatrix} x \\ y \end{pmatrix} + \begin{pmatrix} e \\ f \end{pmatrix}$$

keeping only some of the past data is adequate because there are many paths to each point (to the limit of any real display). When a point is encountered that has already been plotted, we know that we have been within a pixel of this location before. If the iterated function system is defined using *contraction mappings*, knowing we have been within a pixel is enough to know that we need not follow this branch of the depth first search any further. This is because the images of all points after a contraction mapping are closer together than the original points. So if two points were within a pixel, their images under the contraction mapping will be within a pixel. If some of the IFSs functions are not very contractive, some small inaccuracies can be introduced, since although two points may lie in the same pixel and their images under the mapping are within a pixel width, the images might lie in adjacent pixels rather than both in one pixel. On a reasonably high-resolution display, the inaccuracies are noticeable only to a sharp-eyed observer who knows what to look for.

Suppose you have discovered a fractal relating to your business, as in the market share model. Chances are you'd like to apply this to making more money.

A market share model such as the one discussed might be applied to any company making products that are partial substitutes for each other—sports cars and pickup trucks, for instance, or peanut butter and tuna. Perhaps the corporate strategists have decided they value the peanut butter market share three times as much as the tunafish market. Then they want to maximize 3P + T, where P is peanut butter market share and T is tuna market share. But they can't get all the market for either product. They are constrained to points on a fractal like in figures 8.4 and 8.5. Once we've defined a peanut-butter/tuna market share fractal, we will need a way to find the point on it which maximizes 3P + T. We'll also want to know how to get there (using a series of product introductions or advertising campaigns) from where we are today.

To solve this problem, I developed another software tool. The two-dimensional version of the tool finds a polygon that encloses the fractal defined by an iterated function system and then iteratively finds other polygons lying within all previous polygons and also enclosing the fractal. In this way, it finds a polygon that approximates the boundary of the fractal's convex closure—the shape you would get if you wrapped a string tightly around the fractal. Finding a polygon approximating the shape of the tightly wrapped string reduces the tuna/peanut-butter maximization problem to one for which many solution methods are available. It turns out that the solution is applicable to another problem too, that of scaling a fractal for display so that it fills the display screen nicely without clipping. Basically, you need a tight upper bound on the extent of the fractal in the coordinate directions. The only published explanation of this tool currently available is in the *IBM Technical Disclosure Bulletin.*[12]

It is worth noting here that rescaling of an iterated function system is simple once you determine what scaling you wish to apply. Here is what you do. Suppose your IFS consists of the mappings $f_1, f_2, \ldots f_n$, and you wish to apply the rescaling function g to the whole fractal. Simply replace each function f_i with $gf_i g^{-1}$. (Note: g^{-1} is the inverse function for g.) This new iterated function system, $gf_1 g^{-1}, gf_2 g^{-1}, \ldots gf_n g^{-1}$, is like the original except it has been stretched or squished by g. This works because the application of g^{-1} transforms a point back to the "old" coordinates, where we apply f_i to it and then transform the result back to the "new" coordinates with g. Put that all together and we have $gf_i g^{-1}$. Readers might wish to investigate the effect of using a different g for each f_i. In particular, using a different g for each f_i might be a good approach to the galactic interactions model.

In the next two decades I expect many more tools like those described in this chapter will be developed. More and better tools will make fractal modeling easier and more accessible to both experts and less knowledgeable people. Fractals must be accessible to nonexperts who wish to apply them to modeling. Model builders probably aren't trying to learn about fractals; they are just trying to use the best techniques possible to help with their own investigations. Fractal models are likely

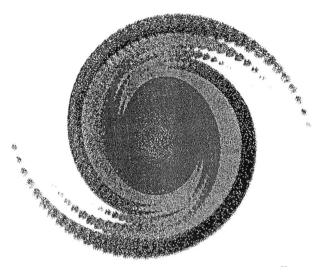

Figure 8.6   The fractal galaxy from table 8.3.   Is there an efficient way to manipulate such fractals, to simulate gravitational interactions between two such galaxies?

to be the best techniques in many circumstances, and as they get easier to create and manipulate, we should see more of them.

To close, I'd like to offer the reader a dream, a fractal model that doesn't exist yet. One of the fascinating aspects of fractals is that things that look quite complex can be defined with just a short description. Often shortening a description assists us in understanding and manipulating. For instance, we can write 2 + 2 + 2 + 2 + 2 + 2 + 2 + 2, but it's easier to understand and deal with if we write 8∗2. And, depending on the circumstances, it might be even more revealing if we write 2^4. In modeling, shortening descriptions is so important that we usually are willing to and indeed expect to discard information in order to get a more tractable model.

To model galactic interactions, we must discard information. One galaxy will contain billions of stars. In principle, the collision of two galaxies can be modeled by simply tracking the locations of the stars as they tug each other with gravity (and possibly electromagnetically). But the problem is too big. No computer will ever be able to process so much data, at least not in a human lifetime. But we can't just sit and watch nature play itself out either, we don't live nearly long enough. An often-used method of modeling galaxies is to pretend they are made up of a relatively small number of objects much bigger than the real objects.[13] Wouldn't it be great if we could describe a galaxy, with all its different sizes of objects—open clusters, planetary systems, stars, asteroids, dust, molecular hydrogen—with just a few numbers? Sounds like a job for a fractal! I attempted this a few years ago, producing images like the one shown

---

Table 8.1.
IFS Coefficients for Figure 8.4

This and following tables give the iterated function system coefficients used to generate figures 8.4 to 8.6. Each line provides the coefficients a-f of one mapping of the form

$$\begin{pmatrix} x \\ y \end{pmatrix}_{new} = \begin{bmatrix} a & b \\ c & d \end{bmatrix} \begin{pmatrix} x \\ y \end{pmatrix} + \begin{pmatrix} e \\ f \end{pmatrix}$$

The collection of mappings in each table defines the IFS for the corresponding figure.

Comments	a	b	c	d	e	f
Convert product y users to product x	0.4	0.5	0	0.5	0.1	0
Get more product x users	0.7	-0.1	0	0.8	0.25	0
New x product bombs	0.5	0.1	0	0.6	0.1	0
Convert product x users to product y	0.6	0	0.6	0.3	0	0.1
Get more product y users	0.8	0	0.1	0.6	0	0.25
Advertising campaign (sell more of both x and y)	0.76	0.095	0.095	0.76	0.1	0.1

---

in figure 8.6. The IFS codes for this fractal are in table 8.3, so you can use the Senile Depth First Search algorithm to display it yourself and try some modifications.

But what about galactic interactions? Everything about the "galaxy" in figure 8.6 is embodied in table 8.3. Perhaps it is possible to work directly on these numbers, and the corresponding set for another galaxy with which it interacts, to see how the galaxies change as they pass near or through one another. Probably another set of numbers needs to be added for velocity, since figure 8.6 represents only positional data. The nonlinearity caused by the inverse-square relation for gravitational attraction makes the problem less than straightforward, at least when we are trying to stick with affine transformations like those in the tables. The interested reader might wish to consider whether there are other, better fractal representations of a galaxy, ways to intermix nonlinear mappings with affine

Table 8.2.
IFS Coefficients for Figure 8.5

Comments	a	b	c	d	e	f
Get more product x users	0.7	-0.1	0	0.8	0.25	0
New x product bombs	0.5	0.1	0	0.6	0.1	0
Get more product y users	0.8	0	0.1	0.6	0	0.25
Advertising campaign 1	0.5	0.2	0.2	0.5	0.2	0.2
Advertising campaign 2	0.76	0.095	0.095	0.76	0.1	0.1

Table 8.3.
IFS Coefficients for Figure 8.6

Comments	a	b	c	d	e	f
5 degree twist, tiny shift North	.97627	-.085413	.085413	.97627	0	2.0
5 degree twist, tiny shift WSW	.97627	-.085413	.085413	.97627	-1.732	-1.0
5 degree twist, tiny shift ESE	.97627	-.085413	.085413	.97627	1.732	-1.0
Medium shrink, no twist	.75175	-.27362	.27362	.75175	0	0
Asymmetrical shrink, no twist	0	-.5	.6	0	0	0
Western outlying cluster	0	-.02	.02	0	-297	0
Eastern outlying cluster	0	.02	-.02	0	297	0

mappings that might help, and how best to integrate mass and position over the fractal galaxy to determine its net gravitational attraction to objects within and around it. By doing so you will be developing applications and inventing techniques we can expect to use with fractals in the future.

Notes

1. B. Mandelbrot, *The Fractal Geometry of Nature* (New York: W. H. Freeman and Co., 1983), and C8-C15. See also pp. 71-72.

2. N. Goel, L. Knox, and J. Norman, "From Artificial Life to Real Life: Computer Simulation of Plant Growth," *International Journal of General Systems* 18 (4) (Fall Semester, 1991): 291-319.

3. J. Kaandorp, "Modelling Growth Forms of Sponges with Fractal Techniques," in *Fractals and Chaos,* eds. A. Crilly, R. Earnshaw, and H. Jones (New York: Springer-Verlag, 1991), 71-88.

4. L. Knox, "Reconstruction of L-System Triplets Using General System Problem Solving Techniques," Final Project for *SS-510, System Problem Solving* (Binghamton University, Fall semester 1989). See also note 2.

5. K. Falconer, *Fractal Geometry* (Chichester: John Wiley & Sons, 1990), 267-272.

6. M. Batty, "Cities as Fractals: Simulating Growth and Form," in *Fractals and Chaos,* eds. Crilly, Earnshaw, and Jones, 43-69.

7. T. Greer, "A Computer Model of Underground Nest Construction by Ants," M.S. thesis, University of Illinois at Urbana-Champaign, 1980.

8. D. Hillel, *Soil and Water* (New York: Academic Press, 1971), 16.

9. M. Barnsley, *Fractals Everywhere* (San Diego: Academic Press, 1988), 82. See also pp. 2-3 and 90.

10. T. Greer, "Two Fractal Models for Operations Research," Society of Industrial Engineers *Envoy* 1 (3) (January/February 1992): 3-5.

11. T. Greer, "Senile Depth First Search Applied to Iterated Function Systems," *IBM Technical Disclosure Bulletin* 34 (8) (January 1992): 275-277.

12. T. Greer, "Method for Scaling Fractal for Display by Finding Upper Bound on Extent," *IBM Technical Disclosure Bulletin* 36 (8A) (June 1993): 489-491.

13. A. Peratt, "Plasma Cosmology," *Sky & Telescope* 83 (2) (February 1992): 136-140.

9

Fractal Holograms

Douglas Winsand

High-resolution fractals are attractive because of their hypnotic visual intricacy. If their resolution is increased sufficiently, fractals can become diffraction gratings splitting incident white light into delicate spectra. Extending the resolution even further will create a fractal hologram. In this chapter diffraction and interference, the wave phenomena underlying holograms, are described. Fractals well suited for creation of holograms are discussed. Many practical problems must be solved in order to create fractal holograms. These problems and possible solutions are introduced.

SCULPTING LIGHT

Since their discovery, fractals have fascinated many of us because of their visual intricacy. As computer graphics and processing power have improved, fractals have been generated with ever increasing detail. For example, fractals have been generated on grids of up to 4,096 by 5,120 pixels on a 7.5" by 8.5" plot.[1] These detailed fractals are even more hypnotic than the lower-resolution fractals. Since fractals are mathematical objects, there is no real limit to their resolution. If the resolution of a fractal is increased enough, the detail will literally become microscopic. If the resolution is increased further, a fractal can act like a diffraction grating splitting incident white light into spectra. Increase the resolution even further, and a fractal can act as a hologram. In the near future, the technology will exist to allow artists to create holograms using very high resolution fractals etched in gold on glass. The first of these fractal holograms may appear in large company lobbies or art museums, but as the technology improves, fractal holograms as art objects will become available to the general public.

What would a fractal hologram look like? The repeating structures at different scales will cause various parts of the fractal to interact with light in different ways. Large structures in the fractal would act like normal macroscopic objects reflecting or absorbing light depending on their reflectivity and color. Very fine structures in the fractal would split incident white light into spectra, and structures at scales near the wavelength of light could cause enough diffraction and interference of incident light to create the illusion of three-dimensional structures, lenses, or other optical devices. An approximation of the image visible from a position perpendicular to the hologram can be generated by applying a fourier transform to the fractal. Thus the original intricacy of the fractal would be mixed with delicate spectra and possibly three-dimensional structures. Fractal holograms will allow future artists to separate, mix, and sculpt light waves themselves. To understand how this is possible, it is necessary to review some basic principles of light propagation.

LIGHT AS A WAVE PHENOMENON

As first postulated by Christian Huygens in 1678 and later proven by Thomas Young (1801-1804) and Augustine Fresnel (1815-1826), light consists of waves. Like other waves, light waves can be described by their wavelength, amplitude,

phase, and direction of propagation.[2] In one dimension, monochromatic light can
be described as a function of these quantities in the following equation:

$$y = A \cos \left(\left(2 \frac{\pi}{\lambda} \right) x - \Phi \right) \tag{9.1}$$

where A is the amplitude of the wave, λ is the wavelength, and Φ is the phase of
the wave.

Typical wavelengths for light waves vary between 7 by 10^{-7} meters
(3/100,000 of an inch) for red light and 4×10^{-7} meters (1.6/100,000 of an inch)
for violet light. Young was the first to demonstrate the wave nature of light by
observing the pattern of dark and light fringes behind an illuminated slit. Figure
9.1 shows a side view of a narrow slit with parallel light waves incident from the
left. The intensity of the light at large distances to the right of the slit can be found
by summing the contributions of light waves emanating from every point across
the slit. The result for points at large distances from the slit at an angle from the
original direction of the incident light is:

$$I = I_0 \; \frac{\sin^2 \left((\pi \, a / \lambda) \; \sin\theta \right)}{\left((\pi \, a / \lambda) \; \sin\theta \right)^2} \tag{9.2}$$

where I_0 is the intensity of the incident light and I is the intensity of diffracted
light.[3] Three important properties of this equation relate to fractal holograms.

First, notice how the intensity of the diffracted light varies with the angle
θ. The $\sin^2$ function in the numerator is sinusoidal and always positive. It is
modulated by the squared function in the denominator, which causes the
intensity to fall off rapidly with increasing angle away from the direction of the
incident light.

Second, the equation tells us how the intensity of the light that has passed
through the slit varies as a function of the slit width a. Figure 9.2a shows the

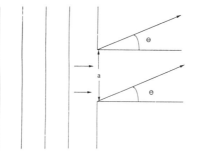

Figure 9.1 Parallel waves of light traveling to the right toward a slit indicated by the
vertical line near the center of the figure. Light waves are diffracted by the slit of
width a at angles measured with respect to the original direction of the light waves.

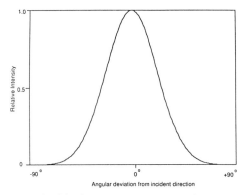

Figure 9.2a A graph of the function in equation 9.2 for a slit of width a = λ. The vertical axis represents the intensity of the diffracted light relative to the intensity of the incident light. The horizontal axis is the angle θ measured relative to the original direction of the incident light. (See figure 9.1.) Notice that light is diffracted at large angles.

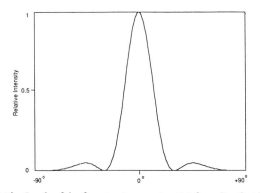

Figure 9.2b Graph of the function in equation 9.2 for a slit of width a = 2 λ. Far less light is diffracted at large angles and there are angles where no light appears.

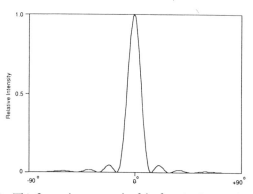

Figure 9.2c This figure shows a graph of the function in equation 9.2 for a slit of width a = 4 λ. Very little light is diffracted away from the original direction of the incident light.

intensity as a function of angle for a slit that is one wavelength wide. Notice that although light appears directly behind the slit a substantial fraction of light is diffracted to the sides, and some is even sent at right angles to the incident direction. Figure 9.2b shows the intensity as a function of angle for a slit two wavelengths wide. In this case, the intensity of the light is high directly behind the slit but falls off rapidly at angles to the sides of the slit. Figure 9.2c shows the intensity as a function of angle for a slit four wavelengths wide. As the slit width increases relative to the wavelength of the incident light, the pattern approaches the macroscopic case where almost all the light appears directly behind the slit with extremely narrow bands of light immediately surrounding the bright central band. Holograms work through diffraction and interference of incident light. If we want to create a fractal hologram that has a significant diffractive effect on incident light, some of its structure must be on the order of a wavelength of light in size.

Last, equation 9.2 depends on the wavelength or color of the incident light. If figure 9.2b shows the intensity of light as a function of angle for red light through a slit width of one wavelength, figure 9.2a shows the intensity function for violet light passing through the same slit. As you can see, there are angles where red photons are scattered but no violet photons appear. Thus, a fixed slit width will diffract light of different colors differently. This explains how small slits can split white light into its component colors.

We have examined the effects of narrow slits on light. In reality, fractal holograms will be constructed from round or square pixels. Equations can be derived for the diffraction of light from round or square holes. However, the fundamental relations just discussed still apply.

More complicated effects occur when light passes through closely spaced multiple slits. The two-slit case is the most commonly studied. The equation for the intensity of light at a great distance from two slits is:

$$I = I_0 \frac{\sin^2 ((\pi a/\lambda) \sin\theta)}{((\pi a/\lambda) \sin\theta)^2} \cos^2 (\frac{\pi}{\lambda} (a+b) \sin\theta) \qquad (9.3)$$

where a is the slit width as before and b is the distance between the two slits.[4] As for the previous single-slit case, the intensity pattern depends on the wavelength and on the slit width. In addition, the distance between the slits, b, also affects the pattern of scattered light. If the slits are very far apart, the light from the two slits is less likely to interfere due to the weak intensity of light diffracted at large angles and also due to the large distances involved. Conversely, light from closely spaced slits will interfere strongly. Holograms also strongly interfere with incident light.

The multiple slit case has been studied intensively for the design of diffraction gratings. The principles we have discussed also apply to diffraction gratings. The most efficient gratings have small slit widths that are closely spaced. As the number of adjacent slits is increased, light of different colors is split more efficiently.

HOLOGRAMS

Holograms are recorded interference patterns between light beams. In the classical case, a coherent beam of light interferes with light reflected from an object that has been illuminated by the same coherent beam of light. (See figure 9.3) A light-sensitive emulsion placed where the two beams meet records the interference pattern between the two beams of light. The interference pattern recorded in the emulsion is the hologram. If that hologram is later illuminated with a coherent beam of light at the same frequency and from the same direction, the light wave that had originally come from the object will be reconstructed. To the viewer, it will appear as if the original object has reappeared. It is important to note that holograms result from the interference pattern between a reference beam of light and a light beam reflected from an object. Notice that there is an implied notion of an object in a hologram. Fractal holograms, in contrast, do not result from the interference between light rays from an object and a reference source. The notion of object is missing in a fractal hologram.

A close examination of a hologram will reveal an extremely fine, intricate pattern of light and dark areas. In fact, the interference pattern of light and dark areas can be mathematically decomposed into a superimposition of a large number of diffraction gratings of various spacings and orientations.[5]

There are many kinds of holograms. Holograms created by interfering beams of light in a photographic emulsion are actually volume holograms. The recorded interference pattern is a three-dimensional structure since, relative to the wavelength of light, a photographic emulsion is thick. Two-dimensional planar holograms are also possible, and fractal holograms fall into this category. There are two types of planar holograms: reflection and transmission holograms. In reflection holograms, light is reflected from the surface and interferes with neighboring light waves from nearby reflecting points. Transmission holograms interfere with and diffract light passing through the hologram surface. All of the holograms discussed in the remainder of this chapter are planar holograms.

In 1966 Barry Brown and Adolf Lohman proposed that the patterns in a physical hologram could be calculated ahead of time and used to generate artificial holograms.[6] This type of computed hologram is known as a computer-generated hologram. Computer-generated holograms can simulate real holograms of physical objects or nonexistent objects. In addition, computer generated holograms can be designed to simulate optical devices such as lenses. These types of computer-generated holograms are called holographic optical elements (HOEs). Like fractal holograms, there is no implicit object in an HOE. Fractal holograms are a new type of computer-generated hologram. Computer-generated holograms have traditionally been converted to physical holograms by printing the computed pattern on a large plotter and photo-reducing the resultant plot. By photo-reducing the pattern, a much higher effective resolution can be obtained. The photo-reduced film serves as the hologram.

FRACTALS SUITED FOR HOLOGRAMS

Now we can begin to determine which fractals would be best suited to the creation of aesthetically striking holograms. Several factors will affect the quality of the resulting fractal holograms. These include size of the pixels used, structure of the fractal, method used to create the fractal, and optical properties of the materials used in printing.

Pixel size affects the quality of fractal holograms in several ways. The smaller the pixels, the more likely that the final fractal will have structures with sizes close to the wavelength of light. However, very small individual pixels will be impossible to see, and if a fractal consists of sparsely spaced pixels, very little light will be reflected, absorbed, or transmitted. A bright hologram requires that a large proportion of light be diffracted. Also, very small pixels will require much larger computation times to generate the complete fractal.

The structure of the fractal will have a strong effect on the quality of the resulting hologram. Fractals with little small scale structure will not make good holograms. Fractals with diffraction grating-like structures or those with closely spaced points will be particularly effective. (See figures 9.4 and 9.5.)

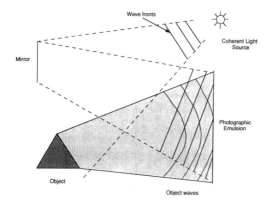

Figure 9.3 Coherent light waves reflected from an object interfere with waves from the source to create an interference pattern in the photographic emulsion.

I'll now discuss how these ideas relate to the famous fractal object known as the *Mandelbrot set*. In the traditional method of representing this fractal, points in the complex plane are colored black or white depending on their long-term behavior when iterating $z = z^2 + c$. The set consists of a central black region with an intricate boundary and a surrounding white region. The boundary, upon closer examination, turns out to be extremely detailed with wildly twisted fingers and myriads of smaller copies of the central black region. Would this fractal be a good hologram if the resolution were high enough?

Let's imagine that the Mandelbrot set were calculated to a very high resolution and printed using a perfectly opaque substance onto a perfectly transparent substrate. The central opaque region will appear exactly as we are used to seeing it. Many of the smaller boundary structures also will appear normally. However, very close to the boundary, there are "microscopic" structures that will be small enough to strongly diffract the light and cause interference between diffracted beams. The region outside the visible boundary of the set is dense with smaller copies of the set and small scale structures capable of diffracting light. Thus the boundary of this extremely high resolution Mandelbrot set will be edged with spectra and three-dimensional structures, making it a good candidate for a hologram.

Another way that the Mandelbrot set is often represented is by contour mappings where all regions with the same escape rate are colored identically. Often the values of the escape rates are alternately colored black and white. This version of the Mandelbrot set would have a very convoluted diffraction grating at the boundary. Thus this version would also be a good hologram along the boundary of the set.

Each point in the Mandelbrot set corresponds to a fractal called a *Julia set*. Julia sets created from points within the boundary of the Mandelbrot set form

Figure 9.4 A Julia set generated from a point closer to the boundary of the Mandelbrot set. This fractal has fine structure along its boundary and would be a candidate for a fractal hologram.

Figure 9.5 Fractal from *Computers, Pattern, Chaos, and Beauty* by Clifford Pickover (New York: St. Martin's Press, 1990), figure 12.4. The figure is rich in fine detail and would make an excellent fractal hologram.

connected fractals with complex boundaries. Those Julia sets created using points near the center of the Mandelbrot set have "smooth" boundaries and would not make good holograms. Julia sets created from points within the Mandelbrot set but near the boundary have very intricate boundaries and are likely to be the best candidates for fractal holograms. Julia sets created from points outside the Mandelbrot set are disconnected and are often called dusts. Nearest to the boundary of the Mandelbrot set, the corresponding Julia sets will be very densely scattered dusts. If we calculate the point density at various points within a disconnected Julia set, there will be areas where the points are dispersed at average distances close to the wavelength of light. The light diffracted from these nearby points will interfere and thus present good areas for holographic effects. In areas where the points are widely spaced, interference effects will be negligible.

Another type of fractal is typified by the *Attractor of Henon*.[7] This fractal is generated using an iterated functional mapping on points in the plane. The algorithm for creating these fractals is:

1. Select an initial point (x_0, y_0) in the plane.
2. Generate a new point (x_1, y_1) using the equations:

$$x_{n+1} = f(x_n, y_n)$$
$$y_{n+1} = g(x_n, y_n) \tag{9.4}$$

3. Plot the new point and iterate.

Depending on the functions f and g that are selected, the successive points may diverge toward infinity or remain within a bounded region. Divergent point sets are unlikely to make good fractal holograms. Convergent and bounded point sets will make good fractal holograms depending on the point density. The following are examples of bounded and convergent point set iterated function mappings. Suggested initial values for x and y are included.[8]

$$x_0 = 0; \ y_0 = 1$$
$$x_{n+1} = y_n + 1.4x_n^2 - 1$$
$$y_{n+1} = x_n^2 - 0.3x_n \tag{9.5}$$

$$x_0 = 0; \ y_0 = 0.58$$
$$x_{n+1} = y_n + 1.4x_n^2 - 1$$
$$y_{n+1} = x_n^2 - y_n^2 + 1 \tag{9.6}$$

$$x_0 = 1; \ y_0 = 0.5$$
$$x_{n+1} = y_n - \cos^{2.5}(x_n)$$
$$y_{n+1} = 0.01x_n^2 - x_n + 1.125 \tag{9.7}$$

$$x_0 = 0.01; \ y_0 = 0.001$$
$$x_{n+1} = 2y_n \sin(x_n) - 1$$
$$y_{n+1} = 3 \cos(y_n) - 3 \sin(x_n) \tag{9.8}$$

$$x_0 = 0.001; \ y_0 = 0.001$$
$$x_{n+1} = 1.4y_n \sin(y_n) - x_n$$
$$y_{n+1} = 1.4 \cos(y_n) - 0.3 \sin(x_n) \tag{9.9}$$

$$x_0 = 0.75; \ y_0 = 0.5$$
$$x_{n+1} = y_n - |\cos(x_n) - \sin(x_n)|^{1.25}$$
$$y_{n+1} = 0.01x_n^2 - x_n + 1.5 \tag{9.10}$$

Iterated function systems are similar to the point set functions just described and were introduced by M. Barnsley and S. Demko in 1985.[9] These systems have the added property that they can be guaranteed to converge and that arbitrary shapes can be encoded. Most often the constituent function mappings are chosen to be affine. In the plane, affine mappings can be expressed as a 2 by 2 matrix plus a 2 by 1 vector. An iterated function system may have multiple affine mappings. To guarantee that the system converges, the determinants of each of the constituent 2 by 2 matrices must be less than 1. In geometrical terms, this means that the area enclosed by a set of points shrinks after the points are mapped by the matrix.

Because of their ability to encode arbitrary shapes, iterated function systems could prove to be especially valuable for the creation of fractal holograms. If the desired pattern of points to create a particular holographic effect is known, then, in theory, an iterated function system that encodes that particular pattern could be created.

Like the previously described point sets, iterated function systems often start with an initial seed point in the plane. This seed point is mapped by the constituent functions to new points, which are plotted and the process is iterated. A later development allows the use of entire patterns instead of single points. These patterns, called condensation sets,[10] are mapped and plotted in a manner similar to that for single points. To create a fractal hologram, a diffraction grating pattern could be used as the initial condensation set. Since a plane hologram can be decomposed into a constituent set of overlapping diffraction gratings, this method would allow the creation of a large number of holograms. The spacing between the grating lines would decrease in the course of the mapping process because each of the mappings must be contractive (determinant less than 1). To prevent the spacing from becoming too small too quickly, the determinants of the constituent matrices would have to be slightly less than 1.0. These iterated function systems would converge slowly since the convergence rate is determined by the largest of the determinants. However, since entire diffraction grating patterns are being mapped, a small number of iterations would be sufficient to create an interesting hologram.

PRACTICAL CONSIDERATIONS

Fractal holograms are clearly theoretically possible, but in practice their creation raises a host of problems. Fractal holograms will have to be computed accurately, stored, and then printed onto a physical surface using a technique capable of the extremely fine resolution necessary to diffract light.

What resolution would be required to generate a clear, distinct fractal hologram? Hologram resolution is usually expressed in terms of line pairs per millimeter. A line pair consists of two parallel lines separated by a small distance. The highest-quality visual holograms have resolutions as high as 5,000 line pairs per millimeter (~125,000 lines pairs/inch). A line pair would, at a minimum, have to be three pixels wide (one on, one off, and one on). If we assume that 500 line pairs per millimeter would be sufficient for a lower-quality hologram, then we would have to calculate 1,500 pixels per millimeter of the final hologram. To cover an 8½ by 11-inch sheet of paper, a fractal hologram would need 323,850 pixels across the sheet and 419,100 pixels down the sheet. The total number of pixels that would need to be calculated for the entire sheet would be about 1.357×10^{11}, or 135.7 billion pixels.

How much time would be required to calculate this fractal hologram? Assuming it would require an average of 6 microseconds per pixel (six clock cycles per floating point operation, 100 floating point operations per pixel) on a 100 Mhz personal computer, it would take approximately 9.8 days of computer time to calculate the hologram. Integer calculations could reduce this time substantially. Assuming that we store the hologram pixels as single bits (black or white), we would need about 17 gigabytes of storage. These are large numbers but not outside the bounds of present technology, particularly with modern parallel computers or special-purpose hardware.

Computer-generated holograms have been created with much lower resolutions than 500 line pairs per millimeter, although they are not as distinct as high-resolution holograms created with lasers. If we assume 100 line pairs per millimeter, the numbers become more manageable. It would only require 679 megabytes of storage for the entire 8½ by 11-inch fractal hologram and would require 9.4 hours of computer time.

Methods could be used to reduce the computations needed. For example, most of the interior of the Mandelbrot set does not need to calculated. A low-resolution calculation could identify parts of the interior of the set that will remain black at higher resolutions. In addition, more advanced methods can bound the interior and exterior of the Mandelbrot set.[11]

Another problem that would have to be considered would be numerical accuracy. Depending on the resolution, the limited precision of floating point calculations on some computers could introduce large errors in the finest details of the hologram. For many fractals, extended precision floating point calculations would be required.

Another practical problem will arise for fractals consisting of large numbers of scattered points. Exceedingly small pixels will be invisible, so areas of a fractal consisting of widely scattered points may not be visible to the human eye. For these fractals, a postprocessing step may be necessary. The density of points across the fractal could be calculated. Points in sparse areas could be enlarged or filled in enough to make them visible. Although this would represent a distortion of the original fractal, the final visual effect would be more pleasing. Another benefit of calculating the density of points across the fractal is that this density also indicates roughly how much light will be reflected or transmitted from that area.

Fractals are traditionally printed in black or colored inks on paper. For holographic applications, other methods will have to be used. Fractal holograms can be either transmissive or reflective. As the names imply, transmissive holograms interfere with light passing through them, while reflective holograms interfere with light reflected from their surfaces. Transmissive fractals can be etched out of an opaque material overlaid on a transparent substrate or out of a high-index-of-refraction material overlaid on a material with a low index of refraction. Transmissive holograms constructed from two materials with different indices of refraction will be the most efficient since they will absorb the smallest amount of light. Reflective holograms can be constructed in several ways as well. The fractal can be printed in a highly reflective material overlaid onto an absorbent material or vice versa. The choice would be made based on the total area covered by the fractal.

The simplest method for creating a fractal hologram is to generate and print a very large fractal on paper,[12] photograph and reduce it, and view the reduction in laser light or direct sunlight. In an experiment, I have created a large Julia set in sections on a home computer, printed and pasted the sections together, and photographed the final fractal image with a 35 mm camera. The final printed, sectioned Julia set, when pasted together, ended up measuring two and one-half feet by three and one-half feet. After photographing this large fractal, the resultant 35 mm negative was used as the fractal hologram since the negative had as much fine detail as the original and the image of the fractal was reduced by a factor of 30. I also generated a diffraction grating by printing parallel lines on a sheet of printer paper, mosaicing several such sheets together, and photographing that. When viewed in direct sunlight, the diffraction grating 35 mm negative does successfully split the light into spectra although they are not widely split or vivid. The Julia set film did not split light as hoped. The small gaps in this fractal were not quite small enough or numerous enough to act as efficient gratings. I believe that if I had made the original fractal larger or reduced it further, it would have started to split light. With dot matrix or laser printers, printer dots are designed to overlap. A laser printer with a 300 dot per inch mode will not print 300 distinct and separate dots within the space of one inch.

A method for generating a fractal hologram with higher resolution would be to write a fractal hologram onto a CD or compact disc.[13] CDs split incident light quite well, as you can see when viewing them in almost any light source but especially in sunlight.

The highest-resolution fractal holograms might borrow technology used in the manufacture of integrated circuits. In integrated circuit manufacturing, engineers design the patterns to be laid onto the chips. These patterns are plotted on large format plotters, and photo-reduced to generate masks used in the chip manufacturing process. Masks are created for an entire silicon wafer, which measures several inches across. A fractal hologram could be created using the same techniques. Of course, the goal is not to create silicon with different electrical properties but a surface with opaque and transparent areas, reflective and absorbent areas, or areas with different indexes of refraction. A standard process in the manufacture of chips is laying down a conductive metal layer. Such a layer, if made thick enough, will be opaque and reflective. If this process could be adapted for work on flat sheets of glass, then a transmissive fractal hologram could be created. If a reflective metal such as gold or silver were used, the same surface could act as a reflective hologram. A factor that must be considered is how much total light is transmitted or reflected. A fractal with large areas of opaque material would make better reflective holograms; one with small opaque areas would be better suited for a transmissive hologram. To create large wall-size fractal holograms or any other computer-generated holograms, existent technologies will have to be extended beyond their present capabilities. Integrated circuit manufacturing techniques are designed for silicon-wafer-size formats.

The best transmissive fractal holograms will be ones written with a high index of refraction material overlaid on a material with a low index of refraction or vice versa. This type of hologram operates somewhat differently from the holograms constructed of metal overlaid on glass. Instead of reflecting or blocking incident light, this type of hologram causes phase shifts in the incident light that passes through the different materials. Phase shifts between nearby light sources causes interference similar to that discussed previously. Another way of thinking about this type of hologram is that each area with overlaid high-index-of-refraction material acts like a lens. The advantage is that more light is passed through the hologram, resulting in a brighter, more distinct hologram. Possible materials suited to this type of hologram are diamond deposited over glass. Recent advances in deposition of thin diamond films may make creation of this type of hologram possible. Of course, many technical issues would have to be resolved to make this technique practical.

CONCLUSION

The holographic methods discussed in this chapter may be used for exhibiting large-scale patterns in massive data sets. More and more often, scientists are encountering massive sets of data that must be analyzed. Writing these data sets in holographic formats might reveal patterns in the data that are invisible when looking at small portions of the data. Is there some hidden pattern in the digits of pi? Pi has recently been calculated to over a billion digits, which is beginning to get much too large for traditional methods of examination. Would a pattern appear in the digits of pi if it was expressed as a binary number and written as a hologram?

As reproduction techniques improve, counterfeiting is becoming a serious problem. Color copiers are now capable of duplicating many valuable documents, such as certificates, stocks, and currency. Since there are as many potential fractal holograms as there are fractals, individual document series could have dedicated holographic emblems. Embedding a fractal hologram into a valuable document could make reproduction very difficult or impossible. Of course, an inexpensive method to write a fractal hologram onto paper or plastic would have to be found.

As microminiature technologies improve, computers get faster, and storage methods improve, creation of fractal holograms will become easier. Artists will have a new medium to express themselves, scientists will have a new tool for discovering patterns in very large data sets, and counterfeiters' careers will become less profitable.

Notes

1. H. O. Peitgen and D. Saupe, *The Beauty of Fractals* (New York: Springer-Verlag, 1986).

2. F. W. Sears and M. W. Zemansky, *University Physics* 3rd ed. part 2, (Reading, MA: Addison-Wesley, 1964).

3. C. Curry, *Wave Optics—Interference and Diffraction* (London: Edward Arnold Ltd., 1957).

4. Ibid.

5. Howard M. Smith, *Principles of Holography* (New York: Wiley-Interscience, 1969).

6. G. Tricoles, "Computer Generated Holograms: An Historical Review," *Applied Optics* 26 (20) (October 1987).

7. D. Hofstadter, "Metamagical Themas," *Scientific American* (November 1981).

8. D. Winsand, *COMPUTE!'s Second Book of Atari* (Greensboro, NC: COMPUTE! Books, 1982), pp. 67-75.

9. M. Barnsley, *Fractals Everywhere* (San Diego: Academic Press, 1988).

10. Ibid.

11. H. O. Peitgen and D. Saupe, *The Science of Fractal Images* (New York: Springer-Verlag, 1988).

12. D. Casasent and A. Lee, "Computer Generated Hologram Recording Using a Laser Printer," *Applied Optics* 26 (1) (January 1989).

13. T. Yatagai, J. Camacho-Basilio, and H. Onda, "Recording of Computer Generated Holograms on an Optical Disk Master," *Applied Optics* 28 (6) (March 1989).

10

Boardrooms of the Future: The Fractal Nature of Organizations

Glenda Eoyang and Kevin Dooley

———

A fractal organization is characterized by its self-similarity across functions and organizational levels and by its complex boundaries. It also is influenced by fluctuations and tensions. We explore the future reality of the fractal organization by following Barbara and Jack on a fictional journey as they start their new virtual company.

———

INTRODUCTION

Business organizations often are described in terms of images and metaphors. Today the metaphors that we use are drawn from the best of *yesterday's* scientists and mathematicians. Images of hierarchical boxes and lines, or two-dimensional flow charts, flood the white boards and annual reports of progressive, thoughtful corporations. Metaphors of machines, linear mathematics, and causes with their predictable effects, dominate our corporate "communications" meetings and our marketing literature.

What would happen if we turned our attention to the images and metaphors of *today's* scientists and mathematicians? What if we heeded the works of those who investigate the behavior of nonlinear physical and mathematical systems? How would the visions and actions of our corporate selves be transformed? We might fill our flip charts with fractal images and our computational forecasting gear with cellular automata. We might turn our predictive efforts toward creative anticipation and our control resources toward adaptation.

When we communicate in our organizations, we transmit information from one context to another. Today, following the lead of yesterday's scientists, we consider only spatial contexts; we focus our communication on moving information from one place to another. Improvements in technology increase speed and reliability, but communication still means only movement of information in space. Today's students of nature define an infinite number of variables that can define an informational context—relationships, power, autonomy, culture, entropy, linguistic communities, and so on. If we allow ourselves to think about our businesses and our business transactions in such a world of rich, unlimited variables, perhaps we would transcend the limits of habit and resources that tend to bound our financial and organizational decisions.

Come with us into an imaginary world in which the movers and shakers of business depend on the images and metaphors of nonlinear mathematics and the mechanisms of tomorrow's technology to help them make decisions and explain their actions.

FOUNDING A FRACTAL ORGANIZATION

The meeting had been a long and grueling one. As Barbara's eyes adapted to the light, she tried to focus on the desk clock. Could it possibly be only 13:24? She had moved into the virtual reality (VR) state at 9:00 to meet Jack and discuss their possible business venture. This was by no means her longest encounter with Jack, but it was surely the most productive. Barbara was physically, mentally, and emotionally electrified with what she remembered of their work. She was eager to review the results of the meeting, so she carefully removed the electrodes from her temples and turned to retrieve the disk on which their thoughts and images had been recorded. A careful review of the data would help to confirm her intuition: Their business would be incredibly innovative and successful.

Barbara and Jack had known each other for only about a year. Both graduate students in the Gates Virtual Business University, they had stumbled onto one another in a real-time seminar on customer analysis. The content was old hat to both of them, so they wandered away from the seminar to investigate their own theories of the multidimensional analysis of wants and needs. At some point, it became obvious that they should become business partners, so that they could use essential reality to test the theories they had created in virtual reality.

Today's session was the culmination of their adventures, but a great deal of work lay ahead of them both. Jack's assignment was to move into the net and search for signs of any organization that shared their fundamental principles. Barbara was to design and run simulations to test the viability of their plans. On Friday, they would join again in the virtual reality to share their findings and prepare to file their plans with his government and with hers. By Monday, if no unexpected political, economic, or electronic surprise interfered, they would open the electronic doors of their business.

Jack slipped his disk into the reader and started the search utility. He wanted to review and summarize the fundamental principles they had defined. This summary would help him set up efficient search criteria as he scanned the net for signs of similar business enterprises.

He moved to the module "Generative Algorithm," where he and Barbara had first simulated some of the likely future paths of their organization. Switching quickly to REVIEW mode, he was able to relive this part of the conversation in depth. The digitized audio version of their conversation formed the voiceover for the high-resolution visual display: Her comments on the right, his on the left, and the computer's linguistic interpretation superimposed over the two. Icons on the bottom of the screen made other sources of input available: data summaries relevant to topics under discussion; geographical, economic, and demographic maps of areas of interest; bibliographic and electronic references related to their discussion; and parallel translation services, including cultural, gender, and linguistic translation options. After running through this segment of the conversation, Jack opened his documentation buffer and summarized their findings.

Jack knew that any generative algorithm required three interdependent components: (1) initial conditions—the starting context of the company; (2) control parameters—the physical and social resources of the company; and (3) defined operators—the communication and decision-making structure.

Once values for these three components were specified, the generative algorithm would transform the conditions and parameters into complex images of reality. Recent research had confirmed that complex images that evolved from such generative algorithms correlated closely to objective reality. Both Jack and Barbara believed that if they built their business on a reliable generative algorithm and tested that algorithm in virtual space, they would be able to create a business that would adapt to changing conditions while it maintained the stability required for longevity and financial success.

The initial conditions they chose must be appropriate as both inputs and outputs of their business processes. Under this heading, Jack carefully typed three items:

- identified needs
- client characteristics
- perception in the marketplace

Jack recognized that each of these initial conditions was multidimensional because they all included various independent, dependent, and nonlinear interdependent variables. He did not take time to decompose each of the initial conditions into their constituent dimensions because the repository already included detailed analyses of each. Rather, he opened the reference icon and linked his list of initial conditions to the Data Dictionary he considered to be seminal and reliable.

He turned his attention next to the category of control parameters. This component of the algorithm would be somewhat trickier to define. He knew that their selection of control parameters would be critical, since they would dictate the boundaries and the dynamical evolution of the company. All previous studies that were documented in the Data Dictionary had considered only a single control parameter, ignoring the multidimensionality of control, which the competition usually considered highly proprietary.

Both Barbara and Jack were convinced that a single control parameter seldom determined the behavior of any system, much less that of a complex business entity. Their first research project investigated failed entrepreneurial endeavors, and their findings revealed a strong pattern: Focus on a single control parameter (such as market penetration) in ignorance of others (such as resource availability) could lead to rapid but uncontrolled growth, followed by complete collapse. So, Jack took a deep breath and wrote under the title of control parameter:

- values
- technology
- resources

As no one had documented the component variables of these control parameters, Jack had to depend on the decomposition that he and Barbara had derived during their morning's discussion. After searching the disk, he reviewed their discussion and defined the components of each of the control parameters. Barbara had pointed out that in the market segment they were targeting, the resource control parameter was going to be greatly influenced by vendor selection and physical plant location. Predator-prey simulations based on ecological population dynamics were useful in showing that while local competitors would tend to be cooperative and thus predictable, a few of the global competitors could wage a price war and make their supply chain rather chaotic. Barbara's simulations would define boundaries and optimal values for each of the contributors to the control parameters.

That finished the biggest step of the work. Jack's final task of defining the operators, or business processes, for their corporation was easy in comparison. For the sheer joy of the experience, he replayed that portion of their meeting. A smile came to his lips as he spoke into his mike: "Operators: Listen and adapt." As these words appeared on his screen, he felt a happy chill. Jack recognized that this business process design was the competitive advantage he and Barbara were embedding in their nascent corporation.

Jack's work provided the raw materials Barbara would need for her simulations. So he quickly transmitted his summary to her along with a holographic image of a "thumbs-up" sign and moved on to his list of other tasks.

While Jack documented the algorithm, Barbara had rerun their morning conversation. She evaluated the conversation in terms of what she remembered of their generative algorithm. She asked and answered what she considered to be a critical question: Did she and Jack apply their generative algorithm as they discussed their business endeavor? In a quick, qualitative review, Barbara assured herself that they had. By the time she had completed the review, she received the priority message from Jack. This basic information would allow her to build a computer model that would show her some likely futures resulting from their generative algorithm.

She began by simulating some of the scenarios that she and Jack had recorded during the morning. One set of images showed that their control metaparameter "technology" would be able to adapt to any new restrictions—both evolutions looked remarkably similar! She continued to run other studies and looked at these images to identify the similarities across the scenarios. She measured the dimensionality of the images to ensure that their fractal dimensions were similar. She tested the likely ranges of values for the control parameters to identify the critical values and correlated those to business situations that would be especially challenging or difficult.

After she was satisfied with the design of the simulation, Barbara loaded in a range of values for each of the dimensions of each of the initial conditions and control parameters, shipped the shell and data to her supercomputer link, and went to the kitchen for a cup of tea. She returned to a long list of files, each containing an image that had resulted from the generative algorithm. She opened and ran her

SS Utility, which tests for similarity and self-similarity in sets of complex images. It was useful to look for similarity between two images because this helped quantify the impact when a parameter or initial condition was changed. She could also look for self-similarity in a single image, which indicated the relative constancy of values, purpose, power, and knowledge throughout the company. As she expected, the utility identified two classes of images: those with a high degree of similarity, those with no mathematically discernible similarity to others.

She focused first on those that were similar, plotting the variables and values that had resulted in the similar images. She checked the extremes of each value and, to her delight but not amazement, discovered that they would have reasonable and likely correlates in her real business future. She also quantified the nature of self-similarity using a parameter known as the *fractal dimension* and found signs of sufficient internal similarity to indicate corporate stability.

Then she turned her attention to situations in which the images showed no similarity to the others. Dissimilarity was especially evident when time to market was varied by more than 20 percent over a 12-month period. Such inconsistencies required special attention and adjustments to initial conditions, control parameters, and rates of iteration. She discarded images in which the values were highly unlikely in the real world. She knew, for example, that their employees would value diversity and commitment to family. She toyed with the simulations until she was sure that all situations they anticipated could be made to be commensurate with the fundamental vision of the organization at its most stable.

Finally, she shifted the resolution on her computer monitor and the granularity of her mathematical models to simulate interactions at the level of individual, team, organization, industry, and economy. Again, she was pleased to find elements of self-similarity regardless of the scale of the image generated by the simulation. This would help ensure coherence of communication across all levels of the hierarchy.

After she had satisfied herself that their algorithms worked with their anticipated scenarios, she began to simulate a variety of scenarios based on random input. She examined each of the resulting images and found that they all shared self-similarity with the completed set. Barbara's next message to Jack was clearly ecstatic: "I think we've got it!" She copied both Jack and their banker on the final findings. Their banker, being well versed in applications of complexity theory to business, would recognize in their work a strong, solid, and dynamic beginning for a business of the future.

BUSINESS IMPLICATIONS

In this section, we compare the visual characteristics of a fractal image and the characteristics of a functioning organization at a given point in time; we also compare the generation of fractals with the dynamical processes of human corporate systems.

Granted, it is difficult for us to imagine a simple iterated system like the generative algorithm resulting in the complexity and variety required for a successful corporation. But it is equally difficult to imagine that the basic algorithms that drive a cellular automaton result in its complex behavior, or that a simple nonlinear equation can, upon iteration, generate a beautiful and complicated fractal image, or that something as simple as a dripping faucet generates a strange attractor with both random elements and surprising structure. In this world of complex, nonlinear phenomena, a very simple algorithm can generate functional and beautiful complexity. Might a simple algorithm generate a functional and profitable corporate entity?

Several authors in the organizational literature have been taken with the creative imagery of fractals and have explicitly used the concept of fractals in their writings to describe the nature of organizational systems. Zimmerman and Hurst[1] have characterized the fractal organization as:

- Having its growth influenced by small fluctuations and tensions—creativity can arise from conflict, and opposite viewpoints can be reconciled by implicitly new strategies.
- Exhibiting self-similarity—skills, knowledge, values, and behavior are replicated at different levels and functional divisions of the organization, so as to create a more holistic entity and increase communication.
- Having complex boundaries—the external boundaries between the organization and its customers, vendors, and competitors are fuzzy, and cooperative strategies dominate; likewise, the internal boundaries between management and labor, between divisions and work groups, and between functional areas become fuzzy as the organization is managed more systemically.

In a fractal organization, boundaries are the most interesting, creative parts of the organization.[2] No organizational line is unambiguous—within each boundary hides an infinite variety that reflects basic self-similarities. This may mean that the most similar parts of the organization may not be functionally or geographically the closest, and conversely, "near neighbors" may be surprising different. Divisions between people and groups depend on more on perception than formal organizational structure—the closer you look, the more complexity you will find. The complexity of boundaries adds to the diversity of the organization, and diversity maximizes learning potential.

Applications of the fractal metaphor generate innovative strategies for organizational development interventions.[3] M. Wheatley says that fractal boundaries "suggest the futility of searching for ever finer measures of discrete parts of the system." [4]

R. Stacey discusses fractal boundaries as complexity which can only arise when components of the organization create tension by pulling in contradictory directions.[5]

When organizational development is seen as self-organization, Zimmerman and Hurst state that the fractal organization will create high redundancy of function, as opposed to high redundancy of parts, through cross-training and cross-functional work teams. This functional redundancy results in a more streamlined and efficient, but fragile, structure.[6] H.-J. Warnecke uses the word "fractal" to refer to self-directed, autonomous work groups within the factory.[7] He states the important principles behind the fractal work group include self-similarity of goals between team and company, autonomy, cooperation instead of competition, and increased networking to reduce losses of information and knowledge across boundaries.

If a fractal image can be used to describe and influence the behavior of organizations, then the process of generating a fractal might be analogous to the process of creating a new corporation. The Barbara and Jack of our story were applying this principle. What was Barbara looking for in the graphic images generated by her business simulator? She was looking for features that are characteristic of fractals. Those include:

- Heterogeneity that appears in patterns which are not identical but self-similar across many different levels of scale—assuring an organizational constancy of purpose.
- Patterns that are recognizable, but not predictable from one part of an image to another or from one generation to another—assuring diversity and creativity.
- Boundaries between different areas that are not clean, straight lines but are instead infinitely complex, allowing for a rich mixing of transition from one side of the boundary to the other—assuring maximum flexibility in communication and knowledge transfer.

These characteristics in the simulations assured Barbara, Jack, and their banker that the corporation they plan to build will meet the fundamental requirements for business success in the future. Those characteristics include:

- Ability to integrate diversity into some comprehensible whole.
- Balancing stability with adaptation within the same corporate structure.
- Corporate identity that gives cohesion to all tasks but also allows for sufficient differentiation to allow for local variations when called for by the market.
- Organizational boundaries that encourage ambiguity and transformation, as parts of the organization share information and resources to the best advantage of all.

- Recognition that variables are not manipulated independently but relate to each other in complex nonlinear ways.
- Assurance that the behavior of all individuals and groups is fundamentally consistent with the behavior of the corporation and that the behavior of the corporation is synchronized with changes in the industry and the economy as a whole.
- Ability to comprehend the nature of the whole by understanding its basic principles and iterative behavior, not by isolating components in hopes of gaining predictability and control.

These characteristics form the basis of the organizational structure that will be both self-reflective and flexible enough to transform to meet the challenges of the future. The metaphors that support them belong to the sciences of complexity and the mathematics of chaos, not to the linear, Newtonian images of the past. In a fractal organization, the management answers may sound familiar, but the questions are innovative and enlightening. This is what we mean when we say businesses should pay attention to the images and metaphors of *today's* brightest scientists and mathematicians as we plan for tomorrow's organizations.

Notes

This work has been partially funded by 3M Engineering Systems and Technology Laboratory, and Honeywell Solid State Electronics Center. We also wish to thank Dr. Pickover for his helpful editorial comments.

1. B. Zimmerman and D. Hurst, "Breaking the Boundaries: The Fractal Organization," *Journal of Management Inquiry* 2 (4) (1993): 334-355. B. Zimmerman, "Strategic Planning at FEDMET," in *Strategic Management and Organizational Dynamics,* ed. Ralph Stacey (London: Pitman, 1989).

2. G. Eoyang, "Organization as Fractal," in *Proceedings of the Annual Chaos Network Conference,* ed. Mark Michaels (1992): 59-61.

3. S. Guastello, K. Dooley, and J. Goldstein, "Chaos, Organizational Theory, and Organizational Development," in *Chaos Theory in Psychology,* eds. A. Gilgen and F. Abraham (Westport, CT: Greenwood Publishing Group, 1995), 267-278.

4. M. Wheatley, *Leadership and the New Science* (San Francisco: Berrett-Koehler, 1992), 129

5. R. Stacey, *Managing the Unknowable* (San Francisco: Jossey-Bass, 1992).

6. R. Ulanowicz, "Growth and Development: Variational Principles Reconsidered," *European Journal of Operations Research* 30 (1992): 173-178.

7. H.-J. Warnecke, *The Fractal Company* (Berlin: Springer-Verlag, 1993).

IV

Fractals in Music
and Sound

11

Fractals in Music

Manfred Schroeder

Certain number-theoretic sequences with fractal properties have been found to generate appealing melodies and rhythms when the numerical data is transformed into musical notes and rhythmic patterns. Fractal designs have been used in composition, leading to melodies that, although purely mathematical in origin, sound surprisingly baroque, that is, like a piece of baroque music. Some of the musically most interesting mathematical sequences can be generated by recursion, such as the sequence 0 1 1 2 1 2 2 3 . . . , which counts the number of ones in the binary representation of successive integers beginning with zero. Another number-theoretic sequence that has been put to good use in music, albeit rhythmic rather than melodic, is the rabbit sequence 1 0 1 1 0 1 0 1 . . . (so called because of its relation to the procreation of rabbits as envisioned by the medieval Italian mathematician Fibonacci). The analysis of music in terms of fractal patterns may lead to new insights in the classification of music. New musical instruments with fractal design features, such as drums with fractal perimeters, await exploration.

A BIT OF NUMBER THEORY

Fractal sequences of numbers make excellent raw material for melodies and musical rhythms. Many of the musically most interesting fractal sequences come from number theory. A prime example of such a sequence is the Morse-Thue sequence and its close relatives.

For starters, consider the positive integers, 0, 1, 2, 3, 4, 5, 6, 7, . . . , and write them down in the binary number system:

$$0, 1, 10, 11, 100, 101, 110, 111, \ldots \qquad (11.1)$$

Now count the number of ones for each integer. This yields the sequence

$$0, 1, 1, 2, 1, 2, 2, 3, \ldots \qquad (11.2)$$

This "ones-counting sequence" also can be generated recursively by the following simple rule.[1] Consider the first 2^m terms of the sequence. The next 2^m terms are obtained by adding one to each term. Each recursion will thus double the number of known terms.

Starting with a single 0, the first four recursions will yield

```
0
0 1
0 1 1 2
0 1 1 2 1 2 2 3
0 1 1 2 1 2 2 3 1 2 2 3 2 3 3 4
```

The reason why this recursion works is, of course, the fact, for $2^{m-1} \leq k \leq 2^m$, the integer $k + 2^m$ has exactly one additional 1 in its binary notation. For example, for $m = 3$ and $k = 5$ (that is, binary 101), we have $k + 2^3 = 13$, which equals 1101 in binary.

Before continuing, let us note an important property of the ones-counting sequence: It is self-similar, that is, certain parts of the infinite sequence contain the entire sequence. In fact, underlining every second term in equation 11.2, beginning with the initial 0, reproduces the sequence:

$$\underline{0}, 1, \underline{1}, 2, \underline{1}, 2, \underline{2}, 3, \ldots.$$

This process of taking every second term (or, more generally, every mth "atom") is called deflation in physics. Deflation plays an important role in renormalization theories that earned Kenneth Wilson the Nobel Prize. But this self-similarity works both ways: Whenever we encounter deflation, inflation is not far behind.[2] Indeed, our ones-counting sequence also can be generated by "inflating" every single term—that is, by writing directly behind it a new term that is the original term augmented by one. Thus each 0 turns into the pair 0 1, each 1 turns into the pair 1 2, and so on. This inflation rule mimics, on a "microscopic" level, the "macroscopic" recursion that always considers blocks of 2^m terms together. Conversely, the block recursion rule may be said to have "inherited" the basic inflation rule $a \rightarrow a, a + 1$.

This self-similarity ("parts equal the whole") is an important aspect of fractals and contributes to their beauty in geometrical designs. Indeed, self-similarity is a widespread attribute of nature and—like mirror symmetry—one of the most fundamental symmetries of the laws of physics. In science this symmetry is also called scaling invariance. For example, Newton's law of gravitational attraction is, to use a favorite modern locution, "true on all scales," from the smallest dimensions (apples falling from trees) to the entire universe. The reason our little ones-counting sequence has this respectable property borders on the trivial: Taking every second term means multiplying the underlying integer by 2, which shifts the binary representation one place to the left without changing the number of ones. For example, 23 equals 10111 in binary and 2 x 23 = 46 equals 101110; both 23 and 46 have four ones.

If we use the ones-counting sequence (equation 11.2) as exponents of 2:

$$2^0, 2^1, 2^1, 2^2, 2^1, 2^2, 2^2, 2^3, \ldots$$

we obtain another interesting sequence:

$$1, 2, 2, 4, 2, 4, 4, 8, \ldots \tag{11.3}$$

which I have called the "Dress sequence" in honor of the German mathematician Andreas Dress, who used equation 11.3 to model the chemical reaction rate of a catalytic process by a cellular automaton.[3] The Dress sequence, also, can be generated recursively: Consider the first 2^m terms: The next 2^m terms are obtained by multiplying each term by 2. Starting with a single 1, the first three recursions yield

1								$\sum_0 = 1$
1	2							$\sum_1 = 3$
1	2	2	4					$\sum_2 = 9$
1	2	2	4	2	4	4	8	$\sum_3 = 27$

Interestingly, the partial sums $\sum_m$ of length $L_m = 2^m$, $m = 0, 1, 2, 3, \ldots$, of the Dress sequence equal an integer power of 3, namely 3^m. This follows directly from its recursive construction and the starting value $\sum_0 = 1$. In terms of the lengths L_m, the partial sum $\sum_m$ follows a familiar fractal law:

$$\sum_m = L_m^{\log 3/\log 2} \qquad (11.4)$$

The exponent log 3/log 2 is the reciprocal of the Hausdorff dimension, log 2/log 3 = 0.63 . . . of the famous triadic *Cantor set*, an early fractal obtained by eliminating middle-thirds from a straight-line segment. The growth law (equation 11.4) is reminiscent of the Sierpinski triangle, see figure 11.1, whose Hausdorff dimension equals D = log 3/log 2 = 1.58 . . . because a doubling of a linear dimension (such as the altitude or base of the Sierpinski triangle) triples its area. (For a solid triangle, doubling of a linear dimension would, of course, quadruple the area.) The Sierpinski triangle is closely related to Pascal's triangle of binomial coefficients:

$$
\begin{array}{ccccccccc}
 & & & & 1 & & & & \\
 & & & 1 & & 1 & & & \\
 & & 1 & & 2 & & 1 & & \\
 & 1 & & 3 & & 3 & & 1 & \\
1 & & 4 & & 6 & & 4 & & 1 \\
 & & & & \cdots & & & &
\end{array}
\qquad (11.5)
$$

Taking the entries in equation 11.5 modulo 2 yields

$$
\begin{array}{ccccccccc}
 & & & & 1 & & & & \\
 & & & 1 & & 1 & & & \\
 & & 1 & & 0 & & 1 & & \\
 & 1 & & 1 & & 1 & & 1 & \\
1 & & 0 & & 0 & & 0 & & 1 \\
 & & & & \cdots & & & &
\end{array}
\qquad (11.6)
$$

which can be considered a discrete version of the Sierpinski triangle (1 corresponding to black and 0 to white). Note that the sums of successive rows are identical with the Dress sequence 1, 2, 2, 4, ...

If we use the ones-counting sequence as exponents not of 2 but of $2^{1/12}$ (the semitone frequency interval of a well-tempered musical scale), we obtain a "melody" defined by the sequence of relative frequencies

$$2^0 = 1, 2^{1/12}, 2^{1/12}, 2^{2/12}, 2^{1/12}, 2^{2/12}, 2^{2/12}, 2^{3/12}, \ldots \qquad (11.7)$$

Figure 11.1 Sierpinski triangle, a geometric fractal related to the binary representation of the integers.

Another way of converting the ones-counting sequence into a melody, in the
C-major scale, say, is effected by the following transcription suggested by my
student Lars Kindermann:

$$1 = C, 2 = D, 3 = E, \text{ etc.} \qquad (11.8)$$

which (ignoring the initial 0) yields the musical "score"

$$CCDCDDE. \ldots \qquad (11.9)$$

(See the fourth row of figure 11.2.)

To introduce an element of rhythm into this tune, we may decide to sound
a repeated note only once and hold it until the next different note. Thus our little
tune (equation 11.9) would turn into

$$C \ DCD \ E \ldots, \qquad (11.10)$$

an appealing if not very moving ditty.

Baroque Integers															
	1	2	3	4	5	6	7	8	9	10	11	12	13	14	15 ⋯
Binary	1	10	11	100	101	110	111	1000	1001	1010	1011	1100	1101	1110	1111 ⋯
Number of 1's	1	1	2	1	2	2	3	1	2	2	3	2	3	3	4 ⋯
C Major	C	C	D	C	D	D	E	C	D	D	E	C	E	E	F ⋯
Every 3rd Note			D			D			D			C			F ⋯
Melody			D									C			F ⋯

Figure 11.2 The ones-counting sequence and a melody derived from it.

Melodies constructed in this manner become decidedly more engaging when, following Kindermann, we "undersample" certain sequences obtained from number theory. Specifically, keeping only every third term of the ones-counting sequence (figure 11.2, fifth row) results in a much improved melody (figure 2, sixth row):

$$D\ CF.\ldots \qquad\qquad (11.11)$$

Dramatic results have been obtained by taking every sixty-third term, which, in C major, gives

$$BA^{63}GABACABAD.\ldots \qquad\qquad (11.12)$$

(See figure 11.3.) The notation A^{63} means that the letter A is repeated 63 times. With this convention, the corresponding musical note is sounded just once followed by 62 silent intervals (a curious beginning for a tune that then takes off on a capricious melodic itinerary).

Why do melodies based on such simple number-theoretic ideas sound so attractive? Of course, beauty lies in the ears of the listener. But on examining the numerical evidence, one discovers interesting patterns: Small melodic structures repeat within larger structures that are in turn repeated within even larger structures—the telltale token of fractal forms: whorls within whorls within whorls in turbulence; galaxies within clusters of galaxies within superclusters of galaxies in the design of (our) universe. The reader is invited to search for such patterns in figure 11.3.

```
BAAAAAAAAAAAAAAAAAAAAAAAAAAAAAAAAAAAAAAAAAAAAAAAAAAAAAAAAAAAAAAAAAAAAGAB\
ACABADABACABAEABACABADABACABAFABACABADABACABAEABACABADABACABAGGAABBAACCA\
ABBAADDAABBAACCAABBAAEEAABBAACCAABBAADDAABBAACCAABBAAGFGACBCADCDACBCAEDE\
ACBCADCDACBCAFEFACBCADCDACBCAEDEACBCADCDACBCAGGGGAAAABBBBAAAACCCCAAAABBB\
BAAAADDDDAAAABBBBAAAACCCCAAAABBBBAAAAGEFEGABADBCBDABAECDCEABADBCBDABAFDE\
DFABADBCBDABAECDCEABADBCBDABAGGFFGGAACCBBCCAADDCCDDAACCBBCCAAEEDDEEAACCB\
BCCAADDCCDDAACCBBCCAAGFGEGFGADCDBDCDAEDECEDEADCDBDCDAFEFDFEFADCDBDCDAEDE\
CEDEADCDBDCDAGGGGGGGGGAAAAAAAABBBBBBBAAAAAAAAACCCCCCCCAAAAAAAAABBBBBBBAAA\
AAAAAGDEDFDEDGABACABAEBCBDBCBEABACABAFCDCECDCFABACABAEBCBDBCBEABACABAGGE\
EFFEEGGAABBAADDBBCCBBDDAABBAAEECCDDCCEEAABBAADDBBCCBBDDAABBAAGFGDFEFDGFG\
ACBCAEDEBDCDBEDEACBCAFEFCEDECFEFACBCAEDEBDCDBEDEACBCAGGGGGFFFFGGGGAAAACCC\
CBBBBCCCCAAAADDDDCCCCDDDDAAAACCCCBBBBCCCCAAAAGEFEGDEDGEFEGABAECDCEBCBECD\
CEABAFDEDFCDCFDEDFABAECDCEBCBECDCEABAGGFFGGEEGGFFGGAADDCCDDBBDDCCDDAAEED\
DEECCEEDDEEAADDCCDDBBDDCCDDAAGFGEGFGDGFGEGFGAEDECEDEBEDECEDEAFEFDFEFCFEF\
DFEFAEDECEDEBEDECEDEAGGGGGGGGGGGGGGGGGG
```

Figure 11.3 Every sixty-third term of the ones-counting sequence translated into the C-major musical scale.

FRACTAL MUSIC FROM BACH TO LIGETI

The appearance of fractal patterns in music has a long and respectable history. K. J. Hsu and A. Hsu discovered fractal structures in J. S. Bach's "Invention No. 1 in C Major" and other baroque and classical works.[4] Just as in the just-described decimating procedure for number-theoretic sequences, the Hsus have reduced several of Bach's Inventions to as little as one sixty-fourth of the original composition while preserving its basic structure! What would happen to Bach if you took every sixty-third note?

Other fractal aspects of existing music were discovered by R. F. Voss and J. Clark at the IBM Research Center in Yorktown Heights, New York.[5] They found that the Fourier spectra of the pitch and the loudness of successive notes of Bach's First Brandenburg Concerto (and many other compositions) show a $1/f$-like dependence on frequency, which is characteristic of many fractal phenomena, including $1/f$ noise. This is not surprising, as Voss points out, because a $1/f^2$ dependence (also called "brown noise" after Brownian motion) would sound rather dull while a $1/f^0$ law ("white noise") would sound excessively random.[6] Thus the ubiquitous $1/f$ dependence ("pink noise") appears as an excellent compromise between too little and too much "surprise" in a musical composition, in agreement with the theory of aesthetic appeal propounded by the American mathematician George David Birkhoff (1884-1944).[7]

The modern Hungarian-born composer and former physicist Gyorgy Ligeti, who lives in Hamburg, Germany, has been inspired by self-similar fractal images and has used fractal designs deliberately in his compositions, such as his "Etudes for Piano." Charles Wuorinen has collaborated directly with Mandelbrot and incorporated fractal ideas in his compositions.

Noises with self-similar power-law spectra $1/f^\beta$ have the curious property, first pointed out by Mandelbrot,[8] that they will sound the same no matter at what speed they are reproduced because frequency scaling does not change the exponent β. This is in stark contrast to most other acoustic signals, which sound quite different when speeded up or slowed down. Speech, for example, assumes a "Donald Duck" –like quality when speeded up by a factor of two.

But self-similar designs in music hold still other surprises in store. Fractal waveforms, originally proposed by Roger N. Shepard and Jean-Claude Risset, can continually rise in apparent pitch without getting any higher in frequency content.[9] Even more astounding, there even exist fractal waveforms (see figure 11.4) that sound lower in pitch when all their frequencies are raised. The waveform consists of 11 pure tones (sinusoids) spanning the audio frequency range from 10 Hz to 18,284 Hz, each component being higher in frequency by a factor $2^{13/12} \sim$ 2.119 than the one below it. Thus increasing all frequencies by a factor $2^{13/12}$ will not change the auditory percept. But raising all frequencies by a factor of 2 (one octave), rather than 2.119, will lower the perceived pitch by one semitone.[10]

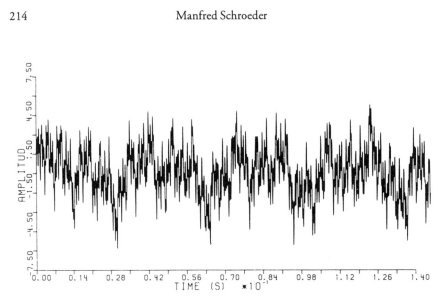

Figure 11.4 Fractal waveform that, paradoxically, sounds lower in pitch when speeded up. Although the waveform itself looks unremarkable, it has a self-similar Fourier spectrum with a frequency scaling similarity factor of $2^{13/12}$.

THE FRACTAL NATURE
OF NUMBER-THEORETIC SEQUENCES

Let us backtrack to our original ones-counting sequence (equation 11.2) and rewrite the sequence modulo 2 (omitting the commas):

$$0\ 1\ 1\ 0\ 1\ 0\ 0\ 1\ldots \qquad (11.13)$$

This is the famous Morse-Thue sequence, which was a first example of a discrete sequence that completely characterizes the trajectory of certain nonlinear dynamic systems in what physicists call phase space (an abstract mathematical space in which the oscillations of a swing, for example, become an ellipse). The astonishing fact that an entire *continuous* trajectory in a multidimensional space could be fully represented by a *discrete* sequence of *discrete* symbols was discovered by the Princeton mathematician Marston Morse in 1922; it still astounds many an expert.

An instance of a dynamic system in which the Morse-Thue sequence occurs is the "logistic parabola," also called "quadratic map," which underlies the famous Mandelbrot set:

$$z_{n+1} = z_n^2 + c. \qquad (11.14)$$

As simple as this equation looks, it gives rise to the breathtaking visual complexities, well known from the depictions, in black and white (or better yet in color) of the Mandelbrot set.[11]

What could be simpler than squaring a number and adding a constant to it as in equation 11.14? Yet beginning at $c = 0$ and decreasing it to $c = -1.4011 \ldots$ (with the starting value $z_0 = 0$) leads to a cascade of period-doublings from period length $P = 1$ (for $c = 0$) via $P = 2$ (for $c = -1$) and $P = 4$ (for $c = -1.3107 \ldots$) to complete chaos (for $c = -1.4011 \ldots$, the so-called Feigenbaum point).

If we characterize successive values of z_n (the itinerary or *orbit*) after $z_0 = 0$ by their algebraic signs, we obtain, at the Feigenbaum point, the chaotic-looking sign sequence:

$$- + - - - + - + - + - - - + - -. \ldots \qquad (11.15)$$

Actually, this sequence is anything but chaotic; it is obtained by a simple recursion.[12] To obtain the first 2^m repeat the first 2^{m-1} terms and change the last sign. Beginning with a single minus sign this yields

$$
\begin{aligned}
&- \\
&- + \\
&- + - - \\
&- + - - - + - + \\
&- + - - - + - + - + - - - + - -.
\end{aligned}
$$

Not surprisingly, the sequence is self-similar: Take every second term and the sequence turns into its own negative

$$+ - + + + - + -. \ldots$$

and repeating the process (of taking every second term) actually reproduces the sequence:

$$- + - -. \ldots$$

Even before its appearance in dynamic systems, the sequence in equation 11.15 was introduced in 1906 by the Norwegian mathematician Axel Thue as an example of an aperiodic, recursively computable string of symbols.

But what is the relation between the Morse-Thue sequence and fractal music? Compute the running product of sequence in equation 11.15 beginning with a single + sign:

$$+ - - + - + + - + - - + - + + - +. \ldots \qquad (11.16)$$

Replacing each + by 0 and each - by 1, we indeed obtain the Morse-Thue sequence (equation 11.13).

The Morse-Thue sequence also boasts a simple recursive construction: Starting with a single +, keep appending the negative of what you already have:

                        +
                        + -
                        + - - +
                        + - - + - + + - . . . .

While taking every second term turns the sequence into its negative, taking every third term has some strange consequences. The decimated sequence starts with a lot of + signs before a first lonely - sign appears:

            + + + + + + + - + + + + + + + + - + + + + + + + + +. . . .

In fact, there seems to be a marked preponderance of + signs. As a result, the partial sums—in contrast to those for the Morse-Thue sequence itself, which keep hovering around zero—keep getting bigger and bigger in an appealing self-similar manner. (See figure 11.5.)

Is the average growth linear, as figure 11.5 might suggest? No, it isn't; see figure 11.6, which shows the partial sums up to 10^4 on a log-log plot. To this curve a straight line can be fitted with a slope of 0.79 . . ., which mathematical theory shows to equal $s = \log 3/\log 4$, a value reminiscent of the Hausdorff dimension of the original triadic Cantor set $D = \log 2/\log 3 = \frac{1}{2}s$. Thus the decimated Morse-Thue sequence is revealed as a bona-fide fractal, having a growth according to a power law with a fractional exponent! Similar fractal behavior has been

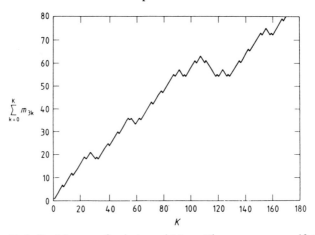

Figure 11.5 Partial sums of a decimated Morse-Thue sequence, a self-similar fractal.

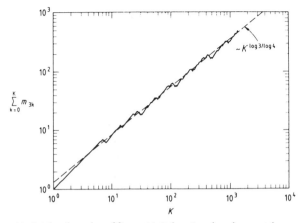

Figure 11.6 A log/log-plot of figure 11.5 showing that the partial sums grow with a fractional exponent log 3/log 4.

observed for other decimating processes, such as taking every fifth, seventh, thirty-first, or sixty-third term. Figure 11.7 shows a bar graph obtained by taking every sixty-third term of the ones-counting sequence and turning it into a melody, according to the Kindermann convention. I encourage readers to generate this and similarly constructed fractal melodies—on the piano, on a computer, or simply by whistling—to convince themselves of the strange "baroque" quality of such tunes.

Interestingly, there is also a recursive formula for the decimated Morse-Thue sequence. My student Gerriet Müller first conjectured and then proved the

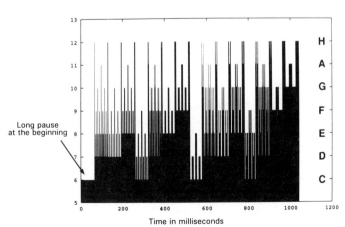

Figure 11.7 Graphical representation of fractal melody obtained from decimating the ones-counting sequence.

following (not exactly simple) recursion for every third term x_k. Let $x_0 = y_0 = +1$, then

$$x_k = x_{k-1}, \hat{x}_{k-1}, y_{k-1}$$

$$y_k = y_{k-1}, (-\hat{y}_{k-1}), x_{k-1} \qquad (11.17)$$

Here $\hat{x}_{k-1}$ is the sequence x_{k-1} read backward. The commas signify concatenation. Thus, $x_1 = + + +$, $y_1 = + - + + +$, $x_2 = + + + + + + + - + + +$, and so on. The sequence y_m may be musically interesting in its own right or in combination with the sequence x_m as a kind of "duet."

Similar recursions are believed to exist for other decimations of the Morse-Thue sequence but have not been proved.

RHYTHM FROM RABBITS

Another number-theoretic sequence that is a treasure trove of astounding applications is a binary sequence that I have called a rabbit sequence. It is related to the Golden Mean and the famous Fibonacci numbers that made their appearance in A.D. 1202. In that year, toward the end of the Dark Ages, an Italian mathematician by the name of Leonardo da Pisa, better known as Fibonacci (son of Bonacci), published a book, *Liber Abaci,* in which he considered, among numerous other fertile problems, the multiplication of rabbits, which are well known for their prolific procreation. Fibonacci's reasoning was based on what we would now call a mathematical model of the breeding process—an outrageously simplified one, to be sure, just like many a modern mathematical model. Fibonacci assumed that in each generation a mature rabbit pair produced one young rabbit pair, which itself grew to maturity in one generation. Also, rabbits never died; they stayed alive forever. How will their numbers grow; how will they multiply?

Designating a young rabbit pair by 0 and a mature pair by 1, Fibonacci's rule implies the following recursive mapping: $0 \rightarrow 1$, $1 \rightarrow 10$. Starting with a single young pair, the first six generations can be represented as follows:

(1)	0									1
(2)	1									1
(3)	1	0								2
(4)	1	0	1							3
(5)	1	0	1	1	0					5
(6)	1	0	1	1	0	1	0	1		8

Another method of generating the next generation of the rabbit sequence is to append to the last generation the one before it. Thus the fifth generation (10110) is obtained from the fourth generation (101) by appending the third generation (10) to the fourth generation.

The numbers in the rightmost column are the numbers of rabbit pairs in each generation, the so-called Fibonacci numbers F_n, which obey the recursion $F_1 = F_2 = 1$, $F_n = F_{n-1} + F_{n-2}$ and occur in innumerable contexts. They also can be calculated directly by the surprisingly simple formula

$$F_n = \left[\frac{\gamma^{-n}}{\sqrt{5}} \right] \qquad (11.18)$$

where $\gamma = (\sqrt{5} - 1)/2 = 0.618 \ldots$ is the Golden Mean and the funny-looking brackets mean "round down to the nearest integer."

The ubiquitous Fibonacci numbers have been employed as parents for musical tunes by an author who calls her (him)self by the transparent pseudonym U. Phonious in a brief note entitled "Playing the Numbers."[13] The euphonious trick is to reduce the Fibonacci numbers modulo some well-chosen integer, say 51. This produces periodic sequences of whole(some) numbers that can sound rather appealing, especially in combination with other sequences that define both melodies and rhythms.

For the mathematically inclined reader, it may be challenging to guess and then prove a formula for the period lengths L_M for a given modulus M. Here are some *numerical* results: $L_2 = 3$, $L_3 = 8$, $L_4 = 6$, $L_5 = 20$, $L_6 = 24$, $L_7 = 16, \ldots, L_{10} = 60, \ldots, L_{14} = 48, \ldots, L_{25} = 100, \ldots, L_{29} = 14, \ldots, L_{50} = 300$, $L_{51} = 72, \ldots$, $L_{98} = 336, \ldots$. Looks quite random, doesn't it? But, of course, the L_M cannot be random.

After squeezing the Fibonacci sequence 1, 1, 2, 3, 5,8, . . . a bit, we now turn our attention to the underlying binary bit sequence, the *rabbit* sequence:

$$\{r_k\} = 1\ 0\ 1\ 1\ 0\ 1\ 0\ 1\ 1\ 0\ 1\ 1\ 0. \ldots \qquad (11.19)$$

Like the Morse-Thue and the ones-counting sequences, the infinite rabbit sequence is self-similar: Part of the sequence equals the entire sequence. Replace every 10 by a 1 and every 1 not followed by a 0 by a 0 and the infinite sequence reproduces itself:

$$1\ 0\ 1\ 1\ 0\ 1\ 0\ 1. \ldots \qquad (11.20)$$

The rabbit sequence is also fundamental in the construction of quasi-periodic lattices that underlie newly discovered quasi-crystals. Quasi-crystals are a new state of solid matter that are neither periodic in space (like crystals of rock salt) nor amorphous (like graphite and glass). The rabbit sequence also appears in the symbolic dynamics of the critical circle map that describes the mode-locking between coupled oscillators, such as planets and moons. The rabbit sequence also can be generated directly, without recursion, by means of the sheer inexhaustible Golden Mean :

$$r_k = [(k + 1)\, \vartheta] - [k\, \vartheta]. \tag{11.21}$$

(As with equation 11.18, the bracket symbols in equations 11.21 and 11.22 indicate to round down to the nearest integer.) We can even say in what positions k all the 1s and 0s are:

$$r_k = 1 \text{ for } k = [n\, \vartheta^{-1}] \tag{11.22}$$

$$r_k = 0 \text{ for } k = [\, n\, \vartheta^{-2}\,]$$

where n runs through all the natural number n = 1, 2, 3, 4. . . . Equation 11.22 shows that, asymptotically, the ratio of the number of 0s to the number of 1s equals the Golden Mean. The pair of integer sequences generated by equation 11.22, called Beatty sequences:

$$1\ 3\ 4\ 6\ 8\ 9 \ldots$$

$$2\ 5\ 7\ 10\ 13\ 15 \ldots \tag{11.23}$$

exhaust all the positive integers and define the winning positions in the game of Fibonacci Nim.[14]

An alternative way of generating this pair of Beatty sequences (equation 11.23) is to start with 1, add 1 to it, and write the sum (2) below the 1. Then write to the right of the 1 the smallest positive integer that has not already appeared (3), add 2 to it, and write the sum (5) below the 3. The next smallest unused number is 4, add 3 to it and write the sum (7) below the 4, and so forth, always adding the next larger number.

There is also a geometric version of Fibonacci Nim, the board game Corner the Lady, and a geometric construction of the winning positions.[15]

BEATS FROM BITS

With all these multifarious properties, it is perhaps small wonder that the rabbit sequence has been put to musical use. Since the Golden Mean is an irrational number, it follows from equation 11.21 that the rabbit sequence is aperiodic. However, certain patterns repeat in a rabbit kind of fashion. For example, the pattern 101 (in equation 11.18) occurs again and again with successive spacings (between each final member of the pattern [equation 11.1] and the next first member [Equation 11.1] of the next occurrence of the pattern) being equal to 1, 0, 1, 1, . . . , which is the rabbit sequence. The same is true for any other patterns: Each partial rabbit pattern repeats in a manner that mimics the rabbit pattern itself!

My students Wolf Dieter Brandt, Rainer Wilhelm, and Holger Behme have converted the rabbit sequence into a rhythm: a short beat for each 0 and long or double beat for each 1. The result is an astounding rhythmic design. No matter what subpattern we listened to, it never repeated itself periodically but followed the same rabbit rhythm, including its own pattern: a rhythm within a rhythm within a rhythm. . . . Readers are encouraged to experiment with rabbit rhythms and to invent their own variations on this theme, using generalized rabbit sequences based not on the Golden Mean but one of the Silver Means as described in my book: *Fractals, Chaos, Power Laws: Minutes from an Infinite Paradise*.[16] This book also contains numerous other recipes for self-similar fractal designs, including paper-folding sequences and self-similar chain codes (which are important in computer graphics).

In addition to periodic, quasi-periodic, and fractal sequences, random and pseudorandom sequences also may be useful for musical purposes. The best known and most widely used pseudorandom sequence are those that are variously called shift-register, maximum-length, or Galois sequences because they are obtained from finite (Galois) fields in number theory.[17] These sequences are distinguished by having flat Fourier spectra, which makes them good candidates for precision measurements at extremely low signal-to-noise ratios, such as the travel time of radar pulses to the planet Mercury when it is behind the sun as seen from Earth (a sensitive test of Einstein's General Theory of Relativity) or acoustic measurements in a concert hall during an actual performance without disturbing the audience. In another application to acoustics, number-theoretic sequences, including Galois sequences and quadratic-residue sequences, have been used in the design of diffusers for sound waves, diffusing them into broad patterns and thereby improving the acoustical quality of concert and lecture halls, recording studios, churches, and even private homes.[18] (A quadratic-residue sequence is obtained by squaring successive integers, dividing by a [prime] number, and taking only the remainder. Thus, for the prime number 7, the quadratic-residue sequence is 0, 1, 4, 2, 2, 4, 1; 0, 1, 4, and so on, repeated periodically with a period length of 7.)

Beyond the pseudorandom sequences derived from Galois fields, which are completely predictable, there lurk other sequences, one of which is the binary Gambler's Fallacy Sequence in which each bit at the end of a given string is the opposite of what one would expect from the past.[19]

As a further variation on our scheme, fractal number-theoretic rhythms might be combined with fractal melodies. More generally, the periodic, quasi-periodic, pseudorandom, and fractal sequences mentioned here should be considered

but samplings of possible building blocks or "strands" to be woven into compositions of artistic merit by the musically gifted.

CAN YOU HEAR THE SHAPE OF A DRUM?

Traditional musical instruments do not only exist in Euclidean space; their designs
are Euclidean and the sounds they produce travel in Euclidean space. The
geometry of a musical instrument largely determines the sound spectrum it
produces (although think of violins: The wood, the string tension, and the bow
pressure are also important ingredients of musical quality).

But does the Fourier spectrum of sound completely determine the geometry? This is a deep mathematical question that Mark Kac once asked.[20] Unfortunately, it does not, or we could calculate from a human speech sound the shape of
the vocal tract and the position of the tongue. This would be a great boon for
automatic speech recognition and "visible speech" to aid the deaf to understand
speech by means of visual cues beyond lip reading. Nevertheless, in a well-defined
sense, the Fourier spectrum does provide half the geometric information in
Euclidean designs.

But suppose we leave our comfortable Euclidean cottage and start thinking
along fractal lines (if that is possible). What would a trumpet with a fractal flare or
drum with a fractal perimeter sound like? Michael Berry gave a preliminary answer,[21]
and the problem was finally resolved by M. S. Lapidus and J. Fleckinger-Pellé, who
showed that the *Minkowski* (not Hausdorff) dimension is the determining fractal
parameter. But while the theoretical questions have been answered, broad exploration
of fractal designs for musical instruments is still waiting in the wings. Music is never
immobile: For musical instruments as well as rhythm and melody, fractals are bound
to affect the future.

Notes

1. M. Schroeder, *Fractals, Chaos, Power Laws: Minutes from an Infinite Paradise*
    (New York: W. H. Freeman, 1992).
2. Ibid.
3. A. W. M. Dress, M. Gerhardt, N. I. Jaeger, P. J. Plath, and H. Schuster, "Some
    Proposals Concerning the Mathematical Modelling of Oscillating Heteroge-
    neous Catalytic Reactions on Metal Surfaces," in L. Rensing and N. I. Jaeger,
    eds., *Temporal Order* (Berlin: Springer, 1985), 67-74.
4. K. J. Hsu and A. Hsu, "Self-similarity of the 1/f-noise Called Music," *Proceedings
    of the National Academy of Science* 88 (1991): 3507-3509.

5. R. V. Voss and J. Clark, "1/f Noise in Music: Music from 1/f Noise," *Journal of the Acoustical Society of America* 63(1978): 258-263.

6. M. Gardner, "White and Brown Music, Fractal Curves and 1/f Noise," *Scientific American* 238 (April 1978): 16-32.

7. Schroeder, *Fractals, Chaos, Power Laws,* 107-112.

8. B. B. Mandelbrot, *The Fractal Geometry of Nature* (New York: Freeman, 1983).

9. R. N. Shepard, "Circularity in Judgments of Pitch," *Journal of the Acoustical Society of America* 36 (1964): 2346-2353. J. C. Risset, "Paradoxes de hauteur," *7th International Congress of Acoustics,* Budapest, paper S10 (1971): 20.

10. M. R. Schroeder, "Auditory Paradox Based on Fractal Waveform," *Journal of the Acoustical Society of America* 79 (1986):186-189. M. R. Schroeder, "Is There Such a Thing as Fractal Music?" *Nature* (London) 325 (1986): 765-766.

11. Mandelbrot, *Fractal Nature of Geometry.*

12. M. R. Schroeder, *Number Theory in Science and Communication. With Applications in Cryptography, Physics, Digital Information and Self-Similarity,* 2nd enlarged ed. (Berlin: Springer, 1990).

13. U. Phonius, "Playing the Numbers," *American Mathematical Monthly* 97 (10) (December 1990): 889.

14. Schroeder, *Fractals, Chaos, Power Laws.*

15. Ibid.

16. Ibid.

17. Schroeder, *Number Theory in Science and Communication.*

18. Ibid.

19. A. Ehrenfeucht and J. Mycielski, "A Pseudorandom Sequence—How Random Is It?" in "Unsolved Problems," R. Guy, ed., *The American Mathematical Monthly* 99 (April 1992): 373-375.

20. Mark Kac, *Enigmas of Chance* (New York: Harper & Row, 1985).

21. M. V. Berry, "Diffractals," *Journal of Physics* A12 (1979): 781-797.

12

Using Strange Attractors to Model Sound

Jonathan Mackenzie

This chapter explores the possible application of fractals and chaos to the computer modeling of sound. Both are recent developments with enormous potential; the former because of the understanding it offers about the nature of complex phenomena, and the latter because of its power to represent and manipulate sound for creative purposes. This chapter presents a number of ideas, techniques, results and speculations to provide insight into these topics.

INTRODUCTION

This chapter is about applying science and technology to the arts. In particular, the science is that of fractal geometry and chaos theory, the technology is the computer, and the medium of interest, sound. Fractals and chaos are recent developments that are revolutionizing our understanding of the complex and irregular nature of the world. Their relevance and use is currently spreading through a diverse range of subjects. A number of developing areas of interest are characterized by the overlap of both scientific and artistic concerns. Although fractals and chaos are essentially scientific phenomena, their reliance on computer-generated images and the accessible and revelatory nature of some of the concepts have inspired this interdisciplinary fusion. In particular, two subjects have emerged that have considerable popularity: visual art and music. Both combine fractal and chaotic models with computer technology to provide powerful tools for artistic experimentation. The aim of my work is to seek a parallel to this, but involving sound.

Consider the images shown in figure 12.1. These are examples of the power of fractals and chaos. Using only very simple models it is possible to create images that can be either complex abstract forms or realistic replicas of natural objects. The question is: Can the same be found in the acoustic domain? For example, could a complex, naturally occurring sound be represented with a simple model? Does there exist an aural equivalent of the Julia set?

Figure 12.1 A synthetic cloud, fern, and Julia set. These were generated with the public domain software *fractint* using the *plasma*, *ifs*, and *julia* options.

Interest in fractal music has concentrated on the arrangement of sequences of notes with reference to fractal or chaotic models. Although the end product is audio, the actual sounds used are conventional natural or synthetic ones.[1] The time scale on which fractals and chaos are being used, then, is different from that of the sounds themselves. Musical fluctuations range from thousandths of Hertz up to several Hertz. Audio fluctuations, however, range from hundreds of Hertz to tens of thousands. An important discovery that supports the use of fractals and chaos for music composition is that, when analyzed, music from a wide range of cultures and historical periods is found to have fractal properties.[2] It has been suggested by Mandelbrot, though, that such properties should not extend beyond the musical structure to the sounds themselves, as these are governed by different mechanisms.[3]

But why should this necessarily be the case? What about the complex and irregular side of musical sound, for example the hiss of a breathy saxophone or the crash of a cymbal? Also, what about nonmusical sound? All around us there are complex and irregular sounds generated by our environments: a burbling brook, splashing water, the roaring of the wind, the rumble of thunder and the variety of screeching, scraping, buzzing, and humming noises made by machinery. Is it, perhaps, that these sounds represent an aural equivalent to the shapes found in nature that have been neglected by Euclidean geometry and then rediscovered with fractals? Compare these two quotes, one from the contemporary composer Iannis Xenakis on the limitations of the conventional Fourier approach of modeling musical sound and the other from Mandelbrot in the introduction to his *The Fractal Geometry of Nature*: "It is as though we wanted to express a sinuous mountain silhouette by portions of circles . . ."[4]; "Clouds are not spheres, mountains are not cones, coastlines are not circles, and bark is not smooth, nor does lightning travel in straight lines."[5]

So, the aim of my chapter is to explore the possibility of developing computerized tools that would allow control over such sounds for creative uses. Apart from the interest in this as a research topic, the potential applications for such tools include computer music composition and the generation of sound effects for film and television.

MODELING SOUND

To begin, it is necessary to give an overview of what a sound model is and what its desired features are. With this background, it then will be possible to look to the theories of fractals and chaos to see how they may be applied.

Figure 12.2 shows a diagram of what is meant by a sound model. It requires that the sound be digitized so that it can exist, so to speak, within a computer. Sound is a fluctuation in time of air pressure that, via the digitization process, may

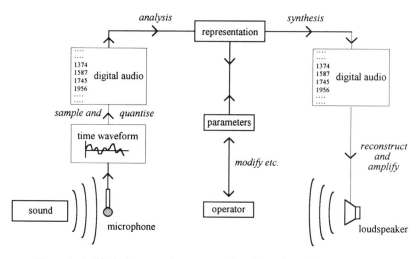

Figure 12.2 Block diagram of a computer-based sound model.

then be described by a stream of numbers known as *digital audio*. This, in turn, may then be reconstructed to form sound again. Digital audio is an enormously flexible way of representing sound. A naturally occurring sound may be modified by processing the numbers that represent it, or numbers may be generated from algorithms and turned into synthetic sound. Any such manipulation may be done with the unlimited flexibility of a computer program. However, digital audio in its raw form may be overwhelming to deal with. Although there is total control over the sound through control of the stream of numbers, there are a lot of numbers to manipulate. Typically, for high-quality audio, each ten seconds of sound requires about a million bytes of numeric data. Also, it is not entirely obvious how these numbers relate to the perceived quality of the sound. Consequently, some further representation of the sound—that is, the digital audio—is needed to make a good practical model.

Imagine the following idealized model of drum sounds. What might be its attractive features for a composer? The model might make use of the recording of an original drum sound and reproduce it so as to retain its relevant characteristics, discarding any perceptually unimportant information in the process. The model might then allow the sound to be modified in a way related to its physical attributes; for example, to be able to change the sound as if it came from a larger version of the same drum, or one that had a tighter skin and has been struck with a different beater. Furthermore, it might allow drumlike sounds to be generated that would not be possible to create with real instruments.

This example suggests, then, various important features required of the representation. First, there are the analysis and synthesis processes, which transform between the raw digital audio and the desired form of the representation. This transformation must be capable of preserving the perceived qualities of the sound. It also produces a new set of data—the model parameters. The user of the model controls these parameters instead of the raw digital audio. The model is manageable because there should be far fewer parameters than original digital audio data. Also, these parameters should relate meaningfully to the original sound being modeled. If the user manipulates these parameters, the sound should be modified in an interesting or useful way. For more detail on the methodology of sound models and conventional representations, see M.V. Mathews, J.C. Risset and D.L. Wessel, and F.R. Moore.[6]

In the next two sections some of the theory of fractals and chaos are reviewed before I return to the subject of modeling sound to discuss how the two might be combined.

FRACTALS AND CHAOS

Chaos theory is about a new understanding of dynamics, the way in which systems behave through time. It concerns the realization that deterministic systems that obey fixed laws can exhibit unpredictable behavior. This runs contrary to the established viewpoint, dating back to Newton, that the behavior of deterministic systems can be calculated for all future time. Also, chaotic behavior, characterized by being irregular and complex, may be found in very simple systems. This, again, is apparently contradictory according to traditional scientific expectations that complex behavior arises only in complex systems.

The theory of fractals, however, provides a new understanding of geometry. It is based on a realization that there exists a large class of geometric objects not encompassed by the traditional Euclidean geometry of points, lines and circles, or the forms of differential calculus—that is, smooth curves. Fractal objects have properties unlike those of their traditional counterparts because of the way they fill space. For example, they typically have dimensions that are not integers. Shapes with infinite circumference can surround a finite area. Curves may be everywhere continuous, but nowhere differentiable. Many have the same form when viewed at different scales, a property known as self-similarity. Like chaos, it is also possible to construct complex fractal forms using only simple rules.

Of greatest importance, perhaps, is that both chaos and fractals can represent natural phenomena with uncanny accuracy. Advancements in abstract theory have been paralleled with discoveries of real-world phenomena that confirm the relevance and usefulness of chaos and fractals.

STRANGE ATTRACTORS

An attractor is a geometric object arising in the theory of dynamical systems. In chaotic systems, the attractor has strange properties that relate to it having a fractal structure. This section provides background regarding strange attractors before suggesting their use as sound models.

The study of dynamical systems is often based on the concept of a system's state. Typically this is a set of numbers—for example, a vector—that completely describes the condition of a system at a single moment in time. Geometry enters the theory in the form of state space—not a physical space, but an abstract one. This space is configured such that the state of the system is represented by a single point somewhere within it. In dynamic systems, the state changes with time and consequently the point in state space moves. With deterministic systems, this movement is governed by fixed laws or rules. This may be written with a simple equation:

$$x_{n+1} = F(x_n) \tag{12.1}$$

This indicates that the state at the nth+1 time step is determined, uniquely, by the previous state at time n and the fixed rule F, which is known as a mapping. The path then taken by the state x or, equivalently, the point in state space, is called a *trajectory*. This is shown in figure 12.3.

Now, an *attractor,* as its name suggests, is an object in state space to which trajectories are "pulled," or attracted, in the long term. Irrespective of the starting state of the system, after enough time has passed, the state is guaranteed to be found moving along a trajectory somewhere on the attractor. It is therefore a description of the long-term dynamics of the system. Attractors come in a variety of different shapes. Traditionally, only Euclidean forms of attractor were known (for example, points, circles, and torii—see figure 12.4). For a chaotic system, however, there exists a 'strange' attractor to which trajectories are pulled and which has a

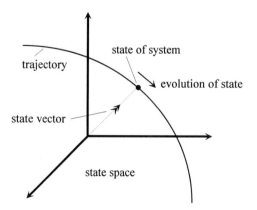

Figure 12.3 Diagram showing the state and its trajectory in state space.

non-Euclidean (fractal) geometry. An example is shown in figure 12.5. This is the attractor of the Lorenz system, constructed by iterating the system of equations,

$$x_{n+1} = x_n + \dot{x} \, \partial t$$
$$y_{n+1} = y_n + \dot{y} \, \partial t$$
$$z_{n+1} = x_n + \dot{z} \, \partial t \qquad (12.2)$$

where

$$\dot{x} = \sigma(y_n - z_n)$$
$$\dot{y} = Rx_n - y_n - x_n z_n$$
$$\dot{z} = x_n y_n - bz_n \qquad (12.3)$$

and

$$\sigma = 10.0, \, R = 28.0, \, b = 2.67, \, \delta \, t = 0.0 \qquad (12.4)$$

while plotting the x variable against the y variable.

Notice how the trajectories never cross one another or meet to form a closed loop. They have a banded arrangement that similarly occurs on a range of scales when the attractor is successively magnified. These geometric properties are typical of fractal objects and correspond to the complex and irregular nature of the system's dynamic behavior.

To summarize, a strange attractor is a geometric object that embodies the dynamical behavior of a chaotic system and is itself a fractal. It is defined by the system rule, F, and may be complex and intricate despite F being simple. This

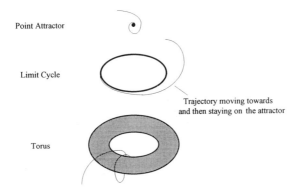

Figure 12.4 Three traditional attractors that have Euclidean geometry.

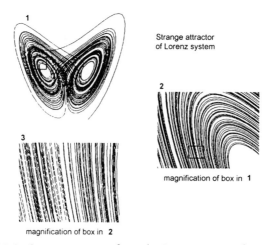

Strange attractor
of Lorenz system

magnification of box in **1**

magnification of box in **2**

Figure 12.5   Strange attractor, from the Lorenz system, showing its fractal properties.

therefore presents a powerful way of manipulating complex fractal objects via simple models. Both the fern and Julia set in figure 12.1 are examples of strange attractors that occur for dynamical systems defined by simple rules. Moreover, for some systems, a slight change in the mapping F, defined by a small number of parameters, results in a small change in the strange attractor. An example of this is shown in figure 12.6. In other words, the chaotic system model provides not

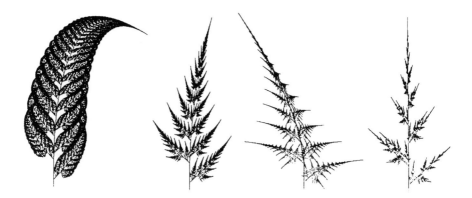

Figure 12.6 An illustration of the power to control strange attractors by modifying the mapping that defines them. Shown are a number of variants of the fractal fern made by slightly altering some of the 28 parameters that define the mapping.

only complex fractal objects from simple systems but a means to control the generation of similar forms.

Perhaps strange attractors have the right kind of properties to be used as a representational basis for sounds. They are simple, controllable systems exhibiting complex behavior that can represent nature or are beautiful in their own right. In order to proceed with this idea, it is necessary to ask a number of questions.

DIAGNOSING NATURALLY OCCURRING FRACTAL SOUND

Having proposed that strange attractors be used as the basis of a sound model, it is now necessary to ask two questions: Are any sounds the product of chaotic systems that possess (fractal) strange attractors? Do any other features of sound have fractal properties? For example, are their corresponding time waveforms fractal? The answer to both these questions is yes—there is evidence showing that some sounds do have chaotic and fractal properties. The following are brief outlines of some examples.

A number of researchers have found that woodwind instruments, such as the recorder, oboe, clarinet, and saxophone, can display a phenomenon known as *bifurcation*—see, for example, Gibiat and Lindenman.[7] Bifurcation describes the way in which a nonlinear dynamical system radically changes its behavior as some parameter of the system is changed. Viewed in state space, this corresponds to a change in form of the attractor. It has already been established that woodwinds rely on a nonlinear mechanism in the mouthpiece to produce the raw excitation tone that then drives the resonating part of the instrument to produce the sound. In woodwind instruments, bifurcation occurs as the blowing pressure at the mouthpiece—the parameter of the system—is increased. For low pressures, nothing is heard and the attractor in state space is a fixed point. Then a bifurcation occurs at the point in which the blowing pressure is sufficient to cause a tone. The attractor is now a simple closed loop. This is the standard regime in which such instruments are usually played. For higher pressures, however, the sound can alter in a variety of ways involving the introduction of new frequencies. The form of the attractor changes to a twisted closed loop, or a torus. For even higher pressure, the sound becomes very rich and noisy. This is found to be a bifurcation to chaos, and again the attractor changes its form and becomes a fractal object.

The sounds of gongs and cymbals are typically complex and irregular despite the instruments themselves having a relatively simple shape. It has been suggested that physical nonlinearities in these instruments allow chaotic dynamics to exist that are responsible for the sounds they make. In this case the nonlinearities are their rims and domes. Experiments have shown that bifurcation sequences occur when a gong is excited with a sinusoidal vibration of variable intensity.[8]

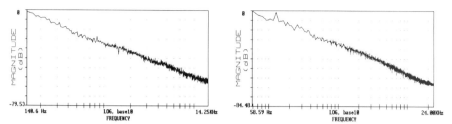

Figure 12.7 Power spectrum for both wind noise (left) and the ambient sound of a large room with ventilation fans (right). Both show an approximately $1/f$ shape over the audio band of frequencies, which is seen as a straight line on logarithmic scales.

These two examples suggest that chaotic dynamics are responsible for certain sounds, but what about the existence of other fractal features apart from state space attractors? The most obvious place to look is at the time waveform of the acoustic signal. One means of diagnosing a fractal structure is to estimate the geometric dimension of the object in question. If this is not an integer quantity, as it is for Euclidean forms, then this is a strong indication that the object is a fractal. Such dimensional analysis has been applied to sound time waveforms. These have included mainly speech sounds but also a variety of animal noises.[9]

Finally, in the course of my own work, I have discovered that the sound of the wind has a strongly fractal nature. This is borne out by spectral analysis of the time waveform, which reveals a $1/f$ form to the power spectrum—see figure 12.7. This means that the power, or intensity, of the signal is distributed over the audio spectrum such that it is inversely proportional to the frequency. Consequently, the sound has a strong low-frequency element to it, which can be heard as a low, deep rumble. A $1/f$ relationship such as this implies that the time waveform is statistically self-affine. This means that if a portion of the waveform graph is magnified, it will have similar average properties to the original. Something like a $1/f$ spectrum can also be found for the ambient sound of rooms where the main sources of noise are ventilation fans. Signals with this pattern to their spectra, known generically as $1/f$ noises, are common in a wide variety of situations including, as mentioned earlier, music. Despite this, there is no single explanation of their origins, although they often are associated with complex nonlinear dynamical systems.

These examples provide strong positive evidence that some natural sounds do have chaotic and fractal features. This places the idea of a strange attractor model for sound on a firm physical basis. In any case, it also is possible to generate abstract synthetic sounds directly, without any reference to the real world. How, then, can strange attractors be used as part of a sound model? The examples suggest two approaches: modeling the fractal attractor in state space and modeling the fractal time waveform.

A synthetic strange attractor can be used in both cases. In the second case, the strange attractor must be in the form of a time waveform—a single valued function of time. Such objects already exist in the form of Barnsley's fractal interpolation functions, or FIFs, discussed in the following two sections. The first of these explores a technique for creating synthetic sounds using a small number of parameters. The second considers how FIFs may be used to represent natural sound in an analysis/synthesis model.

SYNTHESIZING ABSTRACT SOUNDS WITH FIFs

This section describes my own technique for generating synthetic fractal sounds using fractal interpolation functions. Both an outline of the theory and a program code example are provided to encourage experimentation by the reader. Of particular interest for computer music composition are the class of sounds in which both the rhythm and timbre derive from the same fractal structure.

Fractal interpolation functions (FIFs) are a class of iterated function systems (IFS), a general scheme for describing and manipulating fractal attractors. The fern pictures shown previously, for example, are IFS attractors. IFS provide a well-understood, easily implemented, and manageable methodology for working with complex fractal attractors defined with simple chaotic systems.[10] FIFs are a special type of IFS with restrictions made on the parameters that define the attractor to make it of a specific form. As the name suggests, the attractor is a function that passes through a given set of points on a graph. In particular, FIFs are single-valued functions of one variable; since sound waveforms are functions of the same form, FIFs can be used directly to represent them.

FIFs are the strange attractors of chaotic systems generated by

$$x_{n+1} = F(x_n) \tag{12.5}$$

where the state space is two-dimensional and represents the x-y plane on which a graph resides. F is composed of a set of linear mappings having a special form and arrangement to guarantee the attractor is a single-valued function. These mappings are derived from the position of the interpolation points.

Figure 12.8 shows an example. Three linear maps, labeled, w_1, w_2, w_3, are defined by the four interpolation points (the (x,y) pairs) and three associated vertical scaling factors, labeled d. The diagram shows the effect of each mapping on the area A. It is the inverses of the ws that form the mapping F. The combination of the ws (let this be called W) may be used, however, to geometrically construct the attractor by repeatedly applying it to the graph of any function in the x-y plane. As long as all the ds are less than 1 in magnitude, the repeated application, or

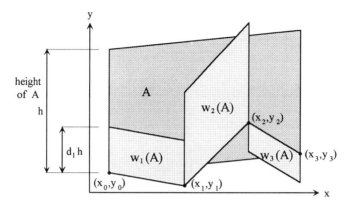

Figure 12.8 The interpolation points, corresponding linear mappings, and their effect on the area A.

iteration, of W to any starting function will generate a fractal function that interpolates the points (x,y). This is shown in figure 12.9.

In this case, 16 interpolation points are specified, their values having been taken from regularly spaced points along a sine wave. The sequence shows the initial function, in this case $y = 0$, which is a straight line along the x-axis, and the effect of repeatedly applying W to it. The first iteration reveals the position of the 16 interpolation points, and after only three iterations, the resulting fractal function appears.

This example illustrates how the FIF generation technique works. The resulting function may then be turned into sound with, for example, a sound card or digital-to-analog converter attached to a personal computer. In this case it does not produce a particularly interesting sound, just a steady-state tone. Figure 12.10 shows the result from using interpolation points that are also derived from the sine wave but which are exponentially placed along the x-axis (that is, at x values 1, 2, 4, 8, 16 and so on as opposed to points 10, 20, 30, 40 . . .). Now the resulting function is much more complex. Its fractal, self-similar structure is revealed by a magnification of one large ripple, which consists of a smaller version of the whole waveform. The waveform contains 65,536 samples and, at a sample rate of 48kHz, lasts about one and a half seconds when turned into sound. It consists of a complex pulsing sequence of percussive tones. Both the tones fall in pitch, and the rhythm slows down, a consequence of the same structure existing on different time scales but being perceived differently. This is an example of the function having an audible as well as visual fractal structure. The fractal structure also is exhibited by playing the sound at a different speed, which can be done using a variable speed tape player, changing the sampling rate, or using a sampling instrument. Played at half the speed, for example, the sound retains approximately the same perceived qualities but lasts twice as long.

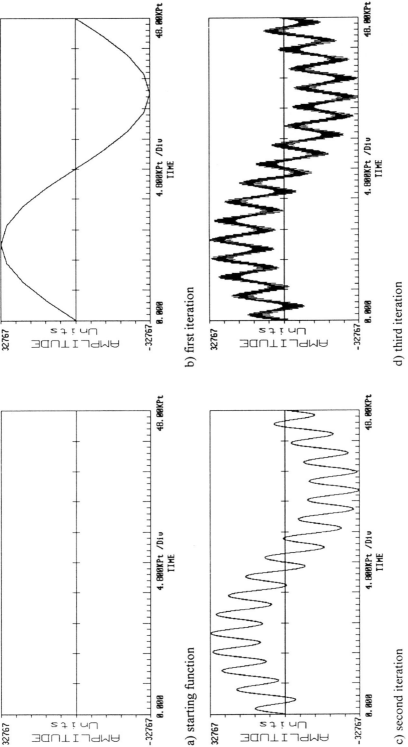

Figure 12.9 Starting function and sequence of three iterations of the composite mapping W, showing how an FIF is produced. This example used 16 interpolation points taken from a single cycle of a sinewave.

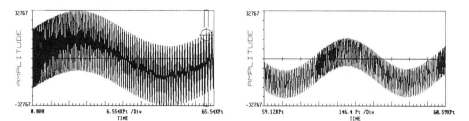

Figure 12.10 An FIF created with interpolation points taken from a sinewave but with exponential *x* spacing. On the left is the whole waveform and on the right a magnification of a small part of it, illustrating the function's fractal structure.

A wide variety of complex sounds may be created by choosing a small set of interpolation points, and this provides a powerful and easily manageable synthesis technique. Program code 2.1 shows how this might be implemented on a personal computer to generate an array of numbers that can then be turned into sound using any available sound card.

Program code 12.1

Generating Digital Audio with Fractal Interpolation Functions

---

This program begins with a set of interpolation points, (x,y) and vertical scaling factors, d, specified by the user, and produces an array containing an FIF which may be treated as a digital audio sequence and converted into sound using a sound card.

---

arrays and variables
x[nmaps+1], y[nmaps+1] - interpolation points
nmaps - number of linear mappings
d[nmaps] - vertical scaling factors
nsamps - length, in samples, of resulting FIF
a[nmaps], c[nmaps], e[nmaps], e[nmaps] - mapping parameters
numits - number of iterations
array1[nsmaps], array2[nsmaps] - arrays containing consecutive
                                 iterations of the FIF

---

format of input data
    x[0] y[0] ---   note no scaling factor here, as there is only one
                         per pair of interpolation points
    x[1] y[1] d[1]
    ....
    x[nmaps] y[nmaps] d[nmaps]   so that there are nmaps+1
                                  interpolation points
x[n]>x[n-1]   - the interpolation points must be in sequence
                          along the x-axis
-1<d[n]<1   - all scaling factors of magnitude less than 1

---

some trial input data to produce a rhythm/timbre
    nmaps=8
    numits=4
    nsamps=32768

Program Code 12.1 (continued)

```
The duration of the sound will then depend on the sample rate used
to play it
x            y           d
0            0
8192         1.0         0.8
16384        0           0.7
20480        -0.5        0.6
24576        0           0.5
26624        0.25        0.4
28672        0           0.3
30720        -0.125      0.2
32767        0           0.1.
```

---

subroutines

A set of nmaps mappings of the form $w\begin{pmatrix} x \\ y \end{pmatrix} = \begin{pmatrix} a & 0 \\ c & d \end{pmatrix}\begin{pmatrix} x \\ y \end{pmatrix} + \begin{pmatrix} e \\ f \end{pmatrix}$ are created

form the input data with the subroutine *calcmaps*, which are then used by
determalg, to generate the FIF

---

calcmaps
```
    for n=1 to nmaps
    {
        b=x[nmaps]-x[0];        /* an intermediate product */
        a[n]=(x[n]-x[n-1])/b;
        e[n]=(x[nmaps]*x[n-1]-x[0]*x[n])/b;
        c[n]=(y[n]-y[n-1]-d[n]*(y[nmaps]-y[0]))/b;
        f[n]=(x[nmaps]*y[n-1]-x[0]*y[n]-d[n]*(x[nmaps]*y[0]-
              x[0]*y[nmaps]))/b;
    }
```

---

determalg
```
    for i=1 to numits      /*once for each iteration */
    {
        for j=1 to nmaps      /* once for each linear mapping */
        {
            for k=0 to numelems-1
            {
                x=k;
                y=array1[k];
                newx=a[j]*x+e[j];           /* Map every value in
                                               array1 to */
                newy=c[j]*x+d[j]*y+f[j];/* a new one */
                array2[int(newx)]=newy; /* and place it in
                                           array2. */
                                        /* Note that int()
                                           returns the integer
                                           value */
            }
        }
        for k=0 to numelems-1           /*copy array2 into array1 for
            array1[k]=array2[k];          next iteration */
    }
```

---

output
arrayl now contains the FIF which can be scaled and formatted
appropriately for the sound card being used.

Finally, figure 12.11 shows two examples of how sounds having both a rhythm and timbre may be constructed. Starting with a rough idea of a desired rhythm, a small set of interpolation points is constructed. The FIF generation routine then adds complex detail to this crude shape to produce the final sound. The degree of detail may be controlled with the value of the vertical scaling factors, d. Typically, the results are complex percussive rhythms with a number of layers of detail that combine to produce very interesting sounds. I believe these form a new class of sound as yet unheard. They have been generated with a simple method and very little data. They present a situation opposite to that considered by Mandelbrot because both the musical and timbral structure follow the same fractal scaling laws, which shows that there is an aural equivalent to abstract fractal images.[11]

MODELING NATURAL SOUNDS WITH FIFs

The last section presented a technique for the direct synthesis of abstract sound with FIFs. As this is a successful way of managing complex sounds with simple systems, the question arises: What about using FIFs as part of an analysis/synthesis model? A natural sound then could be represented with an FIF and modified by altering the interpolation point parameters. In order to achieve this, an analysis algorithm is needed to solve the inverse problem for FIFs. That is, we need a means of generating a set of interpolation points from a given waveform such that the resulting FIF matches the original. This has been attempted by using an algorithm devised by David Mazel that systematically searches for a set of interpolation points.[12] The algorithm has been modified to cope with the larger amounts of data required for sound and to improve its performance. Such an algorithm may be evaluated by comparing the degree of data compression obtained to the quality of the result. The degree of data compression is found by comparing the amount of data required to store the original sound waveform to that required to store the interpolation points and vertical scaling factors. This is measured as a compression ratio. The quality of the result is found by measuring the difference between the original waveform and the FIF generated with the analysis data. This is expressed as a *signal-to-noise ratio,* or SNR. Typically, for an increase in the compression ratio, there will be a decrease in the SNR.

The modified Mazel algorithm allows the compression ratio to be adjusted to requirements. For certain sounds the results are good when compared with requantiztion—the process of throwing away data to give a coarser representation of the original waveform. For a particular amount of degradation to the original sound, for example, a certain value of SNR, the corresponding amount of compression obtained with the algorithm is about three times higher than if the least significant data of the original had been disposed of to give the same SNR.

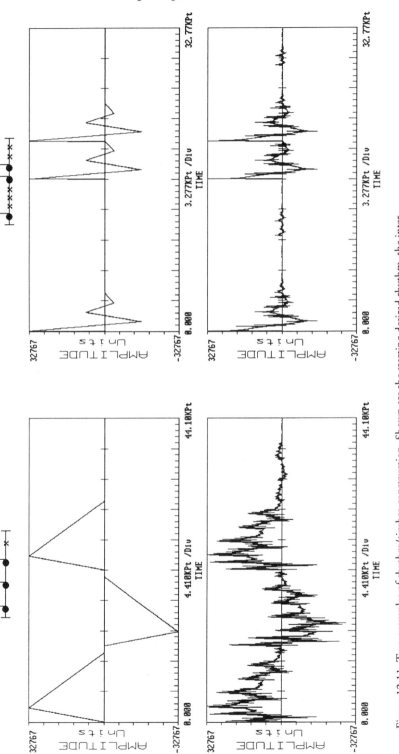

Figure 12.11 Two examples of rhythm/timbre construction. Shown are the starting desired rhythm, the inter-polation points chosen to represent this pattern, and the resulting FIF.

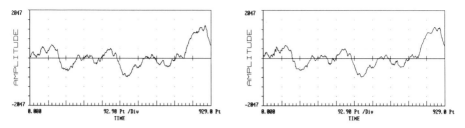

Figure 12.12   Original waveform of wind sound (left) and an FIF created by analysis and then resynthesis (right).

Interestingly, the best results from experiments with a wide range of sounds are obtained for the sound of the wind and the room ambience. These are the sounds that show a $1/f$ power spectrum. This is not surprising—the analysis algorithm is exploiting the fractal structure of the sound waveforms to achieve the relatively good compression performance.

An example of one of the best results found with the modified Mazel algorithm is shown in figure 12.12. The compression ratio is eight to one and the corresponding SNR is 23dB.

Despite this, the amount of degradation for very large compression ratios is unacceptably high. Consequently, it is not possible to reduce the amount of data present in a waveform to a small enough number of interpolation point parameters so that a natural sound may be managed in the same way as with the direct synthesis technique. These results, however, do represent a step in the right direction. Also, it is possible to modify the interpolation points derived by the analysis in order to modify the sound of the resulting FIF. Some interesting time warping and frequency/amplitude modulation effects have been achieved this way.

DYNAMIC MODELING OF NATURAL SOUND

This section discusses a dynamic model for sound. Let us assume that a simple nonlinear dynamical system is responsible for generating a given sound and that this system possesses a strange attractor. The model is then based on representing this attractor as a geometric object. I propose that if a synthetic system is constructed having a similar attractor, then it will have similar dynamics and produce a sound that is similar to the original.

Typically, however, there is no direct access to the original system, its state or attractor. Instead, all that is available is a single function of time that results from observing the original system. This can be considered to be the time waveform (also called a time series) corresponding to the original sound. An

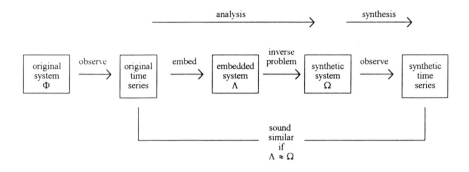

Figure 12.13 Schematic diagram of the dynamic sound model. The Greek letters denote the attractors of the different systems, and the inverse problem describes the process of extracting the rules for the synthetic system so that its attractor approximates that of the embedded system.

important technique exists known as *embedding,* however, which allows the attractor of the original system to be reconstructed in a different state space using only the time series. The technique involves simply constructing a vector of time-delayed values of the time series. This vector may then be treated as the state of the "embedded system." The vector's attractor shares the form and properties of the attractor of the original system. Embedding functions, as a viewing instrument, allow the geometry of the hidden original attractor to be seen. The synthetic system may then be modeled on the embedded one, the goal being to match the geometry of the embedded attractor. This framework is shown in figure 12.13.

The inverse problem is then tackled as follows. A history of the original system is obtained by reconstructing it via the time series in embedded state space. This provides a set of examples of the state transitions

$$x_n \rightarrow x_{n+1} \qquad\qquad (12.6)$$

at a variety of places over the embedded attractor. Because the original system is assumed deterministic, this gives information about the mapping,

$$x_{n+1} = F(x_n) \qquad\qquad (12.7)$$

The set of transitions are examples of F. F can then be estimated by fitting a mapping to the set of example transitions. This idea comes from work on the prediction of time series (for example, by J. D. Farmer and J. J. Sidorowich and M. Casdagli[13]). In this work, however, the emphasis is on accurately predicting a few values of the time series into the near future, not on generating large amounts of a synthetic time series so that it sounds like the original.

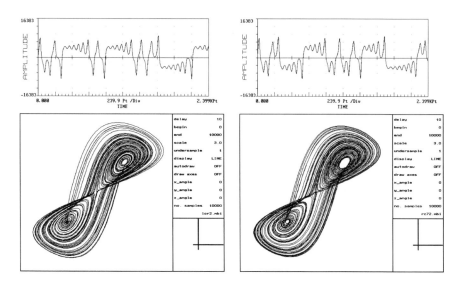

Figure 12.14 Original time series from the Lorenz system and the embedded attractor derived from it (left) and time series and attractor of the synthetic system (right).

Consequently, I have devised a new form of this technique that is suitable for the problem of modeling sound. This technique partitions the embedded state space into a number of domains and fits a linear function to each so that the overall effect is a nonlinear mapping that approximates F. This makes the synthetic system very similar to an IFS—a set of locally linear mappings that together form a nonlinear one and that define a strange attractor. The synthetic system can then be iterated and its attractor compared with that of the embedded system. Finally, the synthetic system can itself be observed to form the synthetic time series that can be turned into sound.

This technique has been tested using the Lorenz system, the simple, synthetic chaotic system discussed earlier. Figure 12.14 shows the original time series derived from this system and the resulting embedded attractor. Note how the topology of the attractor is the same as that shown in figure 12.5 (which is the attractor of the original system) but that its actual shape has been distorted by the process of embedding. Also shown is the attractor of the synthetic system and the synthetic time series. It can be seen how the model accurately preserves the shape of the attractor. Consequently, the synthetic time series is not exactly like the original one, but it shares similar dynamics.

Having confirmed that the modeling technique works for a synthetic input, I have tried it with a number of natural sound time series. One of the sounds I chose was air noise obtained by placing a microphone in a constant stream of air. This complex sound consists of a constant, but irregular, deep rumble. The reason

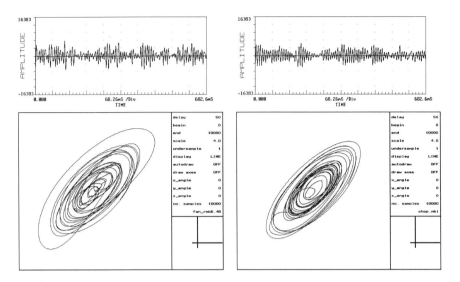

Figure 12.15 Original air noise time series and the embedded attractor derived from it (left) and time series and attractor of the synthetic system (right).

for using this sound is that it is believed to come from a simple chaotic system. (Systems of fluid flow on the edge of turbulence are known to be governed by simple chaotic dynamics.[14] I also used it because of its constant nature, a necessary requirement if an attractor is to be found. Figure 12.15 shows both the embedded and synthetic attractors, and original and synthetic time series.

A certain degree of similarity between the two can be seen, even if it is not as strong as that for the synthetic Lorenz system. Most important, however, is that original and synthetic sounds compare very well. The synthetic version is almost indistinguishable from the original, having the same essential perceived qualities.

This, then, is a very encouraging result. It shows that the perceived qualities of a complex irregular sound can be modeled using a chaotic system by matching strange attractors. I believe this experiment may be the first demonstration of this fact. Also, the synthetic system is relatively simple. Unlimited quantities of the sound can be produced quickly and easily from a system defined by fewer parameters than would be required to store less than half a second of the original. Experiments with a range of other inputs, such as gong sounds, wind noise, and brass instrument tones, also have produced good results, indicating that a chaotic model is capable of preserving essential features of the original signals. For example, a constant tuba tone containing performance irregularities can be very well modeled. Certain statistical patterns of a crashing gong sound can be captured. The results available so far are good enough to conclude that the chaotic model warrants further investigation.[15]

CONCLUSION

This chapter has outlined a number of ways in which chaos and fractals can be applied to model sound by using strange attractors. The main results are that a new class of complex rhythm/timbres may be created with simple systems and that chaotic systems are capable of dynamically modeling naturally occurring irregular sound. While these results are very encouraging, they are only preliminary. This work can be developed in many ways and a large number of experiments could be performed.

On the basis of the results outlined in this chapter, I feel that applying nonlinear dynamics theory to sound modeling has considerable future potential. I imagine that it will provide a basis for modeling complex sound in much the same way that linear theory has for resonant sound. Anyone interested in the creative use of sound could develop a number of powerful computer-based tools and techniques.

Notes

1. For examples, see J. Pressing, "Nonlinear Maps as Generators of Musical Design," *Computer Music Journal* 12 (2) (1988): 35-45; M. Gogins, "Iterated Function Systems Music," *Computer Music Journal* 15 (1) (1991): 40-48; and K. Jones, "Algorithmic Explorations of Juxtaposition and Simultaneity in Computer-Aided Composition," *Proceedings of the International Computer Music Conference, Glasgow*, 305-306.

2. R. F. Voss and J. Clarke, "$1/f$ noise in Music: Music from $1/f$ noise," *Journal of the Acoustical Society of America* 63 (1) (1978): 258-263; K. J. Hsu and A. J. Hsu, "Fractal Geometry of Music," *Proceedings of the National Academy of Sciences USA* 87 (1990): 938-941, K. J. Hsu, and A. J. Hsu, "Self-Similarity of the '$1/f$ noise' Called Music," *Proceedings of the National Academy of Sciences USA* 88 (1991): 3507-3509.

3. B. B. Mandelbrot, *The Fractal Geometry of Nature* (New York: W. H. Freeman & Co., 1983).

4. I. Xenakis, *Formalized Music* (Bloomington, IN: Indiana University Press, 1971).

5. Mandelbrot, *Fractal Geometry of Nature*.

6. M. V. Mathews, "Digital Synthesis of Natural and Unnatural Sounds," *Collected Papers from AES Premier Conference* (New York: Audio Engineering Society, 1982), 239-242; J. C. Risset and D. L. Wessel, "Exploration of Timbre by Analysis and Synthesis," in *The Pyschology of Music*, ed. D. Deutsch (San Diego: Academic Press, 1982), 25-58; and F. R. Moore, *Elements of Computer Music* Englewood Cliffs, NJ: Prentice Hall, 1990).

7. V. Gibiat, "Phase Space Representations of Acoustical Musical Signals," *Journal of Sound and Vibration* 123 (3) (1988): 529-536; V. Gibiat, "From Order to Disorder an Approach of Chaos in Musical Signals," *Proceedings of the Sixth European Signal Processing Conference, Belgium* (1992), 187-190; and E. Lindenman, "Routes to Chaos in a Nonlinear Musical Instrument Model," *84th Convention of the Audio Engineering Society*, paper number 2621 (1988).

8. K. A. Legge, "Nonlinearity, Chaos, and the Sound of Shallow Gongs," *Journal of the Acoustical Society of America* 86 (6) (1989): 2439-2443.

9. C. A. Pickover, *Computers, Pattern, Chaos and Beauty* (New York: St. Martin's Press, 1990); and T. Senevirathne et al. "Amplitude Scale Method: New and Efficient Approach to Measure Fractal Dimension of Speech Waveforms," *Electronics Letters* 28 (4) (1992): 420-422.

10. M. Barnsley, *Fractals Everywhere* (San Diego: Academic Press, 1988).

11. More details on this work may be found in J. P. Mackenzie and M. Sandler, "Fractal Interpolation Functions for Sound Synthesis," *Audio Engineering Society 90th Convention* Preprint 3008; and J. P. Mackenzie, "Using Strange Attractors to Model Sound," Ph.D. Thesis, *University of London*, 1994.

12. D. Mazel, "Using Iterated Function Systems to Model Discrete Sequences," *IEEE Transactions on Signal Processing* 40 (7) (1992): 1724-1734.

13. J. D. Farmer and J. J. Sidorowich, "Predicting Chaotic Time Series," *Physical Review Letters* 59 (8) (1987): 845-84; and M. Casdagli, "Nonlinear Prediction of Chaotic Time Series," *Physica D* 35 (1989): 335-356.

14. See, J. P. Crutchfield, et al., "Chaos," Scientific American 254 (1986): 46; and J. P. Gollub and H. L. Swinney, "Onset of Turbulence in a Rotating Fluid," Physical Review Letters 35 (1975): 927.

15. More detail on this work can be found in J. P. Mackenzie and M. Sandler, "Modelling Sound with Chaos," *Proceedings of the IEEE International Symposium on Circuits and Systems, London* 6 (1994): 93-96; and J. P. Mackenzie, "Using Strange Attractors to Model Sound."

V

Fractals in Medicine

13

Pathology in Geometry and Geometry in Pathology

Gabriel Landini

––––––

In histopathology, diagnosis of normality and disease usually is based on microscopic morphological changes occurring in cells and tissues. However, the analysis and description of such complex and irregular morphologies is difficult and, to date, has been qualitative or subjective. Now this difficulty can be avoided by using fractal geometry, which provides new approaches to objective measurement and understanding of shape complexity.

––––––

The Caterpillar was the first to speak.
"What size do you want to be?" it asked.
"Oh, I'm not particular as to size," Alice hastily replied;
"only one doesn't like changing so often, you know."
"I don't know," said the Caterpillar.
    —Lewis Carrol, *Alice's Adventures in Wonderland*

Before they were even called fractals, many fractal objects were historically regarded as "pathological" or mathematical "monsters" because they did not fit within the frame of Euclidean geometry.[1] The situation has changed dramatically, and through the development of fractal geometry we are able to model and understand many physical and natural processes that previously were considered irregular and seemingly patternless. Paradoxically, in medicine, we can use these "pathological" geometries to quantify morphological features of cells and tissues in order to differentiate "normality" from "pathology."

Histopathology is the branch of pathology that is concerned with the study of the morphological changes (in cells and tissues) during disease; these changes occur at the microscopic (or submicroscopic) level. The purpose of histopathology is to facilitate the interpretation of histological images (samples of tissue) to reach a diagnosis and then decide on the choice of an appropriate treatment. In general, the earlier the diagnosis is made, the greater are the chances of a successful treatment.

Histopathologists are faced daily with the nontrivial task of analyzing complicated patterns in sections of sampled tissues under the microscope. This involves a sequence of comparisons (sometimes made in a nonsystematic way) between features of a sample and features of normal ones. In this sense, diagnosis is recognition, comparison, and quantification. One problem with this approach is that pathological states are dynamic while histological samples are only "snapshots" of the dynamic process, and one must extrapolate in order to diagnose and prognose. The human brain has a remarkable capacity for some types of pattern matching, but in some histopathological problems, pattern matching has been showed to be poorly reproducible both inter- and intraobserver (between different observers and in the same observer at different times).

The idea of using the power of computers to increase accuracy or amount of information is not new. However, it has been shown that in histopathology, although still very promising, image analysis techniques have not yet produced very satisfactory results. One reason is that diagnosis is made through a very complex hierarchical system of decisions. A second reason is that (as in many other sciences) the patterns to be analyzed appear complicated or seemingly irregular. (They are not well described in classic geometrical terms.) As a result, these patterns have been described qualitatively: *abnormal* or *bizarre* cells, *irregular* tumor outlines, *dendritic* ulcers and *geographic* glossitis, among many others. These descriptions, of course, lack objectivity; how bizarre can a cell be? How do we compare two dendritic lesions? Is there a measure for these irregular morphologies? And furthermore, what are the mechanisms underlying the formation of these irregular structures?

FRACTAL CANCER?

When cancer is suspected, the main question for pathologists reduces to: Is the lesion under the microscope malignant or not? Although in most instances the diagnosis can be reached relatively easily, sometimes this is not the case. One problem is that not all histological presentations of the same disease are the same but rather have a set of common features. A second problem, already mentioned, is that some of the features or patterns are difficult to quantify when objective judgments or the use of statistical methods is not possible. For example, irregular structures are not readily quantifiable; yet this is precisely where the methods of fractal geometry are so promising.

Cancer of the mouth (oral cancer) accounts for approximately 5 percent of all diagnosed tumors in Western countries and as much as 60 percent in some South Asian regions. It arises as a result of malignant transformation of the epithelial tissue that lines the mouth. Like many other cancers, its occurrence is thought to depend on a multitude of exogenous (dietary behavior, use of tobacco and alcohol) and endogenous (genetic predisposition) factors. Different parts of the mouth also have different degrees of predisposition to cancerization. The shape of the junction between the epithelial cells and the underlying connective tissue of the normal oral mucosa, called the epithelial-connective tissue interface (ECTI), is slightly undulated to flat. In premalignant lesions such as epithelial dysplasia (called premalignant because they may develop into cancer), the ECTI often becomes irregular due to increased cell proliferation and partial loss of tissue "order" or architectural control. In cases of squamous cell carcinoma (cancer), this effect is more marked, and in addition islands of epithelial cells may invade into the deeper layers of the connective tissue. (See figure 13.1.)

Classically these changes are described as "increased irregularity of the ECTI," but as the shapes are difficult to conceptualize in geometric terms, it has remained a subjective parameter. It is possible to characterize the ECTI by means of the fractal dimension. To estimate the fractal dimension, we superimposed grids of increasing size ε on digitized profiles of the ECTI and counted the number of

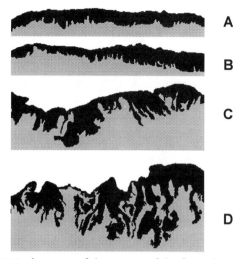

Figure 13.1 Digitized sections of the mucosa of the floor of the mouth with epithelial dysplasia (A, B) and squamous cell carcinoma (C, D). The epithelial-connective tissue interface (ECTI) is the limit between the epithelial tissue (black) and the underlying connective tissue (gray).

"boxes" ($N(\varepsilon)$) that contained any part of the profiles. If a plot of $\log(\varepsilon)$ versus $\log(N(\varepsilon))$ was linear in some range of $-\varepsilon$ (more than one order of magnitude of ε), then the object was considered fractal and the fractal dimension (D) was estimated as $D = -b$, where b is the slope of the linear regression line in that range.

Figure 13.2 shows the fractal dimension including the ECTI in cases of normal, premalignant, and malignant lesions of the floor of the mouth. Using the fractal dimension of the ECTI alone, it is possible to differentiate statistically between normal/mild dysplasia (not likely to become malignant), moderate/severe dysplasia (premalignant lesions with higher chances of transforming into cancer), and squamous cell carcinoma (cancer).[2] The range of self-similarity is small (one to two decades) due to the finite size of the cells which are the structural elements of the tissue. Nevertheless, the fractal dimension gives objective information that is not available otherwise. Self-similar processes involved in growth have been proposed as being responsible for the mosaic patterns in the liver of chimeric rats and in simulated "cell pushing" replicative systems revealing characteristic fractal dimensions.[3] In carcinogenesis, characterization and computer modeling of tumor growth modalities may help to understand local aggressivity, tumor infiltration, and metastatic dynamics (spread of cancer to distant parts of the body).

Another "irregular" feature of cancer tissues that we have been successful in characterizing is the shape of cell nuclei. The nucleus contains the DNA, which controls the cell metabolic machinery. In tumor cells, the function and distribution of DNA are altered, producing loss of control of the regulation of the tissue activity.

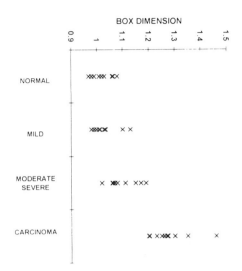

Figure 13.2 Box dimension of the ECTI in normal mucosa of the floor of the mouth, mild epithelial dysplasia, moderate and severe epithelial dysplasia, and squamous cell carcinoma.

As a consequence of these changes, the shape and size of the nuclei are altered. An increase in the variation of shape and size of the nuclei, when compared to normal tissues, is known as *pleomorphism* (which means "many forms"). In an attempt at quantification, image processing techniques and shape factors (which relate areas and perimeters) have been used in the past to characterize nuclear shape. As nuclei are irregular, their perimeters depend on the observational resolution being used (the microscope magnification). As a consequence, nuclear shape measured by these techniques has not been conspicuously successful. Therefore, nuclear pleomorphism has remained a subjective morphological characteristic in diagnosis. (See figure 13.3.)

However, although irregular, nuclei do not seem "strictly fractal" either; if observed at very high resolutions, the nuclear membrane looks relatively smooth (nearly Euclidean), while at lower resolution the irregularity is patent. We have used a semifractal or *asymptotic fractal* model to approach the irregularity of malignant and normal nuclei in transmission electron microscope images.[4] We have applied this methodology, together with a multivariate linear discrimination function, to 672 normal and malignant cells that were then reclassified with 76.6 percent accuracy (84.8 percent of the normal and 67.5 percent of the tumor) using only the perimeter data.

FRACTALS IN YOUR EYES

Many may agree that fractal geometry has a great deal of the visual ingredient. Images of the outside world are captured by an array of cells in the retina of the eye; these

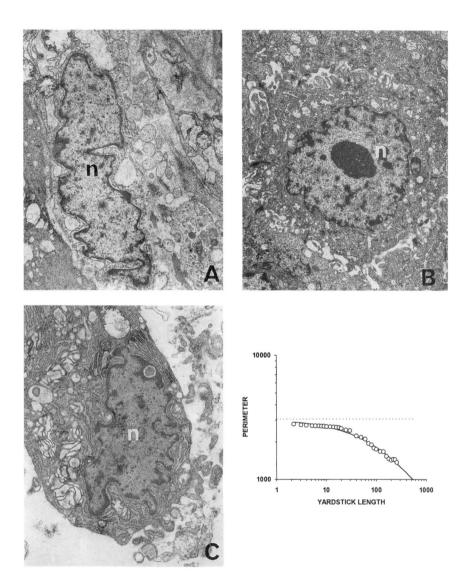

Figure 13.3 Pleomorphic nuclei from a histiocytic tumor (A), an epithelial tumor (B), and a fibrous tumor (C). The graph shows the effect of resolution on perimeter length (for cell C), provides an Euclidean asymptote (dotted line) at high resolution, and a fractal asymptote at low resolutions. The asymptotic fractal formula is $Br = Bm/[1 + (r/L)^c]$, where Br is the perimeter length for a yardstick size r, Bm is the Euclidean asymptote, L is a constant, and c is the asymptotic fractal exponent. Yardstick and perimeter length in pixels (1 pixel = 7.73 nanometers).

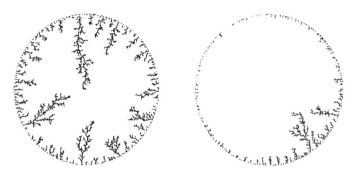

CENTERED LESION   EXCENTRIC LESION

Figure 13.4 Two simulations of corneal neovascularization with a central lesion of the cornea and with an eccentric lesion. Blood vessels invade the cornea from its perimeter. Note the predominance of a large "blood vessel." In vivo, these are called feeder vessels.

respond to a range of frequencies of electromagnetic radiation (visible light) with a change of electrical potential that is transmitted to the brain for further processing. The blood vessels that provide the retinal blood supply arise from a location named the optic disc and distribute on the retina in a branchlike fashion. The vascularization process (the development of the blood vessels) has been a matter of debate. Deterministic models in which the subbranching of vessels obeys certain deterministic or probabilistic rules somewhat similar to the *Lindenmayer systems* have been proposed.[5] However, other mechanisms also have been suggested, such as self-avoiding invasion percolation and nonequilibrium diffusion processes, especially diffusion limited aggregation (DLA).[6] The last was suggested by visual similarities between the retinal blood vessels (which grow in a quasi-two-dimensional space, the retina), and two-dimensional DLA computer-grown clusters. (See figure 13.4.) Furthermore, both share the same fractal dimension ($D \sim 1.7$). Diffusion processes also are known to be involved in the pathological vascularization of the eye cornea following chemical or physical trauma. (These include diffusion of gases, nutrients, angiogenic factors, and cell mediators.) Because normal corneas do not possess blood vessels, neovascularization after an injury may facilitate the resolution of that injury, but it also will be associated with a loss of corneal transparency. In particular, superficial and interstitial types of cornea vascularization can be simulated with DLA models.[7] It has been speculated that since neovascularization may advance toward the stimuli along a diffusion gradient, the same process may be occurring during the vascularization of the retina.

As mentioned, several independent research groups have estimated the fractal dimension of the retinal blood vessels at around 1.7.[8] This has an important consequence: The fractal dimension may be used for objective characterization of the retinal vessels in disease or during treatment monitoring, and it involves only retinal photography, which is a noninvasive method of data acquisition. Fractal characterization may be particularly important in diseases such as diabetic proliferative retinopathy in which proliferation of small vessels may change the architecture (and the fractal dimension) of the blood vessel tree.

CELLULAR AUTOMATA
AND THE FRUSTRATIONS OF A VIRUS

Herpes simplex virus (HSV), a member of the Human herpes virus group of DNA viruses, is characterized by a variable range of host cells, a short growth cycle, neurotropism (special affinity for nerve cells), and rapid spread with destruction of the infected cells. HSV infection may occur in oral or genital mucosas, or skin, producing ulcers. HSV ulcers in the oral mucosa (herpetic gingivostomatitis) are usually rounded to ovoid; sometimes small ulcers coalesce to form a larger one. HSV lesions in the corneal epithelium of the eye (herpes simplex keratopathy) most commonly differ from ulcers in other locations by exhibiting a characteristic arborescent or "dendritic" morphology. Most of these types of lesion remain dendritic and localized, and usually heal within three weeks with antiviral treatment. Occasionally dendritic ulcers can enlarge progressively and change in form to "geographic" or "amoeboid" appearance, often after the inappropriate use of topical corticosteroids (anti-inflammatory drugs). Although still produced by the same virus, amoeboid ulcers have a prolonged clinical course and are more refractory to treatment than dendritic ones. The reasons remain unclear for the resolution of early dendritic lesions and their morphology, the progression to amoeboid morphology, and the clinical severity of large ulcers.

Recently we have shown that the shape of HSV corneal ulcers can be described by the fractal dimension of their outlines, and that the fractal dimension of the outlines has a relationship to the size of the ulcer.[9] There seems to be an inverse relation between the fractal dimension and the maximum diameter of the ulcer. As the virus does not have a metabolism of its own but uses the host cell molecular machinery for its replication, we suspected that the process of *cell infection → viral replication → cell death → ulcer* can be regarded as an algorithmic mechanism that in reality depends on the status of the host cells. For this particular type of problem, cellular automata seem to be the most suitable models available. Cellular automata are mathematical models for complex natural systems that contain a large number of simple identical elements. These elements, called "cells," affect one another by means of local interactions following a set of simple fixed rules that govern the system dynamics.

Table 13.1.
Cellular Automaton Rules for Simulation
of the Herpes Simplex Virus Spread.

Rule 1.

Two types of cells:

permissive (easily infected by HSV),

resistant (not easily infected),

both distributed randomly in the tissue.

The proportion of the permissive cells: $p = \dfrac{Permissive\ cells}{Total\ cells}$

Rule 2.

Permissive → *infected* if 1 or more neighbors are infected (viral infection).

Resistant → *infected* if 5 or more neighbors are infected (viral infection)

or dead (lack of tissue support).

Rule 3.

Infected cell → *dead* cell.

Our model assumes that the corneal epithelium (where the HSV infection takes place) is composed of two populations of epithelial cells with different degrees of permissivity to infection by HSV. (The actual hypothesis is that it depends on the presence or absence of a cell membrane receptor that facilitates HSV infection.) The two types of cells can have three states: alive, infected, or dead. The two populations of cells are distributed randomly, the infection is triggered in a single cell, and the evolution of the infection then depends only on the status of the neighbor cells (spread by contiguity). The rules are summarized in table 13.1. By changing the number of permissive cells in the tissue and the strength of the resistant cells, it is possible to simulate a wide range of lesions from dendritic, to amoeboid, to round, which resemble the range of morphologies common in oral as well as corneal ulcers. (See figure 13.5.) Here the strength of a resistant cell means how many of its neighboring cells must be infected or dead to affect that resistant cell.

For this particular model, when resistant cells are abundant, the ulcers tend to be small and irregular; when they are scarce, then ulcers are larger, with smoother outlines. For simulation of HSV ulcers in the cornea, we found that it was necessary that resistant cells were infected if their neighbors had five or more infected or dead cells (shown in figure 13.5 as the rule ≥ 5). As this particular configuration produces ulcers that have a transition from irregular (dendritic) to large and smooth, we performed 3,600 simulations for different percentages of permissive cells and estimated the fractal dimension of the simulated ulcers with the yardstick method. The method consists of measuring the length of the ulcer outlines $L(\varepsilon)$ with yardsticks of increasing size ε. Then, the fractal dimension (D) was estimated as $D = 1-b$, where b is the slope of the linear part of the regression of log (ε) on $\log(L(\varepsilon))$. The results from these simulations compare directly with the data obtained in vivo regarding the fractal dimension of the outlines and its dependence on the size of the ulcer.[10] Furthermore, this model illustrates the transformation of dendritic ulcers into amoeboid ones when corticosteroids are administered: the drug is depressing the inflammatory response and therefore transforms "resistant" cells into "permissive" ones and allows the infection to percolate through the tissue. In essence, the viral infection can be regarded as a critical percolation phenomenon.

THE FUTURE OF HISTOPATHOLOGY

At this time, analytical and quantitative microscopy is regarded only as an "aid" to the expert, but, the role of semiautomatic analysis may become more prominent in the new methods and techniques that are being considered continuously.

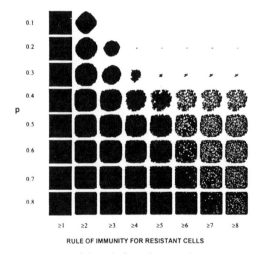

RULE OF IMMUNITY FOR RESISTANT CELLS

Figure 13.5 Variations of the rule for infection of resistant cells and for different proportions of permissive cells. From left to right: rules ≥ 1, ≥ 2, ≥3, ≥ 4, ≥ 5, ≥6, ≥7, and ≥8. Rule ≥ 1 is equal to $p = 1$ or "all permissive cells," since any resistant cell will be infected if it has one or more infected/dead neighbors. From top to bottom: $p = 0.10$ to 0.80 in steps of 0.1 indicates the proportion of permissive cells in the total number of cells. Rule 5 mimics dendritic and amoeboid ulcers. Rules ≥6, ≥7, and ≥ 8 also have such morphology, but islands of resistant cells remain in the ulcer, a feature not found in real HSV infection.

Gradually, systematization of diagnosis is becoming possible; it will continue to develop, not only using the expert system databases approach but also using expert image interpretation systems.

Computer diagnosis? Robot pathologists? It is difficult to imagine machines replacing the expert knowledge of pathologists, at least in the near future, due to the scope and complexity of the subject. However, the word "impossible" is not appropriate. Thanks to computers and fractal geometry, three of the main goals pursued are gradually being achieved—objective measurement, analysis of large amounts of data, and speed—this is a great step forward.

Notes

The author wishes to thank Dr. J. W. Rippin, Oral Pathology Unit, School of Dentistry, The University of Birmingham, U.K. for his constant support and many fruitful discussions.

1. B. B. Mandelbrot, *The Fractal Geometry of Nature* (San Francisco: W. H. Freeman, 1982).

2. G. Landini and J. W. Rippin, "The Fractal Dimensions of the Epithelial-Connective Tissue Interfaces in Premalignant and Malignant Epithelial Lesions of the Floor of the Mouth," *Analytical and Quantitative Cytology and Histology* 15 (2) (1993): 144-149.

3. P. M. Iannaccone, "Fractal Geometry in Mosaic Organs: A New Interpretation of Mosaic Pattern," *FASEB Journal* 4 (1990): 1508-1512; M. K. Khokha, G. Landini, and P. M. Iannaccone, "Fractal Geometry in Rat Chimeras Demonstrates that Repetitive Cell Division Programs May Generate Liver Parenchyma," *Developmental Biology* 165 (2) (1994): 545-555; and G. Landini and J. W. Rippin, "Fractal Fragmentation in Replicative Systems," *Fractals* 1 (2) (1993): 239-246.

4. J. P. Rigaut, "An Empirical Formulation Relating Boundary Lengths to Resolution in Specimens Showing 'Non-Ideally Fractal' Dimensions," *Journal of Microscopy* 133, pt. 1 (1984): 41-54; and G. Landini and J. W. Rippin, "An 'Asymptotic Fractal' Approach to the Morphology of Malignant Cell Nuclei," *Fractals* 1 (3) (1993): 326-335.

5. M. F. Kiani and A. G. Hudetz, "Computer Simulation of Growth of Anastomosing Microvascular Networks," *Journal of Theoretical Biology* 150 (1991): 547-560; H. R. Bittner, "Modelling of Fractal Vessel Systems," in *Fractals in the Fundamental and Applied Sciences*, eds. H. Peitgen, J. Henriques, and L. Penedo (Amsterdam: Elsevier, 1991), 59-70.

6. A. Daxer, "Fractals and Retinal Vessels," *The Lancet* 339 (1992); 618. F. Family, B. R. Masters, and D. E. Platt, "Fractal Pattern Formation in Human Retinal Vessels," *Physica D* 38 (1989): 98-103.

7. G. Landini and G. Misson, "Simulation of Corneal Neovascularization by Inverted Diffusion Limited Aggregation," *Investigative Ophthalmology and Visual Science* 34 (6) (1993): 1872-1875.

8. See G. Landini, G. Misson, and P. I. Murray, "Fractal Analysis of the Normal Human Retinal Fluorescein Angiogram," *Current Eye Research* 12, (1993): 23-27 and references therein.

9. G. Landini, G. Misson, and P. I. Murray, "Fractal Properties of Herpes Simplex Dendritic Keratitis," *Cornea* 11 (6) (1992): 510-514; G. P. Misson, G. Landini, and P. I. Murray, "Size Dependent Variation in Fractal Dimensions of Herpes Simplex Epithelial Keratitis," *Current Eye Research* 12 (1993): 957-961; and G. Landini, G. Misson, and P. I. Murray, "Fractal Characterisation and Computer Modelling of Herpes Simplex Virus Spread in the Human Corneal Epithelium," in *FRACTAL 93: Fractals in the Applied and Natural Sciences,* ed. M. Novak (Amsterdam: North-Holland, 1994), 241-253.

10. Landini, Misson, and Murray, "Fractal Characterization," 241.

14

Fractal Statistics: Toward a Theory of Medicine

Bruce J. West

———

In the past decade or so many investigators have applied fractal geometry to anatomy, physiology, and biology. From these studies has emerged a new paradigm of health and the forerunner of a theory of medicine. One of the most successful of these new ideas is the processing of data sets using fractal statistics, in particular Lévy stable distributions. There are some indications of how the newly evolved perspective may influence the future of medicine and indeed what form a theory of medicine may take.

———

INTRODUCTION

In the middle of the 1970s it became apparent to a number of scientists that there was a large class of physical phenomena for which traditional statistical physics was not equipped to describe, much less to explain. The common elements of these phenomena were complexity, nonlinearity, and apparently random fluctuations in space and/or time. I had become somewhat sensitized to these phenomena early on for two reasons: I had done a postdoctoral study in the early 1970s with Elliott Montroll, who held the Einstein Chair of Physics at the University of Rochester, and Elliott and Benoit Mandelbrot were old friends and colleagues, so I had heard a number of Dr. Mandelbrot's talks as early as 1968. From these early lectures and subsequent discussions, it was apparent that the statistical properties of certain systems were often counterintuitive; for example, in some cases the more data you collected, the less reliable the average value of a physical variable could become. The examples I recall from that time were drawn from economics and astrophysics. Rather than being dominated by a narrow band of frequencies, as they would for a simple harmonic system, the spectral properties of systems instead spread themselves into an inverse power law, so that correlations persisted from very short to very long times, no one scale being dominant. This lack of a characteristic scale was a harbinger of fractals. I'll give examples later.

Because of the complexity of natural systems, each time you measure a given quantity, a certain amount of estimation is required. For example, in a simple physics experiment, you estimate the positions of the markings on a ruler or the location of the pointer on a gauge. Thus, if you measure a quantity X, N times say, then instead of having a single quantity, you obtain a collection of quantities $\{X_1, X_2, \ldots, X_N\}$, often called an ensemble. In a similar way the variations in a physiological quantity such as the blood pressure, the beating of the heart, the rate of respiration, and others all give rise to time series that yield various ensembles. In the continuous limit in which the number of independent observations of a quantity approaches infinity, the characteristics of any well-behaved measured quantity is specified by means of a distribution function. The general idea is that any particular measurement has little or no meaning in itself. Nor is it very useful. Only the collection of measurements, the ensemble, has a physical (biological) interpretation, and this meaning is manifest through the distribution function. This function, also called the *probability density*, associates a probability with the occurrence of an event. Usually the average value of the quantity of interest has the greatest probability and deviations from the average are less probable; the greater the deviation from the average, the less the probability. This regularity is observed in even the most random of processes, for example, in the error associated with making a measurement. This regularity in the form of the frequency distribution of errors prompted the nineteenth-century English eccentric Galton to write:

> I know of scarcely anything so apt to impress the imagination as the wonderful form of cosmic order expressed by the "Law of Frequency of Error." The law would have been personified by the Greeks and deified, if they had known of it. It reigns with serenity and in complete self-effacement, amidst the wildest confusion. The huger the mob, and the greater the apparent anarchy, the more perfect is its sway. It is the supreme law of unreason. Whenever a large sample of chaotic elements are taken in hand and marshalled in the order of their magnitude, an unsuspected and most beautiful form of regularity proves to have been latent all along.[1]

Galton was, of course, referring to Gauss' law of error and the normal (Gaussian) distribution.[2] When distributions have finite *moments* to all orders, these moments usually can be used to determine the probability density. In some cases, such as for the normal bell-shaped distribution, the second moment is sufficient to determine the complete behavior of the ensemble. However, when distributions have sufficiently long tails, the first few moments will not characterize the distribution because they diverge. Distributions with infinite moments characterize processes described by noninteger exponents and contain surprises that run counter to our intuition.

Let us consider a number of examples of inverse power law disributions. In figure 14.1a we plot the percentage of authors publishing exactly n papers as a function of n. The straight line on log-log graph paper indicates Latka's Law, which is an exact inverse square law relation. This means that the number of authors publishing a given number of papers $[N(n)]$ varies inversely as the square of the number of papers $N(n) \sim 1/n^2$. This distribution function is an inverse power law and is quite different from the bell-shaped distribution of Gauss. The frequency of appearance of words in English prose is given by Zipf's Law, where if $f(n)$ is the relative frequency of the nth word in the order of the frequency of usage of that word in a typical novel say, then $f(n) \sim 1/n$ is a reasonable fit to the data, as shown in figure 14.1c. Similarly, the frequency of distribution of incomes in western societies is also an inverse power law. The empirical distribution is Pareto's Law where $N(x)$ is the number of people with income x or larger and v is a positive number (Pareto's exponent) then $N(x) \sim 1/x^v$. In figure 14.1b the frequency distribution of income in the United States in 1918 is given.

Jonas Salk and I discussed these three distributions, as well as others, from both the point of view of scaling and the intepretation of the underlying process.[3] To see how different such distributions are from those of Gauss, consider the

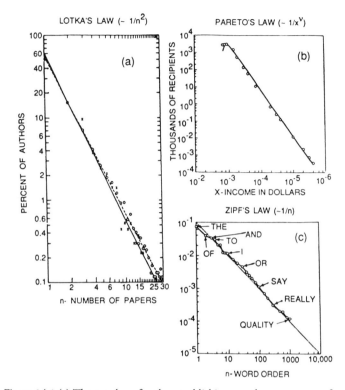

Figure 14.1 (a) The number of authors publishing exactly *n* papers, as a function of *n*. The straight line shows the exact inverse-square law of Lotka. All data are reduced to a basis of exactly 100 authors publishing but a single paper; (b) Frequency distribution of incomes in the United States, 1918; (c) Word frequency as a function of word order for the first 8,728 words in the English language. (From B.J. West and J. Salk, "Complexity, Organization and Uncertainty," *European Journal of Operations Research* (30) (1987): 117-128.

probability that on any given day, you shall meet someone twice your height. Since the distribution of heights in the U.S. population is Gaussian, this probability is very small, although probably not zero. Now consider the distribution of income as given by Pareto's Law. The likelihood of your meeting someone, on any given day, whose income is twice yours is finite. In fact, the probability of encountering someone whose income is ten times or 100 times yours is also finite. From this simple comparison it is clear that the underlying process for a bell-shaped distribution is very different from that of an inverse power-law distribution. Much of what we discuss herein is related to these differences.

The geometric notion of a fractal carries over into the domain of statistics in that it is the details of the random fluctuations in the process of interest that are revealed at smaller and smaller scales as the interval of observation is magnified.

Lévy studied the general properties of such processes in the 1920s and 1930s and generalized the *Central limit theorem* (to which the Law of Errors was a forerunner) to include those phenomena for which the second moment diverges. Among the varied and interesting properties of the *Lévy stable distribution* (process) is the fact that it satisfies a scaling law, indicating that the statistical fluctuations in the underlying process are self-similar. Subsequently it has been found that both the inverse power-law spectra and the Lévy distribution, which asymptotically has an inverse power-law form, are a consequence of scaling and fractals.

Now how do these comments relate to a theory of medicine? The *Dictionary of Science* defines a theory as: "an explanation or model based on observation, experimentation, and reasoning, especially one that has been tested and confirmed as a general principle to explain and predict natural phenomena." Thus we shall make the use of the Lévy distribution plausible as a general description of certain classes of biomedical phenomena through a combination of mathematical reasoning and comparison with experiments. We develop the rationale for this theory through a sequence of examples that answer the question "How?" but not the questions "Why?" at least not the *why* in the Aristotelian sense of a final cause.

Herein we briefly review the evolution of these ideas starting with the well-known phenomenon of *diffusion*. Diffusive processes are the backbone of linear descriptions of all natural fluctuating phenomena. Formally we begin with a discussion of a random walk strategy for modeling complex systems since simple random walks faithfully reproduce diffusive phenomena such as Brownian motion. Modified random walks have been applied successfully to the description of biomedical processes such as neuron spike statistics and the clearance of radioactive isotopes from the body. In these cases it is useful to allow the distance traversed by the random walker to represent an instantaneous voltage in describing the neuronal discharge, the level of radioactive isotope tracer in the body as a function of time, or indeed a quantitative measure of any other process that has both a random and deterministic component.

The next topic in the order of complexity is *anomalous diffusion*, where the anomaly can be either the time dependence of a process' *variance* or the statistics of the random variable. The variance, denoted by σ^2, can be expressed as $\sigma \propto t^{2H}$ where t is the time over which the process has evolved and H is a positive parameter. Normal diffusion, which is to say Brownian motion, occurs when $H = 1/2$, and the variance increases linearly with time, as found by Einstein in 1905. It is worth pointing out that these anomalous processes have inverse power law spectra, $1/f^{2H+1}$, where f is the frequency. Note the relation between the exponents in the variance and the spectrum. We find a large number of biomedical processes have such spectra: the interbeat interval of the heart,[4] the spectrum of a single *QRS pulse*,[5] and the fetal heart rate,[6] to name just a few. It is possible that such processes are described by *fractional Brownian motion (fBm)*, but we shall find that often there is a change in statistics rather than a change in the variance, and a Lévy stable distribution is required to describe the process. Both these types of anomalies arise in the modeling of DNA sequences[7] and ion-channel gating.[8]

These examples are presented for two reasons: to show some of the successes that have been made in applying fractal statistics to the description of biomedical phenomena and to demonstrate that to obtain a theoretical description of the overall phenomena, a detailed mechanistic model for every process in the human body is not needed. Here we formulate a metatheory, much like the Law of Errors, that is applicable to medicine because of the immense complexity of the systems involved. The distribution function does not provide the detailed reductionistic model of the process that we have sought, but it does provide an overarching description with which every such reductionistic model must be consistent. As we shall see, it may also provide a number of diagnostics for disease. Here we present an overview of how *fractals* have not only entered into the descriptions of biomedical phenomena but have prompted a new paradigm of health for the clinician. For at least five decades physicians have interpreted fluctuations in biomedical processes in terms of the principle of homeostasis: Physiological systems normally operate to reduce variability and to maintain a constancy of internal fluctuations. For example, the apparently regular beating of the heart for an individual at rest, as heard through a stethoscope or at the wrist, has led cardiologists routinely to describe the normal heart rate as regular sinus rhythm. However, more careful analysis reveals that although the average heart rate is 60 beats per minute for healthy, young adults, the heart rate may vary as much as 20 beats per minute every few heartbeats. According to this theory, which reached its full expression with Walter B. Cannon of the Harvard Medical School, any physiological variable should return to its normal steady state after it has been perturbed. As pointed out in Goldberger, Rigney, and West, the principle of homeostasis suggests that these variations are merely transient responses to a fluctuating environment.[9] It might be postulated that the body is less able to maintain a constant process while at rest during disease or aging, so that the magnitudes of the variations would be greater than expected. However, this is not the case. Goldberger and coauthors explain that a different gestalt emerges when the process of interest, say the heartbeat of a normal person is measured again. The time series plot of the normal beat-to-beat variation in heart rate appears ragged, irregular, and, at first glance, completely random. If, however, you concentrate on a few hours of the data, the fluctuations look remarkably similar to structures seen over a number of days. Again, by concentrating on a few minutes of data, the variations resemble those in the few hours' data. The beat-to-beat fluctuations at different time scales appear to be *self-similar*, just like the branches of a geometrical fractal. In other words, the heart rate may fluctuate considerably (even in the absence of fluctuating external stimuli) rather than relaxing to a homeostatic steady state.

DIFFUSIVE PROCESSES

In his 1945 book *What Is Life ?* I. Schrödinger asked: Why are atoms so small?[10] His intent was to understand why atoms are so small relative to the dimensions of the human body. To paraphrase M.V. Volkenshtein,[11] the answer to this question is that the high level of organization necessary for life is possible only in a macroscopic system; otherwise the order would be destroyed by microscopic fluctuations. Thus the strategy for understanding biomedical phenomena may be based on a probabilistic description of complexity, when the phenomenon of interest apparently lacks a characteristic scale.

In the past decade the floodgates have burst open, releasing a torrent of studies on the applications of fractal geometry to physiology and biology. While the irregular structure of natural forms in living organisms has long eluded description, using this new geometry, structures such as the internal membrane surface of a cell or the inner lung surface of the branching trees of the bronchial airway and blood vessels may now be described. E. R. Weibel and others argue that fractal geometry may in fact be a *design principle* for living organisms.[12] Weibel and, independently, A. L. Goldberger and I have suggested that the implicit form of recursion relations used to generate fractals provides insights into the possible strategies used by nature in the genetic programming of living organisms.[13] But these far-reaching conclusions are way ahead of the story. Let us back up and review the simplest of organized processes: diffusion.

Diffusion is a ubiquitous phenomenon in biology: It is important to the cell biologist for descriptions of growth and membrane repair,[14] to the neurophysiologist for descriptions of neuronal activity,[15] and also to the bacteriologist for describing the random motion of motile bacteria.[16] A classic strategy for modeling diffusive processes uses random walks. If a walker takes one step per unit time such that X_j is the step size at step j, with some specified probability, then the total displacement of the walker after N steps is

$$X(t) = \sum_{j=1}^{N} X_j$$

where the total time is $t = N\tau$, and τ is the time required for each step. Here the walker can be a heavy particle embedded in a fluid of lighter particles, as it was in 1829 for the botanist Robert Brown. (Most texts credit the first observation of this phenomenon to him, and that is why it is called Brownian motion.) The actual effect was first described in 1785 by the Dutch physician Jan Ingen-Housz, who observed that finely powdered charcoal floating on an alcohol surface executed a highly erratic random motion.[17] In addition to the actual physical process, random walks can model a variety of natural phenomena. The walker can represent the postsynaptic potential where each step corresponds to an inhibitory or excitatory impulse, or it can be the concentration of chemical species in a chemical reaction or any such process that can be modeled as a time sequence of random changes.

We assume that the statistics of the individual steps are determined by a transition probability density such that $p(x_j)dx_j$ is the probability the walker takes a step of size in the interval $(x_j, x_j + dx_j)$ per unit time. If the likelihood of taking a step of a given size in all possible directions is the same, the probability density is symmetric and the average motion of the walker vanishes $<X(t)> = 0$, where the symbols "$< >$" denote an average of the distribution of step sizes. When the second moments of each of the individual steps are equal the second moment of the random walk, the process in equation 14.1 is given by

$$< X^2(t) > = 2\ Dt. \tag{14.2}$$

The constant D is called the *diffusion coefficient*. It measures the average area covered by the diffusing substance per unit time. This is the familiar result for Brownian motion wherein the second moment increases linearly in time t, an example of which is the growth of an ink blot in water. If we gently put a drop of ink in a stationary fluid, the mean square distance between any two particles in the ink drop increases linearly with time due to thermal agitation of the ink caused by the fluid. The initially small drop of ink spreads out diffusively in such a way that the width of the distribution representing the blob increases linearly with time, as expected for the variance of the distribution.

The probability density for Brownian motion is

$$P(x, t) = (4\ \pi\ Dt)^{-1/2}\ \exp\{-x^2/4Dt\} \tag{14.3}$$

where $P(x, t)dx$ is the probability that the random variable $X(t)$ is in the interval $(x, x + dx)$ at time t independently of where it started. Note that at $t = 0$ the probability density becomes a singular pulse at $x = 0$. Mathematically this is a Dirac delta function $\delta(x)$. It corresponds to a infinitely dense drop of ink placed initially at the origin. Immediately afterward the ink (probability) spreads out in a symmetrical way, generating the profile depicted in figure 14.2. Note that each ink molecule is undergoing the random walk, and $P(x,t)$ represents how an ensemble of such walkers evolves over time.

The scaling properties of $P(x,t)$ can be obtained easily by multiplying x by the constant λ and t by the constant β so that from equation 14.3

$$P(\lambda x, \beta t) = \beta^{-1/2}\ P(x, t) \tag{14.4}$$

as long as $\lambda = \beta^{1/2}$. Thus the distribution for the random variable $\beta^{1/2} X(\beta t)$ is the same as that for $X(t)$. Processes that scale differently in X and t are called *self-affine*.

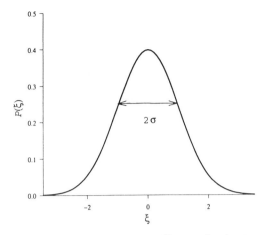

Figure 14.2 The probability density $P(\xi)$ given by the Gaussian distribution equation 14.3 is graphed as a function of ξ, for $\sigma = 1$, with $\xi = x/\sqrt{40t}$

This property means that if you measure a process such as a sequence of heartbeats and determine that the process is self-affine, then the time series on the time scale of minutes would be statistically indistinguishable from that on the time scale of days provided that the amplitudes were properly adjusted. The wave forms in these two cases would be quite similar, but they would not be exactly the same since it is the probability density that satisfies the scaling equation 14.4, not the time series itself.

The random walk model has been used to describe the morphogenesis of complex irregular structures such as the *dendritic branching of neurons* growing in vitro in culture dishes[18] and in the retina of the eye of a cat;[19] random walk models also describe the ramified system of retinal vessel patterns in the developing human eye[20] and the bifurcation patterns of vessels in the cat brain as well as in the human retina.[21] The model used in each of these studies is based on two assumptions: (1) a monomer (single particle) of a given type undergoes a random walk in a two-dimensional space covered by a "checkerboard" lattice; (2) when this monomer comes in contact with another of its species, it sticks to it and ceases moving. To create one of these intricate dendritic structures, first place a seed, (a stationary monomer) at the origin of the lattice. Next release a second monomer from a random point in space far from the origin. This distant monomer is allowed to take steps of equal probability to all adjacent sites at each point in time. Eventually it "diffuses" (random-walks) to the origin where it sticks to the seed-monomer. Once this dimer (double particle) is formed, another monomer is released far from the dimer and diffuses toward the origin, where it eventually sticks to the dimer to form a trimer with a larger number of spatial configurations. This process is repeated again and again until a *diffusion limited aggregation* (DLA) structure emerges. (See figure 14.3.)

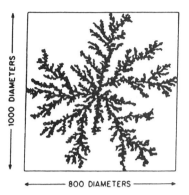

Figure 14.3 A 50,000 particle off-lattice Witten-Sander DLA aggregate.[22]

As pointed out by Feder,[23] among others, the fact that a cluster is random does not necessarily imply that the cluster is fractal. For a cluster to be fractal, its density must decrease as its size increases in a way determined by the exponent in a mass relation. The mass contained in a radius R for a fractal distribution of monomers is

$$M(R) \propto R^d \tag{14.5}$$

where d is called the mass fractal dimension. For example, the mass of a solid sphere, say a beach ball of radius R filled uniformly with sand, is proportional to R^3. In the absence of other knowledge, it usually is assumed that the matter is uniformly distributed in the available space and that d is equal to the Euclidean dimension of the space ($d = 3$). Let us suppose, however, that on closer inspection we observe that the sand is not uniformly distributed, but is instead clumped in distinct spheres of radius R/b, each having a mass that is $1/a$ smaller than the total mass. Thus what we had initially visualized as a beach ball filled uniformly with sand turns out to resemble one filled with basketballs, each of the basketballs being filled uniformly with sand. We now examine one of these basketballs and find that it consists of still smaller spheres, each of radius R/b^2 and each having a mass that is $1/a^2$ smaller than the total mass. Now again the image changes, so that the basketballs appear to be filled with baseballs, and each baseball is uniformly filled with sand. If we assume that this procedure of constructing spheres within spheres within spheres can be telescoped indefinitely, we obtain after n telescoping operations $M(R) = a^n M(R/b^n) \propto R^d (a^n/b^{nd})$. This relation yields a finite value for the total mass in the limit of n becoming infinitely large only if $d = ln\ a / ln\ b$. This fractional value for d is the Hausdorff (fractal) dimension of the mass dispersed throughout the Euclidean volume of radius R.

The mass density $\rho(R)$ is defined in the usual way by the ratio of the mass to volume

$$\rho(R) = \frac{M(R)}{V} \propto R^{d-E} \qquad (14.6)$$

where the volume is given by $V \propto R^E$ in an E-dimensional Euclidean volume. In our beachball example, $E = 3$. Thus we see that the density is constant only if the fractal and Euclidean dimensions are equal. Witten and Sander found the fractal dimension of two-dimensional DLA structures to be approximately 1.7.[24] This means that the mass of the aggregate increases as $R^{1.7}$ and the mass density decreases as $R^{-0.3}$. Family, Masters, and Platt,[25] in their study of blood vessels in the normal human retina, assumed that $M(R)$ is proportional to the total length of all the vessels in a circle of radius R. The fractal dimension d was obtained from the log $M(R)$ versus log R plot shown in figure 14.4, where the slope of the curve gives $d \cong 2.32$, a value smaller than the $E = 3$ dimensional volume in which the vessels are contained. Smith and associates found the fractal dimension of cultured vertebrate central nervous system neurons ranged from 1.14 to 1.6.[26] Caserta and coworkers found the average dimension for 11 cultured chick retina neurons was 1.41, and it was 1.68 for 22 cat retina neurons.[27] Mainster obtained fractal dimensions of 1.63 ± 0.05 and 1.71 ± 0.07 for the retinal arterial and venous patterns, respectively.[28]

A number of these authors speculate that these fractal measures may be able to provide insight into the relationship between vascular patterns and disease. There is a great need for noninvasive methods to diagnose retinal disease and to quantify its severity and degree of progress. However, it remains to be demonstrated that the fractal dimension of retinal blood vessel patterns differ between

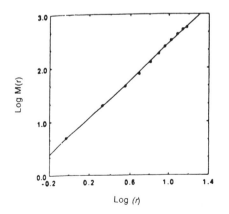

Figure 14.4 A typical plot of the logarithm of the length (mass) of the retinal blood vessels, $M(r)$, versus the logarithm of the radius r. The slope of the line gives $d=1.72 \pm 0.03$.[29]

normal and diabetic, say, or that it can be used to identify infants at risk for retinopathy or prematurity. It is clear however that these suggestions need to be pursued and tested.

Another application of simple diffusion was made by Gerstein and Mandelbrot to model (*GM model*) the spike activity of a single neuron.[30] In neurons, synapses receive both excitatory and inhibitory inputs from presynaptic neurons. In the diffusive model, these inputs are assumed to arrive randomly and to sum linearly over time, so that if the limiting process is continuous rather than discrete the process is Brownian. To apply this model to neuron discharge, they introduced a reflecting and an absorbing barrier and calculated the probability distribution for the voltage to reach the absorbing barrier for the first time. The probability density is determined by starting a large number of random time series between the reflecting and absorbing barriers and recording the time at which each trajectory intercepts the absorbing barrier (threshold potential). The trajectories that reach the threshold potential in a time t represent those neurons in an ensemble that have fired in this time interval. The statistics of the spike discharges are then described by the first passage time distribution, now called the interspike interval distribution

$$Q(a,b,t) = \frac{A}{t^{3/2}} \exp\left[-b\,t - a/t\right] \qquad (14.7)$$

where a is a parameter associated with the difference between the threshold (absorbing barrier) and resting potentials (reflecting barrier). The parameter b is associated with the difference in the arrival rates of excitatory and inhibitory inputs, and A is the

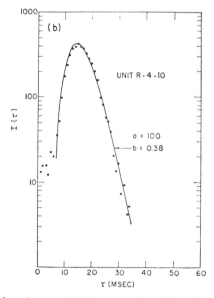

Figure 14.5 A fit to the interspike histograms for the firing of a neuron in the cochlear nucleus of a cat under moderate dial-urethane anesthesia using the GM model.[31]

normalization constant. Thus $Q(a, b, t)dt$ is the probability the neuron fires for the first time in the time interval $(t, t + dt)$. The parameters a and b are adjusted to fit the interspike interval histogram data from various neurons, and the fit to the data in some cases is quite good. (See figure 14.5.) Note that both the peak in the data as well as its considerable variation are accounted for by the distribution equation 14.7.

In a number of neurons the interspike interval distribution has a long tail that could not be accounted for by the GM model. M. E. Wise suggested that the pattern of action potentials apparently consists of mixtures of distributions of the same form but with different parameters for each distribution.[32] This generalization is accommodated in the GM model by making a, b, or both random variables. Consider first the new distribution in which we assume the difference between the arrival rates of excitatory and inhibitory pulses is described by a fractal distribution function, one that does not possess a characteristic scale. B. J. West and W. Deering show that the average interspike interval distribution is

$$<Q (a, b, t) >_b = A\, t^{-w} \exp\{ -a/t \} \qquad (14.8)$$

where the brackets with a b subscript denote an average over this parameter; the value of w is given by

$$w = 3/2 + \mu \qquad (14.9)$$

is the index for the fractal (hyperbolic) distribution for b and A is the normalization constant.[33]

In figure 14.6 the distribution of action potential interspike intervals from a neuron in the somatosensory cortex of a rabbit is displayed on log-log paper. The straight-line fit for large values of the intervals indicates an asymptotic inverse power-law distribution fits the data. The exponent w depends on the type of neuron and experimentally varies in the interval $1 \le w \le 7$, as discussed in Wise.[34]

Similar considerations can be made for physiological radioactive clearance curves, which represent the distribution of rates at which radioactive isotopes are cleared from the body. Here the interval between the absorbing and reflecting barriers is the random variable. Wise discusses in this application, which is intended to represent multiple returns of the radioactive tracer into the blood without being expelled from the body.[35] The new distribution is averaged over the fluctuations in the parameter a,

$$<Q (a, b, t)>_a = A\, e^{-bt} t^{-w} \qquad (14.10)$$

where

$$w = 3/2 - \beta, \qquad (14.11)$$

with β the index for the hyperbolic (fractal) distribution for a and A the normalization constant. The solution given by equation 14.10 is identical to one given by Wise for $\beta \geq 1$ and has been verified for over 180 radioactive clearance curves from human patients and rats yielding $0.3 \leq w \leq 0.52$. (See figure 14.7.) From this figure it is clear that equation 14.10 represents data for times less than one day.

If in this latter context we also consider b to be a random variable over which we average, then averaging over both a and b yields

$$<Q\,(a,\,b,\,t)>_{a,\,b} = A\,t^{-w} \qquad\qquad (14.12)$$

where

$$w = 3/2 + \mu - \beta, \qquad\qquad (14.13)$$

and A is the overall normalization constant. In the long time region $t \geq one\ day$, the distribution takes on the inverse power-law form of equation 14.12 with $0.9 \leq w \leq 1.6$ as discussed by Wise. This behavior is shown in the inset of figure 14.7 for times up to 20 days.

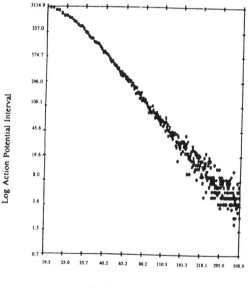

Figure 14.6 Distribution of action potential intervals from a neuron in the somatosensory cortex of a rabbit is plotted versus the logarithm of the interval duration. The asymptotic straight line indicates that the distribution can be represented by an inverse power-law distribution.[36]

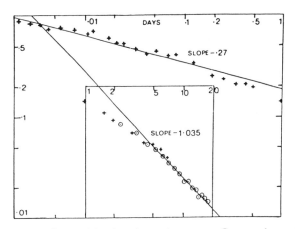

Figure 14.7 Specific activities, in microcuries per gm Ca per microcurie ^{47}Ca injected, + in blood, ⊙ in urine, from 0.02 to 1 day, inset 1 to 20 days after a single injection in a young fit man; replot on 3 x 2 cycle log-log paper of data.[37]

The hyperbolic distributions used in each of the preceding averages are actually asymptotic *Lévy stable distributions*. We point this out because the scaling properties of Lévy processes indicate that there are nested clusters of activity such as we described for the distribution of mass. For example, the spike statistics arise from clusters of activity in which there are embedded smaller clusters, which in turn contain smaller clusters, and so on. This clustering hierarchy shall be discussed more fully when we address Lévy processes explicitly.

ANOMALOUS DIFFUSIVE PROCESSES

We have seen how useful diffusive processes can be for modeling complex phenomena, such as the discharge of neurons and the passage of radioactive isotopes from the body, in a simple and direct way. Let us now consider processes having the same Gaussian statistics but with a variance that is not linear in time. To accomplish this, consider a random walk process $X(t)$ with the correlation function

$$\langle X(t)\, X(t')\rangle = 2\,D\ |t-t'|^{2H} \tag{14.14}$$

This process is simple Brownian motion for $H = \frac{1}{2}$, but is called anomalous diffusion for other values of H. In terms of the random walk model for $H > \frac{1}{2}$, the walker tends to continue in the direction he or she has previously been walking. The higher the value of H, the greater is this tendency to persist in one's direction of motion. For $H < \frac{1}{2}$, however, the walker actively chooses to avoid continuing in a given direction, and changes direction more frequently than in Brownian motion. This behavior has been called antipersistent. The statistics of the random walk process are not changed by the correlation function equation 14.14; they remain Gaussian but their variance changes, yielding

$$P(\Delta x, \Delta t) = (4\pi D\Delta t^{2H})^{-\frac{1}{2}} \exp{-(\Delta x)^2/4 D\Delta t^{2H}} \qquad (14.15)$$

where Δx is the change in the displacement of the walker during the time interval Δt. Equation 14.15 is the probability density for *fBm*.

It is again interesting to scale the displacement by λ and the time interval by β so that from equation 14.15 we obtain

$$P(\lambda\Delta x, \beta\Delta t) = \beta^{-H} P(\Delta x, \Delta t) \qquad (14.16)$$

as long as $\lambda = \beta^H$. Therefore, for $H = \frac{1}{2}$, this scaling relation is just equation 14.4. Thus we see that *fBm* is self-affine, with the random variables $\beta^H X(\beta t)$ and $X(t)$ having the same distribution function. It should be emphasized that as important as the scaling process equation 14.16 is, it does not uniquely determine the probability density. Other probability density functions have the same scaling properties.

The scaling of the correlation function given by equation 14.14 was recently observed in a biological context by applying the random walk model to DNA sequences. Peng and coworkers developed a DNA random walk by assigning the value $X_j = +1$ if the j th element of a DNA sequence is pyrimidine and -1 if the jth element of the sequence is purine in equation 14.1.[38] The four nucleotides (adenine, guanine, thymine, and cytosine, known as A,G,T and C) were analyzed using the DNA walk and the most robust fit to inverse power-law scaling was obtained for the neucleotide class (purine versus pyrimidine) rather than for any specific nucleotide. The "experimental" result of applying the DNA walk procedure to an intron-containing gene is depicted in figure 14.8. An intron-gene is one containing large segments for which there is no discernible coding function; that is, it is junk DNA. The large excursions in the $X(l)$ "landscape" suggests the existence of long-range correlations, where we have replaced the "time" t by the distance along the molecular chain l. In fact, the calculation of the variance of the random walk $\sigma(l)$ shown in figure 14.9 clearly indicates that

$$\sigma_{DNA}(l) \propto l^H \qquad (14.7)$$

where a least square fit to the slope of the log-log curve yields $H = 0.67 \pm 0.01$. One possible explanation of such a correlation would be that the DNA walk is in fact a *fBm* process.

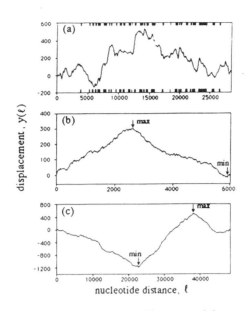

displacement, $y(\ell)$

nucleotide distance, ℓ

Figure 14.8 DNA walk representations. (a) Intron-rich human -cardiac myosin heavy-chain gene sequence, (b) its cDNA, and (c) the intronless bacteriophage DNA sequence. Note the more complex fluctuations for the intron-containing gene in (a) compared with the intronless sequences in (b) and (c). Heavy bars denote coding regions of the gene. So that the graphical representation was not affected by the global differences in concentration between purines and pyrimidines, DNA walk representations were plotted so that the end point has the same vertical displacement as the starting point.[39]

Another interesting property of the DNA sequence emerges from this analysis. There is a difference in correlation between a coded DNA sequence, denoted cDNA, and a gene that has uncoded segments, introns or junk DNA. The same gene is analyzed using the DNA walk but restricting it to the cDNA sequences. This result is shown in figure 14.9 where $H = 0.49 \pm 0.01$, a value consistent with a Brownian process. Thus Peng and associates observed that a gene with and without introns are quite different.[40] Table 14.1 provides a summary of their analysis. Sequences are divided into two groups on the basis of their intron content; it is clear that sequences with introns all have H > ½, and therefore contain long-range correlations, and sequences of cDNA all have H ≅ 0.50 and therefore have at most only short-range correlations. This systematic behavior is shown in figure 14.9, where σ (l) is separately averaged over the two groups in the table yielding overall values of $H = 0.61 \pm 0.03$ for intron-containing genes and nontranscribed genomic regulatory elements and $H = 0.50 \pm 0.01$ for cDNA sequences and genes without introns.

Table 14.1 [+]

Summary of the correlation analysis of 24 sequences selected across the phylogenetic spectrum.

Sequence	Code	Comments	Length (nt)	% Introns	α
	GROUP A				
Adenovirus type 2	ADBDG	Intron-cont. virus	35,937	n.d.	0.56
Caenorhabditis elegans MHC gene 1	CELMY01A	Gene	12,241	51	0.61
C. elegans MHC gene 2	CELMY02A	Gene	10,780	44	0.54
C. elegans MHC gene 3	CELMY03A	Gene	11,621	49	0.61
C. elegans MHC unc 54 gene	CELMYUNC	Gene	9,000	25	0.58
Chicken *c-myb* oncogene	CHKMYB15	Gene (5'-end)	8,200	92	0.60
Chicken embryonic MHC	CHKMYHE	Gene	31,111	78	0.65
Drosophila melanogaster MHC	DROMHC	Gene	22,663	72	0.56
Goat β-globin	GOTGLOBE	Gene*	10,194	n/a	0.58
Human β-globin	HHUMHBB	Chromosomal region	73,326	n/a	0.71
Human metallothionein	HUMMETIA	Gene	2,941	91	0.61
Human α-cardiac MHC	HUMMHCAG1	Gene (N-terminal)	2,366	72	0.65
Human β-cardiac MHC	HUMBMYH7	Gene	28,438	73	0.67
Rat embryonic skeletal MHC	RATMHCG	Gene	25,759	76	0.63
				$\alpha_{mean} \pm (2 s.e.m) =$ 0.61 $\pm$ 0.03	
	GROUP B				
Bacteriophage λ	LAMCG	Intronless Virus	48,502	0	0.53
Chicken *c-myb* oncogene	CHKCMYBR	cDNA	2,218	0	0.50
Chicken nonmuscle MHC	CHKMYHN	cDNA	7,003	0	0.47
Dictiostelium discodium MHC	DDIMYHC	cDNA	6,681	0	0.49
Drosophil melanogaster MHC	DROMYONMA	cDNA	6,338	0	0.47
Human β-cardiac MHC	HUMBMYH7CD	cDNA	6,008	0	0.49
Human dystrophin	HUMDYS	cDNA	13,957	0	0.53
Human embryonic MHC	HUMSMHCE	cDNA (partial)	3,382	0	0.51
Human mitochondrion	HUMMT	Intronless Gene	16,569	0	0.49
Yeast MHC	SCMY01G**	Intronless Gene	6,108	0	0.50
				$\alpha_{mean} \pm (2 s.e.m) =$ 0.50 $\pm$ 0.01	

Sequences are divided into two groups on the basis of their intron content; within each group the sequences are ordered alphabetically. Note that $\alpha > 0.5$ implies the existence of long-range correlations, whereas $\alpha \approx 0.5$ implies only short-range correlations. The second column (code) lists the GenBank names (unless specified otherwise).

nt, Number of nucleotides per sequence.

n/a, not applicable; nontranscribed DNA regions.

*, eta-globin activating region (nontranscribed DNA).

MHC, myosin heavy chain

n.d., no data; exon/intron map not fully known.

**, EMBL name

+ from Peng *et al,* "Long-Range Correlations in Nucleotide Sequences," *Nature* (356) (1992).

It is amazing that coded DNA sequences behave like Brownian motion and junk DNA has long range correlations. Researchers would have guessed just the opposite. Voss finds this analysis incredible, contending that what Peng and coworkers are observing is in fact an artifact of their DNA walk.[41] He argues that the way in which numbers are associated with symbols (amino acids) in the DNA

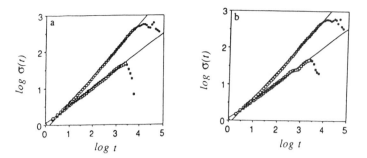

Figure 14.9 The logarithm of the random walk variance $\sigma(t)$ is plotted versus the logarithm of the number of steps taken t. The actual DNA sequences are given in (a) where (0) denotes the HUMHBB (human beta globin chromosomal) sequence and (^) denotes the LINE c sequence. The slopes for the linear fits are 0.72 and 0.49, respectively. The Lévy model sequences are presented in (b) with the same values for the linear fits to the slopes. The solid circles and squares are data omitted from the linear regression fit.[42]

sequence does in fact introduce spurious correlations. While presenting Voss's technique, which is presumably free of these spurious correlations, would lead us too far afield, when his method is applied to DNA sequences, it yields an inverse power-law spectrum $1/f^{\alpha}$ and the corresponding long-range fractal correlations. Averages carried out over greater than 5 time 10^7 bases from over 2.5 times 10^4 sequences in ten classifications (primates, invertebrates, phage, and so on) of the Gen Bank data bank show systematic changes in correlation and spectral exponent with evolutionary category.[43]

In figure 14.10 the spectral density is depicted with the white-noise floor removed (white noise is noise with Gaussian statistics and a spectrum that is constant at all frequencies), and the spectra are averaged over the specific categories in the Gen Bank Release 68 data base as indicated. Systematic changes are clearly seen in the category averages in the figure, $S(f) \sim 1/f^{\alpha}$, with $\alpha \leq 1$ being the signature of fractal (scaling) correlations that extend to the size of the largest sequences (on the order of 10^5 bases). Bacteria and phage show the smallest range of scaling. Voss points out that for the remaining categories, there is a systematic increase with evolutionary status from 0.64 for organelle to the 1.00 (exact $1/f$ noise) for invertebrates followed by a decrease to 0.77 for primates. In the interval $0 < \alpha < 1$, increasing α increases the recovery probability from a DNA error, meaning that more error at the level of DNA can be tolerated without a change in phenotype, but decreases the information content per unit length.

These results from Voss are, of course, not consistent with those of Peng and associates, since they clearly show the power-law correlations for all manner of DNA sequences without regard for their coding structure. They led to some

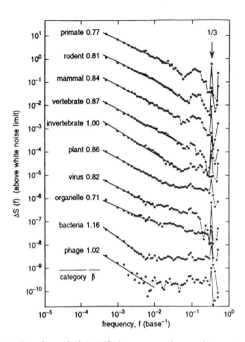

Figure 14.10 Equal-symbol Δ $S(f)$ (spectrum above white noise limit) for categories of DNA sequences from the GenBank Release 68 data bank. Least squares estimates of are tabulated and the resulting fits are shown as solid lines. Δ $S(f)$ offset for clarity.[44]

intense controversy.[45] Although it is not clear which of these models shall eventually be demonstrated to be superior, it is clear that DNA sequences have fractal statistical properties.

Pickover used a related modeling strategy to develop a graphical approach for representing information-containing sequences in DNA or protein sequences.[46] Briefly, the approach involves the mapping of sequence data to a three-dimensional pattern on connected tetrahedra to visualize similarities between, and biochemical properties of, DNA and amino acid sequences. Pickover anticipated the DNA walk in his computer experiments of random RNA (ribonucleic acid) sequences, in which mutations could be injected randomly to simulate molecular evolution. He states: "A provocative avenue of future research is to start with different statistical distributions of bases (*e.g.* Gaussian, Brownian, *etc.*) and determine what folded structures evolve."[47] Now it is time to interleave the phenomenological work of Peng and associates and Voss with the computer modeling of Pickover.

Voss says: "A white noise represents the maximum rate of information transfer, but a $1/f$ noise, with its scale-independent correlations, seems to offer the best compromise between efficient information transfer and immunity to errors on all scales."[48] I made a similar observation in the context of the fractal model of the mammalian lung,[49] arguing that a fractal is very tolerant of the variability in the physiological environment and is in fact preadapted to errors that during morphogenesis would have devastating effects on nonfractal processes. Lefévre also concluded that in an anatomical structure that "efficiency increases with fractal complexity."[50] Thus we see indications from the microscopic structure of DNA to the macrostructure of physiological organs that fractals may be the key to understanding the underlying reason for the way in which biological processes are organized.

NONDIFFUSIVE PROCESSES (LÉVY) STATISTICS

In the preceding section we modified the correlation between the steps in a random walk to obtain *fBm*. Let us now consider what happens when the steps in the random walk become arbitrarily large, resulting in the divergence of the second moment. We can handle such an eventuality by changing the normalization on the random walk displacement series as follows

$$Y_N(t) = \frac{1}{N^{1/\mu}} \sum_{j=1}^{N} X_j \qquad (14.18)$$

so that $X(t) = N^{1/\mu} Y_N(t)$ is the displacement of the random walk after N steps, $t = N\tau$, and μ is a positive parameter. Lévy asked: When is the distribution of the normalized sum $P(y)$ the same as that of the individual steps $p(x)$? Stated differently, this asks when is the character of the whole the same as that of its parts, which is a statistical form of requiring the process to be fractal.

One distribution for which the second moment diverges is the hyperbolic or inverse power-law distribution

$$p(x) = \frac{c}{|x|^{\beta+1}}, \quad 0 < \beta < 2 \qquad (14.19)$$

where c is a normalization constant. Note that this is just the form of the distribution we used earlier for the parameter variations in the spike statistics and radioactive clearance curves. The characteristic function, the Fourier transform of the probability density, corresponding to the fractal distribution equation 14.19 is determined using a Tauberian Theorem to be

$$\tilde{p}(k) \cong 1 - c|k|^\beta \cong e^{-c|k|^\beta} \qquad (14.20)$$

for small k. The statistics of the random walk process can then be determined using the convolution theorem to obtain

$$\tilde{P}(k,N) = e^{-cN|k|^\beta} \tag{14.21}$$

which is the characteristic function for a random walk consisting of N steps carried out with the fractal distribution equation 14.19 for the individual steps. In the continuum limit we can replace equation 14.21 with

$$\tilde{P}(k,t) = e^{-\gamma t|k|^\beta} \tag{14.22}$$

so that the probability density, the inverse Fourier transform of the characteristic function, is

$$P_L(y,t) = \frac{1}{2\pi} \int_{-\infty}^{\infty} e^{iky} e^{-\gamma t|k|^\beta} \, dk \ , \quad 0 < \beta \le 2, \tag{14.23}$$

which is the expression for the centro-symmetric Lévy stable distribution in one dimension. Montroll and I show that for large values of the random variable, the Lévy distribution becomes an inverse power-law distribution

$$P_L(y,t) \sim \frac{1}{|y|^{\mu+1}} \quad 0 < \mu < 2 \ , \tag{14.24}$$

which has the same form as the distribution for the individual steps of the random walk with $\mu = \beta$.[51] Note that equation 14.23 is the appropriate distribution for the random walk equation 14.18 with transition probability equation 14.19.

The scaling property of the Lévy distribution is obtained by scaling the displacement with λ and the time with β to obtain

$$P_L(\lambda x, \beta t) = \beta^{-1/\mu} P_L(x,t) \tag{14.25}$$

as long as $\lambda = \beta^{1/\mu}$. Note that equation 14.25 has the same scaling form as fBm equation 14.16 if the Lévy index μ is the same as $1/H$. Thus the self-affine scaling results from Lévy statistics with $0 < \mu \le 2$ or from Gaussian statistics with a power-law spectrum with $0 < H \le 1$. The scaling relation is the same in both cases and therefore cannot be used to distinguish between the two. Note that equation 14.25 and equation 14.16 are equivalent only for $2 \ge \mu \ge 1$ for which $\frac{1}{2} \le H \le 1$, indicating that Lévy processes give rise to anomalous diffusion that is faster than ordinary Brownian motion. By faster we mean that the variance increases with time as t^α and $\alpha > 1$, whereas $\alpha = 1$ for Brownian motion. When the Lévy index is in the interval $0 < \mu \le 1$, the parameter H is greater than unity, and as unity is the greatest value for H in a diffusive process since it corresponds to ballistic motion, such scaling indicates a Lévy process rather than fBm.

It should be emphasized that there is no differential representation of the evolution of a Lévy stable process. This can perhaps best be seen using the random walk picture. For Brownian motion, the walker steps locally from site to site so that in the continuum limit, the differences in these steps can be represented by derivatives and the evolution of the probability density can be represented by a differential equation in space and time. For a Lévy process, however, the random walk is fractal since the steps at each point in time can be of arbitrary length. Thus steps adjacent in time are not nearby in space, and the best we can do in the continuum limit is obtain an integrodifferential equation to describe the evolution of the probability density. Consider a random walk defined by a jump distribution whose probability of taking a step of unit size is $1/a$, the probability of taking a step a factor b larger is a factor a smaller, . . . , the probability of taking a step a factor b^n larger is a factor a^n smaller. As this walk progresses, the set of sites visited consists of localized clusters of sites, interspersed by gaps, followed by clusters of clusters over a larger spatial scale. This walk generates a hierarchy of clusters, the smallest having about a members in a cluster, the next largest being of size b with approximately a^2 members in the larger cluster, and so on. The parameters a and b determine the number of subclusters in a cluster and the spatial scale size between clusters, respectively. The fractal random walk can best be characterized by the parameter

$$\mu = \ln a/\ln b \qquad (14.26)$$

which is the fractal dimension of the set of points visited during the walk and the set of mass points. Hughes, Shlesinger, and Montroll have shown that the continuous form of this clustered random walk has Lévy statistics with μ being the Lévy index.[52] Thus it is quite clear that a Lévy process has a fractal dimension and also may be called a random fractal process.

Recently, Peng and associates have shown that a scale-invariant biological process, the human heartbeat time series, possesses such Lévy statistics.[53] The data consists of digitized electrocardiograms of beat-to-beat heart rate fluctuations over approximately 24 hours or 10^5 beats recorded with an ambulatory monitor. The time series is constructed by recording the interval between adjacent beats as data, for example, let $f(n)$ be the interval between the n and $(n + 1)$ beat. A great deal of variability is observed in the interbeat interval. The mechanism for this variability apparently arises from a number of sources. The sinus node (the heart's natural pacemaker) receives signals from the two branches of the involuntary (autonomic) portion of the nervous system. One branch is the parasympathetic, whose stimulation decreases the firing rate of the sinus node. The other is the sympathetic, whose stimulation increases the firing rate of the sinus node pacemaker cells. The influence of these two branches results in fluctuations in the heart rate of healthy subjects. Peng and coworkers graphed the histogram for the differences in the beat-to-beat intervals $I(n) = f(n + 1) - f(n)$ and find that this is a stationary Lévy process as shown in figure 14.11.[54] They find that the statistics of healthy and diseased conditions are the same, that being Lévy stable with an index μ = 1.7; however, the spectra for the two cases are quite different. The power spectrum $S_I(f)$ has the form f^{β} where β = 1 - $2H$ and the mean square level of the interbeat fluctuations increases as n^{2H}. Here again H = $\frac{1}{2}$ corresponds to Brownian motion, so that β = 0 indicates the absence of correlations in the time series $I(n)$ (white noise). They observed that for the diseased data set, $\beta \cong 0$ in the low-frequency regime, confirming that $I(n)$ are not correlated over long times. On the other hand they also observed that for the healthy data set, $\beta \cong 1$ indicating a long time correlation in the interbeat interval differences. The anticorrelational properties of $I(n)$ are consistent with a nonlinear feedback system that "kicks" the heart rate away from extremes.

The conclusion is that the different scaling patterns must be a consequence of the ordering of the differences (the correlations are produced by the underlying dynamics) rather than of their statistics. Therefore the fractal dimension can be associated with system complexity, and a reduction of the fractal dimension can be considered indicative of pathology. D. T. Kaplan and M. Talajei note that studies of heart rate using techniques from nonlinear dynamics are consistent with the idea that a loss of complexity is pathological.[55] S.M. Pincus, I.M. Gladstone, and R.A. Ehrenkranz found that dynamical complexity of heart rates was reduced in sick neonates.[56] J. E. Skinner and associates have reported that the correlation diversion of such heart rate series decreased prior to the onset of ventricular fibrillation in pigs;[57] Kaplan and coworkers found that complexity

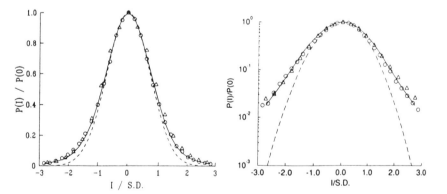

Figure 14.11 The histogram of $I(n)$ for the healthy (circles) and diseased (triangles) subject so $P(I)$ is the probability of finding an interbeat increment in the range $[I - \Delta\ I/2, I + \Delta I/2]$. To facilitate comparison, we divide the variable I by the standard deviation (S.D.) of the increment data and rescale P by $P(0)$. Both histograms are indistinguishable and are well fit by a Lévy stable distribution with μ = 1.7 (solid line). The dashed line is a Gaussian distribution, which is a special case of a Lévy stable distribution with μ = 2.[58]

of both heart rate and blood pressure signals was less in healthy elderly subjects than in healthy younger subjects.[59] These observations are all consistent with the observations of Peng and others[60] as well as with the West-Goldberger hypothesis that the heart rate in healthy people is an aperiodic, chaotic signal and that a "loss" of chaos is indicative of pathology.[61] Note that the present finding of the Lévy distribution has been shown in another context to be the appropriate statistics for chaotic dynamical processes.[62] This may well be the desired connection between dynamics and distributions required to understand these biomedical phenomena.

Just as the first passage time distribution was found to be useful for biological diffusive processes, the *mean first passage time* for Lévy processes is of value in interpreting a number of biomedical data sets. Consider the survival probability of a process that undergoes a change when the random walk displacement achieves a value η. For example, the thermostat in a room switches on when the temperature falls below a preset value and switches off again when the temperature again achieves this threshold. This is the model we discussed for the firing of neurons where the first passage time distribution was interpreted as the interspike interval distribution. Now we relate the first moment of the first passage time distribution to the survival probability. If $T_1(\eta)$ is the mean first passage time to η, then the *survival probability* is

$$F(\eta, t) \cong e^{-t/T_1(\eta)} \qquad (14.27)$$

which becomes exact as $\eta \rightarrow \infty$. Of course, for the processes of interest, η is not merely a single number, because for any complex system it could have a distribution of values. This was the rationale used in the discussion of the firing of neurons. The *mean survival probability* then is obtained by averaging over a distribution of levels $p(\eta)$ to obtain the mean survival probability $F(t)$. The mean first passage time to a level for a Lévy process was found to be $T_1(\eta) \sim \eta^\mu$,[63] so that if $p(\eta)$ is a Gaussian distribution in magnitude of the crossing level, then the mean survival probability is

$$F(t) \sim \exp\{-A t^{\frac{2}{\mu+2}}\} \qquad (14.28)$$

where A is a known constant.[64] Thus the average survival probability has the form of a stretched exponential.

A biomedical application of equation 14.28 may be used to study biomedical applications such as ion channel kinetics. Ions such as sodium, potassium, and chloride can move freely through water but cannot cross the hydrophobic lipids that form cell membranes. However, these ions can interact with proteins in the cell membrane that transport them across the cell membrane. Channels bind to

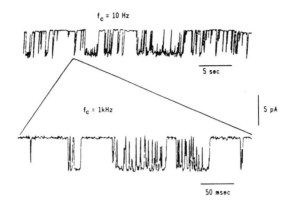

Figure 14.12 The opening and closing of the ion channel is a process that is fractal in time. This recording of a potassium channel in pancreatic cells shows a low resolution display of the data (top), and a small segment of the same data displayed at 100 times greater resolution (bottom). The one closed event in the top trace consists of many closings and openings. That is, this signal is self-similar in time.[65]

many ions at a time and allow them to move down their electrochemical gradient. The channel protein spans the cell membrane and can have different shapes called conformational states. One such state has a central hole, through which electro-chemical gradients can force ions. Pieces of the channel can also rearrange to block the hole and thus block the flow of ions. These open and closed conformational structures differ in energy, and the ambient temperature is high enough to provide sufficient energy to spontaneously switch the channel between conformations that are open and closed to the flow of ions.[66]

Liebovitch and Koniarek mention that over the past 40 years, it has been assumed that the switching of a channel from one conformation state to another can be described by a discrete Markov process.[67] This model assumes that the channel has only a few discrete states, and since the process is Markov, the probability of making a transition depends only on the present state and not on the history of previous states. This picture is not consistent with the time series data depicted in figure 14.12; the current is recorded by the patch clamp technique. These data display a self-similarity from one time scale to the next, at least over an interesting domain—over at least two orders of magnitude there is an inverse power law in time. As the authors stress, this inverse power law over a wide range of closed and open time durations suggests that the channel protein has many very similar conformational states rather than a few discrete ones. Liebovitch and Koniarek suggest that the fractal behavior may be due to the chaotic behavior of a deterministic mechanism rather than having a stochastic origin. We are not so

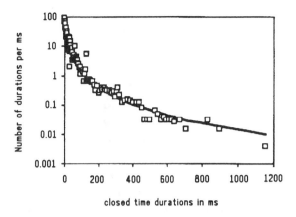

Figure 14.13 Boxes indicate how often the closed times of different durations are found in patch clamp recordings of potassium channels in the corneal endothelium. Curve represents the fractal form derived from the scaling relationship, Eq. 14.29, characteristic of simple fractal scaling.[68]

ready to abandon the statistical interpretations here. Instead, we believe that the Lévy distribution, since it is the most general solution to the Markovian BCSK chain condition may properly describe the ion channel gating statistics.[69] This does not, of course, rule out a dynamical origin for the statistics.

Liebovitch and Koniarek argue that the switching between open and closed configurations is a fractal process in time.[70] Thus they heuristically derive the probability $F(t)$ that the duration of the open (or closed) state is greater than t, and it is given by

$$F(t) = \exp\left\{-A\,t^{2-d/}(d-2)\right\} \qquad (14.29)$$

where d is the fractal dimension of the switching process. This function, the survival probability, was used to fit the closed-time histograms from a potassium channel in the optical membrane of a corneal endothelial cell. (See figure 14.13.)

The stretched exponential form of the probability distribution function equation 14.28 has been obtained in proteins using at least two microscopic models, both of which provide physically plausible descriptions of protein dynamics. Thus this form of the probability density does not require fractal dynamics, but as yet the data have not been used to distinguish unambiguously between alternatives. Dewey and Spencer point out that with the limited experimental data available, it is difficult to statistically distinguish a fractal model of ion channel gating from a complicated conventional model.[71] However, they do mention that an extensive determination of the temperature dependence of the survival probability may be useful in discriminating among models. But this has not as yet been done. They go on to point out that fractal protein dynamics is not a prerequisite for observing a fractal kinetic phenomenon associated with the functionality of the protein. It is conceivable that such effects are being measured in the ion gating experiments.

Comparing equation 14.28 with equation 14.29, we find

$$d = 2 - 2/(2 + \mu) \qquad (14.30)$$

relating the fractal dimension of the switching process d with the index of the Lévy distribution μ. As the Lévy index sweeps the interval $0 < \mu \leq 2$, the fractal dimension goes through $1 < d \leq 1.5$. This result suggests that the fractal nature of the switching process may be a consequence of its being a Lévy stable process.

SUMMARY AND CONCLUSION

The Lévy stable distribution is a general Markov process that satisfies a version of the Central Limit Theorem in which the central moments of the random variable need not be finite. Among its properties is self-affine scaling, so that $X(\lambda t)$ and $\lambda^{1/\mu} X(t)$ both have the same ensemble distribution function. In addition, it has the nested clustering behavior succinctly described by the fractal dimension as given by the Lévy index. Also, the evolution of a Lévy process cannot be described by a differential equation of motion but must rely on an integrodifferential equation. Thus it is the appropriate way to describe processes that abruptly change in time, or start and stop discontinuously in space.

We have indicated a number of complex, biomedical processes that apparently can be described by such Lévy processes and that may therefore be fractal. Among the examples discussed were neuron spike activity, heartbeat intervals, radioactive clearance curves from the body, ion channel gating, and DNA sequences. Others that we have not mentioned here include the architecture of the mammalian lung and the cardiac conduction system.[72] All of these examples taken together compel us to ask the question: Why? Why is the Lévy distribution, or, more generally, the fractal character of statistical processes so prevalent in biomedical processes?

A preliminary answer to this question is related to the fact that fractal processes may be robust against errors introduced by a fluctuating environment. I have presented a simple error model showing that a fractal structure (process) is essentially unresponsive to error compared with the response of structures having a characteristic scale.[73] A fractal is very tolerant of the variability in the physiological environment. This error tolerance can be traced to the broad-band nature of the distribution in scale sizes of a fractal object (or process). This distribution ascribes many scales to each "generation" of the process; therefore, any scale introduced by an error is already present, or nearly so, in the original system. Thus the fractal is preadapted to certain genetic errors and variations in the growth environment. A classical process, on the other hand, has a characteristic scale size; any additional scale introduced by errors or fluctuations has a devastating effect on the process. This is consistent with the findings of West and Deering for a large number of physiological processes.[74]

Thus, in developing a theory of medicine, I have argued that the combination of organization and diversity, observed at all levels of anatomy and control processes in humans, should be described as fractal random processes. Further I assert that Lévy stable statistics captures the essential features of many of these richly structural processes. Therefore, whether you believe that the underlying dynamical behavior of many of these physiological processes is random or deterministic becomes irrelevant, in part due to the recent connections made between dynamical processes and Lévy distributions.[75] These latter studies demonstrate that deterministic dynamical systems with chaotic solutions can be characterized as Lévy stable processes in the asymptotic time domain.

The observation made by E .R. Wiebel[76] concerning the mammalian lung may be reexpressed for many of the physiological processes we have discussed:

> In the case of the lung, the interlocked fractal trees of airways and vessels facilitate by design rapid mixing of air and blood with an efficiency likened to that of high turbulence, but at considerably lower cost: fractal design causes the flow pattern of turbulence to be "frozen into structure," and mixing is made more efficient. . . . Quite clearly, the construct of fractal trees is of high significance for understanding biological structure and its morphogenetic control and perhaps even for understanding how such complex hierarchical structures can be efficiently programmed genetically.

Whereas Wiebel promoted the use of fractal geometry as a design principle for living organisms, we would suggest that Lévy stable statistics are more inclusive and may be even more useful as a design principle. It might be said that we have taken the first step in establishing a theory of medicine, since knowing that a process is, or ought to be, a Lévy stable process may explain the sensitivity an individual has to a particular medication. It may assist in designing a protocol for the testing of a new drug. It may allow researchers to interpret properly extreme fluctuations in data sets. It may enable us to devise new diagnostics for certain diseases. The knowledge that the process under study is designed according to the requirements of Lévy stable processes would strongly influence these and many more physician activities.

Notes

I would like to thank the Office of Naval Research for partial support of this work.

1. F. Galton, *Natural Inheritance* (London: Macmillan, 1989).

2. F. Gauss, *Theoria motus corporum coelestrium* (Hamburg; Dover Engl.: trans., 1809)

3. B. J. West and J. Salk, "Complexity, Organization and Uncertainty," *European Journal of Operations Research* (30) (1987): 117-128.

4. C. K. Peng, J. Mietus, J. M. Hausdorff, S. Havlin, H. E. Stanley, and A. L. Goldberger, "Long-Range Anticorrelation and Non-Gaussian Behavior of the Heart," *Physical Review Letters* (70) (1993): 1343-1346.

5. A. L. Goldberger, V. Bhargava, B. J. West, and A. J. Mandell, "On a Mechanism of Cardiac Electrical Stability: The Fractal Hypothesis," *Biophysics Journal* (48) (1985): 525-528.

6. N. A. J. Gough, "Fractals, Chaos and Fetal Heart Rate," *The Lancet* (339) (1993): 182-183.

7. C. K. Peng, S. V. Buldyrev, A. L. Goldberger, S. Havlin, F. Sciortino, M. Simons, and H. E. Stanley, "Long-Range Correlations in Nucleotide Sequences," *Nature* (356) (1992): 168-170.

8. See, for example, L. S. Liebovitch, J. Freichbarg, and J. P. Koniarek, "Ion Channel Kinetics: A Model Based on Fractal Scaling Rather than Multistate Markov Processes," *Mathematical Biosciences* (89) (1987): 36-68.

9. A. L. Goldberger, D. R. Rigney, and B. J. West, "Chaos and Fractals in Human Physiology," *Scientific American* (262) (1990): 42-49.

10. I. Schrödinger, *What Is Life?* (New York: Macmillan, 1945).

11. M. V. Volkenshtein, *Biophysics* (Moscow: MIR, 1981).

12. E. R. Weibel, "Fractal Geometry: A Design Principle for Living Organisms," *American Journal of Physiology* (261) (1991): L361-L369.

13. B. J. West and A. L. Goldberger, "Physiology in Fractal Dimension," *American Scientists* (75) (1987): 354-364.

14. M. G. P. Stoker, "Role of Diffusion Boundary Layer in Contact Inhibition of Growth," *Nature* (246) (1973): 200.

15. L. M. Ricciardi, *Diffusion Processes and Related Topics in Biology,* Lecture Notes in Biological Mathematics vol. 14, ed. S. Levin (Berlin: Springer, 1977).

16. L. A. Segel, I. Chet, and Y. Henis, "A Simple Quantitative Essay for Bacterial Motility," *Journal of General Microbiology* (98) (1977): 329.

17. J. Ingen-Housz, *Dictionary of Scientific Biology,* ed. C. C. Gillespie, (New York: Scribners, 1973), 11.

18. T. A. Smith, Jr., W. B. Marks, G. D. Lange, W. H. Sheriff Jr., and E. A. Neale, "A Fractal Analysis of Cell Images," *Journal of Neuroscience Methods* (27) (1988): 173-180.

19. F. Caserta, H. E. Stanley, W. D. Elder, G. Doccord, R. E. Hansman, and J. Nittman, "Physical Mechanism Underlying Neurite Outgrowh; A Quantitative Analysis of Neuronal Shape," *Physical Review Letters* (64) (1990): 95-98.

20. F. Family, B. R. Masters, and D. E. Platt, "Fractal Patterns Formation in Human Retinal Vessels," *Physica D* (38) (1989): 98-103; M. A. Mainster, "The Fractal Properties of Retinal Vessels: Embryological and Clinical Implications," *Eye* (4) (1990): 235-241.

21. T. Matsuo, R. Okeda, M. Takohashi, and M. Funota, "Characterization of Bifurcating Structures of Blood Vessels Using Fractal Dimension," *Forma* (5) (1990): 19-27.

22. From T. A. Witten, Jr., and L. M. Sander, "Diffusion Limited Aggregation, A Kinetic Critical Phenomenon," *Physics Letters* (47) (1981): 1400.

23. J. Feder, *Fractals* (New York: Plenum Press, 1988).

24. Witten and Sander, "Diffusion Limited Aggression,"1400.

25. Family, Masters, and Platt, "Fractal Patterns," 98-103.

26. Smith et al., "A Fractal Analysis," 173-180.

27. Caserta et al., "Quantitative Analysis," 95-98.

28. Mainster, "Fractal Properties," 235-241.

29. From Family, Masters, and Platt, "Fractal Patterns."

30. G. L. Gerstein and B. Mandelbrot, "Random Walk Models for the Spike Activity of a Single Neuron," *Biophysics Journal* (4) (1964): 41-68.

31. From Gerstein and Mandelbrot, "Random Walk Models."

32. M. E. Wise, "Spike Distributions for Neurons and Random Walks with Drift to a Fluctuating Threshold," in *Statistical Distributions in Scientific Work,* eds. C. Taillie et al. (Dordrecht-Holland: D. Reidel, 1981), 211-231.

33. B. J. West and W. Deering, "Fractal Physiology for Physicists; Lévy Distributions," *Physics Reports* 246 (1 and 2) (1994): 1-100.

34. Wise, "Spike Distributions," 211-231.

35. M. E. Wise, "Skew Distributions in Biomedicine Including Some with Negative Power of Time," in *Statistical Distributions in Scientific Work,* eds. G. P. Patil et al. (Dordrecth-Holland: D. Reidel, 1975), 241-262.

36. From Wise, "Spike Distributions," figure 4.

37. From Wise, "Skew Distributions."

38. Peng et al., "Nucleotide Sequences," 168-170.

39. From Peng et al., "Nucleotide Sequences."

40. Ibid., 168.

41. R. Voss, "Evolution of Long-Range Fractal Correlations and $1/f$ Noise in DNA Base Sequences," *Physical Review Letters* 68 (1992): 3805.

42. From S. V. Buldyrev, A. L. Goldberger, S. Havlin, C. K. Peng, M. Simons, F. Sciortina, and H. E. Stanley, "Long Range Power-Law Correlation in DNA," *Physical Review Letters* 71 (1993): 1776.

43. H. J. Jeffrey, "Chaos Game Representation of Gene Structure," *Nucleic Acids Research* 18 (1990): 2163.

44. From Voss. "1/ f Noise in DNA."

45. See, for example, Buldyrev et al., "Long Range" and R. F. Voss, "Voss Replies," *Physical Review Letters* 71 (1993): 1777.

46. C. A. Pickover, "DNA and Protein Tetrograms: Biological Sequences as Tetrahedral Movements," *Journal of Molecular Graphics* 10 (1992): 2-17.

47. C. A. Pickover, "Computer Experiments in Molecular Evolution," *Speculations in Science and Technology* 13 (1989): 181-190.

48. Voss, "1/f Noise in DNA," 3805.

49. B. J. West, "Fractals, Intermittency and Morphogenesis," in *Chaos in Biological Systems,* eds. H. Degn, A. V. Holden and L. F. Olsen (New York: Plenum, 1987), 305-314.

50. J. Lefévre, "Is There a Relationship between Fractal Complexity and Functional Efficiency in the Pulmonary Arterial Bed?" *Journal of Physiology* 446 (1992): 5788.

51. E. W. Montroll and B. J. West, "On an Enriched Collection of Stochastic Processes," in *Fluctuation Phenomena,* ed. E. W. Montroll and J. L. Lebowitz, (Amsterdam: North Holland Personal Library, 1979 [1st ed.], 1987/6 [2nd ed.]).

52. B. D. Hughes, M. F. Shlesinger, and E. W. Montroll, "Random Walks with Self-similar Clusters," *Proceedings of the National Academy of Science USA* 78 (1981): 3298-33291.

53. Peng et al., "Non-Gaussian Behavior of the Heart," 1343-1346.

54. Ibid, 1343.

55. D. T. Kaplan and M. Talajic, "Dynamics of Heart Rate," *Chaos* 1 (1991): 251-256.

56. S. M. Pincus, I. M. Gladstone, and R. A. Ehrenkranz, "A Regularity Statistic for Medical Data Analysis," *Journal of Clinical Monitoring* (1991).

57. J. E. Skinner, C. Carpeggiani, C. E. Landisman, and K. W. Fulton, "The Correlation Dimension of Heartbeat Intervals is Reduced in Conscious Pigs by Myocardial Ischemia," *Circulatory Research* 68 (1991): 966-976.

58. From Peng et al., "Non-Gaussian Behavior of the Heart."

59. D. T. Kaplan, M. I. Furman, S. M. Pincus, S. Ryan, L. Lipsitz, and A. L. Goldberger, "Aging and the Complexity of Cardiovascular Dynamics," *Biophysics* 59 (1991): 945-949.

60. Peng et al., "Non-Gaussian Behavior of the Heart," 1343.

61. West and Goldberger, "Physiology in Fractal Dimensions," 354-364; A. L. Goldberger and B. J. West, "Fractals in Physiology and Medicine," *Yale Journal of Biology and Medicine* 60 (1987): 421-435.

62. M. F. Shlesinger, G. M. Zaslavsky, and J. Klafter, "Strange Kinetics," *Nature* 363 (1993): 31-37.

63. V. Seshadri and B. J. West, "Fractal Dimensionality of Lévy Processes," *Proceedings of the National Academy of Science USA* 79 (1982): 4501-4505.

64. K. Lindenberg and B. J. West, "The First, the Biggest, and Other Such Considerations," *Journal of Statistical Physics* 42 (1985): 201-243.

65. From L.S. Liebovitch and J.P. Koniarek, "Ion Channel Kinetics," *IEEE Engineering Medicine and Biology* 11 (1992).

66. For a more complete discussion, see J. B. Bassingthwaighte, L. S. Liebovitch, and B. J. West, *Fractal Physiology* (New York: Oxford University Press, 1994).

67. Liebovitch and Koniarek, "Ion Channel Kinetics," 53-56.

68. From Liebovitch and Koniarek, "Ion Channel Kinetics."

69. Montroll and West, "Enriched Collection of Stochastic Processes" ; P. Lévy, *Théorie de l'addition des variables aléatoires* (Paris: Gauthier-Villars, 1937).

70. Liebovitch and Koniarek, "Ion Channel Kinetics," 36-38.

71. T. G. Dewey and D. B. Spencer, "Are Protein Dynamics Fractal?" *Comments on Molecular, Cellular Biophysics* 7 (1991): 155-171.

72. See, for example, B. J. West, *Fractal Physiology and Chaos in Medicine* (River Edge, NJ: World Scientific, 1990).

73. West, "Fractals," 305-314.

74. West and Deering, "Fractal Physiology for Physicists," 1-100.

75. Shlesinger, Zaslavsky, and Klafter, "Strange Kinetics," 31-37.

76. E. R. Weibel, "Fractal Geometry: A Design Principle for Living Organisms," *American Journal of Physiology* 261 (1991): L361-L369.

VI

Fractals and Mathematics

15

Fractals and the Grand Internet Parallel Processing Project

Jay R. Hill

The Grand Internet Parallel Processing Project was an interesting collaboration of more than four dozen enthusiastic volunteers around the world whose computers are connected by Internet. The collaboration resulted in a sequence of ever closer estimates for the area of the Mandelbrot set, that famous bushy-shaped object in mathematics. Here I report a new lower bound, 1.50585063, derived by summing series expansions for the areas of many thousands of set components. While some would like to have seen simple values such as $3/2$ or $\pi/2$ for the Mandelset area, alas, these are ruled out.

ATTACK OF THE PIXEL COUNTERS

"What is the area of the *Mandelbrot set*?" It was the kind of question sure to start a discussion or "thread" on the Internet, a global computer network linking computers of all kinds in industry, academia, and the home. The thread was going strong when I first saw it on `sci.fractals` (an electronic bulletin board), cross-posted by Jon Noring from FRAC-L (another electronic mail list specializing on fractals). When the question was asked in November 1989, it was quickly answered by a reference to Dave Rabenhorst.[1] Dave had used a "pixel-counting" scheme to obtain an answer of 1.508 to four significant digits. "Pixel counting" means plotting the set on a computer screen with a special color denoting points thought to be inside the Mandelbrot set. By counting these points, or pixels, and noting the total area shown on the screen, the area can be estimated.

The question surfaced again on January 30, 1993, with a reference to a slowly convergent Laurent series.[2] (A Laurent series is similar to a Taylor's series but allows functions to be evaluated in the presence of singularities.) Application of the Laurent series produced a value of 1.7274 as an upper bound for the area. By upper bound I mean a value known to be greater than the actual area. For example, if we could show that all of the set fits inside a circle of radius 2, then the upper bound would be 4π. The Mandelbrot set is defined as the set of points, c, on the complex plane for which the iteration formula $z \rightarrow z^2 + c$ does not diverge. In fact, some of these values of c converge to periodic cycles. Periodic cycles occur in circular or cardioid-shaped regions called components. All points in a component have the same period. Components with periods up through period 12 are illustrated in figure 15.1.

The large cardioid shown in figure 15.1 contains points with period 1. Its boundary has the formula

$$c = (\tfrac{1}{2} \cos \theta - \tfrac{1}{4} \cos 2\theta) + i (\tfrac{1}{2} \sin \theta - \tfrac{1}{4} \sin 2\theta),$$

which can be integrated to give an exact value of $3\pi / 8$ for the area. Recall that i represents $\sqrt{(-1)}$. Iterations with c inside the cardioid converge to $z = \tfrac{1}{2} - \sqrt{(\tfrac{1}{4} - c)}$, which repeats with every iteration.

Figure 15.1 The Mandelbrot set showing components with periods up to 12. Components with periods 1 through 4 are marked.

The largest circle is the period 2 component, whose boundary formula is

$$c = -1 + \tfrac{1}{4} (\cos \theta + i \sin \theta).$$

Iterations with c inside this circle converge to $z = -\tfrac{1}{2} \pm \sqrt{(-\tfrac{3}{4} - c)}$. That is, z alternates between two values (corresponding to the $\pm$). Hence the period is 2. The area of this circle is $\pi / 16$. Adding the areas of this circle and the large cardioid just calculated above gives a value of 1.374, which is certainly a lower bound for the area of the Mandelbrot set. The other smaller components shown in figure 15.1 have period greater than 2 and are not exactly circular, cardioidal, or other simple forms.

Figure 15.1 does not show the filaments, which in fact connect all of the isolated components into one object called the Mandelbrot set. The pixel counters cut the complex plane into many very small squares, and the center of each is used in the basic iteration formula. If the iteration does not diverge after some number of iterations, the pixel is assumed to be in one of the components or on a filament, and the area of the entire square is counted as "inside."

With such differing results for the Mandelbrot set area, 1.508 from pixel counting vs. 1.7274 from the Laurent series, the momentum of the thread could not be stopped. A prediction that the area would turn out to be $\tfrac{3}{2}$ was made. Another argued in favor of $\pi/2$, but without a proof. So Joe Gladstone humorously responded in the tradition of Fermat's last theorem, "I have a wonderfully simple proof of this, but unfortunately the edit buffer of this stupid editor does not have the room . . ." Some mathematicians remarked that the "pixel-counting" scheme is futile, since it cannot measure the filamentary area in the set boundary.

But the pixel counters were undeterred. Jon Noring suggested, "If each of 5,000 people were able to calculate 16,000,000 points, then we would have calculated a total of 80 billion points, all (mostly) free." In other words, idle cycles on Internet workstations could be used to compute the area of the Mandelbrot set. Noring even named it the Grand Internet Parallel Processing Project (GIPPP). Yuval Fisher, a mathematician at the University of California at San Diego, teased the group with his message carefully titled "Computing pixels that are outside of M or near its boundary." He wrote, "Hurry, hurry, hurry. Get your computational pieces now. They are going fast. You'll find out how you can join this waste-of-time by computing some sort of estimate of the area of your very own chunk of M. Join now!"

And GIPPP#1 was off and running. But there soon followed a lively debate between the pixel counters and Fisher after he proclaimed, "Pixel counting is a waste of time, since it doesn't tell you anything about the area. It is not even clear that the limit of counting pixels at progressively high resolutions will converge to the area."

A pixel-counting program with the colorful title "Life, the Universe, and the Area of the Mandelbrot Set" was offered by Ocie Mitchell. Robert Manufo suggested a period identification scheme that speeds up the calculations in the interior of the set. He then used an average of 40 runs of his GIPPP#2 pixel-counting program to obtain 1.50659530 with a standard error of 0.00000209.

The debate raged on. On April 6 Gerald Edgar wrote, "But we need to know whether the entire square is inside or outside or neither (where it is near the boundary). Simple iteration from one starting point will not decide this rigor-ously." He continued, "Certainly you will miss the boundary of the Mandelbrot set, which may conceivably have positive area; but even if we say we are computing the area of the interior of the Mandelbrot set, this is a problem we need to talk about. How can we be sure that the entire square is 'in' the set? Or 'out'? Fisher's 'distance estimator' method will produce some useful data on this, perhaps." Yuval Fisher and I were already working on a program to overcome these kinds of problems. On April 12 I announced to the net our GIPPP#3, which would compute an upper and lower limit to the Mandelbrot set area. Our code was based on code written by Marc Parmet, whose code was based on Yuval Fisher's appendix in *The Science of Fractal Images*.[3] The code estimates the distance from a given point to the boundary of the Mandelbrot set, and it depends on accurate determi-nation of the period of that point when it is in the interior of the set. We therefore were able to determine rigorous upper and lower area bounds and also the interior area as a function of period of the component containing the samples. We used a precise and quick test to determine if sample points are in either the period 1 or 2 component. Since most of the area is in these two components, this shortened the calculation and increased the accuracy. These and additional tests for period 3 and 4 components are discussed in a later section.

Since we know the exact areas of the period 1 and 2 components, we used this exact value rather than the underestimate coming from the GIPPP#3 calculations. As accurate estimates of the area of components for higher periods become available, the lower bound can be adjusted.

A FORMULA AND AREA FOR PERIOD 3 COMPONENTS

A very interesting formula has been derived by Dante Giarrusso and Yuval Fisher for the boundary of the period 3 components.[4] By letting the angle θ go from 0 to 2π, their formula maps a circle to each of the components. They define

$$\omega(\theta) = \mathrm{asinh}\ \{[88 - 27 \exp(i\theta)] / (80\sqrt{5})\};$$

then the boundary point is

$$c = -7/4 - (20/7)\ \{\sinh\ [(\omega(\theta) + 2\pi ki)/3] - 1/\sqrt{80}\}^2 \qquad (15.1)$$

There are two parameters in this formula, θ and k. When θ is zero, one gets a point called the "base," which for a cardioid is the point in the cusp. Circular components are attached to larger components at their "base" point. Equation 15.1 is the solution of a cubic and k selects from its three roots, one for each component. For example, $k = 0$ selects the cardioid whose base is -1.75; $k = 1$ selects the circlelike component whose base is $(-1 + i\sqrt{27})/8$ on the period 1 cardioid. Choose $k = 2$ to select the other circlelike component at $(-1 - i\sqrt{27})/8$.

My first attempt to calculate the areas of the period 3 components was to approximate the component as a many-sided polygon and numerically evaluate the polygon's area. I used 2^{23} points evaluated with equation 15.1 as corners of the polygon and achieved 13-digit accuracy on that estimate.

I'm afraid my secret desire for an exact solution to the total area using "yet to be developed methods" was revealed when I posted my silly poem announcing the period three component area:

My Mandelbrot Flower

She loves me, she loves me not
Will it be no or will it be yes?
From my flower of Mandelbrot
The answer will be anyone's guess.

It's the area of Mandelbrot,
Quadratic, cubic and quartic petal
Whose total is dearly sought.
But each component is transcendental.

1.17809724509205 is so simple,
The period 1 cardioid, you see.
0.19634954084867 for a circle,
0.05654241812768 for period three.

With petal on bud and bud on petal
In an organized infinity.
The solution will never settle
Until the clock strikes eternity.

That is three petals down, infinity to go.

While the debate continued over Riemann sums, Cantor-like sets, and Lebesgue integrals, more than 66 workstations and even a few personal computers began several month's computation. We divided the complex plane near the Mandelbrot set into approximately 1,000 cells. We then determined that 119 of them were "boundary" cells, meaning they were not completely outside or completely inside the set. Volunteers who wished to work on this project were sent a copy of the GIPPP#3 program containing their assigned cell for detailed evaluation.

Forty-four enthusiastic contributors sent in their results. Two folks had to discontinue their calculations after repeated electrical storms caused machine shutdown. The result has been submitted for publication, but I will quote the numbers here.[5] We obtained a lower bound of the area of the Mandelbrot set equal to 1.5031197 and an upper bound of 1.5613027. This eliminated the hopes of both the 3⁄2 and $\pi/2$ enthusiasts.

POLYGONAL APPROXIMATIONS
OF COMPONENT AREAS

By the end of May 1993, Keith Briggs suggested that several formulas by John Stephenson might be useful for calculating areas of components up through period 8.[6] Stephenson's formulas map a circle to these components. Coincidentally, Dante Giarrusso had also derived the same formulas (for periods up to period 7) some years ago.[7] In fact, as Giarrusso pointed out, the formulas for period 3 was published 100 years ago by E. Netto, who was interested in Galois groups of polynomials whose roots can be expressed as rational functions of other roots.[8]

For period 3, let P be a point on the edge of a circle and c be the corresponding point on the component. Netto-Giarrusso-Stephenson's formula is

$$P^2 - (c + 2)\,P + c^3 + 2c^2 + c + 1 = 0 \tag{15.2}$$

where $P = \exp(i\,\theta)/8$ and θ is an angle.

The period 4 mapping formula is a cubic in P and degree 6 in c.

$$P^3 + (c^2 - 3)P^2 - (c^4 + c^3 - c^2 - 3)P - (c^6 + 3c^5 + 3c^4 + 3c^3 + 2c^2 + 1) = 0 \tag{15.3}$$

where $P = \exp(i\,\theta)/16$. Using independent methods, Keith Briggs and I numerically integrated the areas for periods 3 through 5. For each of 10^6 or more values of P, we solved equations 15.2 and 15.3 for c. These form a polygon whose area we evaluated. While I was able to compute estimates through period 5, Briggs used extended precision arithmetic to continue the integration through period 7. On June 17, 1993, he wrote to me, "After a lot of headaches, I have the 'exact' area of the period 7 components of the Mandelset. This is by numerical integration of the regions defined by the roots of Stephenson's polynomial,[9] which in this case is of degree 63. I think this is about as far as we can push this approach." It was indeed a difficult calculation, which prompted me to post my revised poem on the Internet in July.

My Mandelbrot Flower

She loves me, she loves me not
Will it be no or will it be yes?
From my flower of Mandelbrot
The answer will be anyone's guess.

It's the fractal set of Mandelbrot,
With quadratic, cubic and quartic petal
Whose total area is dearly sought.
But each component is transcendental.

The area in period one, 1.17809724509205, is simple.
Since for a cardioid it is exactly $3\pi/8$, you see.
So is 0.19634954084867, period two's bounding circle.
But for the area, 0.05654241812767, in period three

Just sum Giarrusso's form, it's analytical.
The area, 0.02327513769624, in period four

Is found by integrating Stephenson's polynomial.
Find the fifth area, 0.01311506747875, as before.

Now roots seem chaotic, I can't get a fix!
The petals are tiny and hard to integrate.
Keith Briggs computed the area in period six.
Thanks, 0.008934790525 is his estimate.

Briggs found 0.00505792907 in period seven,
And I'm adding 0.00446801353778 for eight.
Of the Mandelbrot area, some make an obsession,
Staying at the terminal until very late.

With roots all a tangle and sums so tiny,
Who can know the area beyond period eight?
Pixel counters and distance estimators a plenty,
With Richardson's method try to extrapolate.

Thanks to Bill Broadley and dozens on Internet
Who helped Yuval Fisher and Jay Hill show
The bounds on the area of the Mandelbrot set
Are 1.5613027 from above and 1.5031197 from below.

Will the net of computers, each plucking a petal,
Someday circumscribe the Mandelbrot infinity?
How near Robert Manufo's 1.506595 will it settle?
We'll all know when the clock strikes eternity.

I had made an estimate for the area in period 8 components. To do this, I used new methods, which I shall discuss later.

EXACT TESTS FOR POINTS
INTERIOR TO COMPONENTS UP TO PERIOD 4

Here is another use for equations 15.2 and 15.3. These formulas provide fast tests for c in period 3 and 4 components. In addition to being useful in the distance estimator method just mentioned, these tests are useful for making plots of the Mandelbrot set. Equation 15.2 can be made into a simple quadratic test. Using the standard solution of equation 15.3 requires a few special sign checks to get the proper cube roots. Let $B = \sqrt{[-(4c + 7)]}$. Then

if abs[8 + 4c (1 - B)] ≤ 1
then c is in the period 3 bud,
Otherwise if abs[8 + 4c (1 + B)] ≤ 1
then c is in the period 3 cardioid.   (15.4)

I will not display my algorithm for period 4 since Dante Giarrusso recently showed me another of his clever formulas.[10] Although equation 15.3 is a cubic, it tests c in all six period 4 components. Here is his test. Let $X = 2 \sqrt{(4c+3)/3}$.

If $X = 0$, c is not in period 4. Otherwise, put
$Y = $ asinh $(-16/X^3)$,
$Z = 2(c + 1)c + 2$,
and

$F = X$ sinh $[(Y + 2 \pi k i)/3]$.

Then if for any k in 0, 1, or 2

abs$[c(F+1)F + Z]$ ≤ ⅛   (15.5)

c is in a period 4 component according to this list:

k	component center
0	-0.156520166833755061799 ± i 1.03224710892283180167
1	-1.310702641336832883563
1	0.282271390766913879697 ± i 0.53006061757852529949
2	-1.940799806529484752232

Just for completeness, I use the following simple test for c in periods 1 and 2. Let $s = [abs(c)]^2$. Then c is in the period 1 cardioid if $(256s - 96)s + 32$ Real(c) ≤ 3, or period 2 circle if $16s + 32$ Real$(c) + 16 ≤ 1$.

A MORE ACCURATE METHOD
FOR THE AREA OF PERIOD 3 COMPONENTS

Although we used millions of samples in the numerical integration of area components, the methods used in the previous section are of limited precision. We are lucky if we get 12- to 15-digit accuracy. To get better results with much less computation, I use a Taylor Series expansion, which leads to a form that can be integrated term by term. The sum will give the desired area. The Taylor series is

$$P(c) = P(r) + (c-r)\, P^{(1)}(r) + (c-r)^2\, P^{(2)}(r)/2! + \ldots \qquad (15.6)$$

Solving equation 15.2 for P in terms of c yields a simple quadratic. This is how I obtained the quick tests, equation 15.4. Next, I evaluate derivatives for Taylor series, equation 15.6. Since there are three components (values of r) to evaluate, let's first take the cardioid at $c = -1.75$.

$$P = 1 + \tfrac{1}{2}\, r\, [1 + \sqrt{(-4r-7)}],$$
$$P^{(1)} = \tfrac{1}{2} - \tfrac{1}{2}\, (6r+7)/\sqrt{(-4r-7)},$$
$$P^{(2)} = -(6r+14)/[(4r+7)\sqrt{(-4r-7)}] \qquad (15.7)$$

and so on. Additional derivatives are evaluated with MathCad™ or your favorite equivalent. The other two period 3 components are at $(-1 \pm i\sqrt{27})/8$ and are obtained by changing the sign of the square root in equation 15.7.

Looking over the results, I find sequences of coefficients such as 6, 12, 72, 720, 10080, 181440, ... and 14, 42, 336, 4200, 70560, 1481760, ... Some sense was made of it by taking second differences, ratios of terms, and so on. The final algorithms can be best expressed in a computer language such as C++. I used C++, since it facilitates complex operations and even many-digit arithmetic by just renaming the data types.

For the cardioid at -1.75 the following algorithm generates the desired series. Let $P^{(k)}$ be the kth derivative of P. Let $f = -4c-7$, $g = \sqrt{f}$, $h = 1/g$, $a = -3$, $b = -7/2$, $k = 0$, then

$$P^{(0)} = 1 + r\,(1 + g)/2, \text{ and } P^{(1)} = \tfrac{1}{2} + (ar + b)\,h.$$

Additional derivatives are found by iteration (here := denotes replacement or updating a variable).

```
begin loop:
    if k >1 then
    P(k) = (ar + b) h,
    h := h/f,
    a := a(-6 + 4k),
    b := b(-6/k -2 + 4k),
    k := k + 1,
end loop when P(k) is small enough.                              (15.8)
```

The other series for components at $(-1 \pm i\sqrt{27})/8$ is identical to this except for changing the sign of g so that $g = -\sqrt{f}$.

The Taylor series (equation 15.8) is expanded about the component "center" or root, r, where $P = 0$. Actually, the series expansions can be evaluated at any point inside the component, but the convergence is best at the center. I used more than 120-digit arithmetic to evaluate the roots of equation 15.2. Setting $P = 0$ we get the polynomial: $r^3 + 2r^2 + r + 1 = 0$. Let

$U = [(25 + \sqrt{621})/54]^{1/3}$
$V = [(25 - \sqrt{621})/54]^{1/3}$
The roots are $r = -(U + V + 2/3)$

$= -1.75487766624669276004950889635852869189460661777279314$
$398928397064608065512808109073822709284225030364837736 79$

and $r = \frac{1}{2}(U + V) - \frac{2}{3} \pm \frac{1}{2} i (U - V) \sqrt{3}$

$= -0.1225611668766536199752455518207356540526966911136 0342$
$8005358014676959672435959454630886453578874848175811 3164359$
$+i\ 0.7448617666197442365931704286043923672401630849068 2457$
$4201847592154415217837839767791143754932964159039252 80487$
$3377.$

An additional factor of $2^3 = 8$ applied to the series (equation 15.6) scales the map to a unit circle. That is, when c is on the edge of a component, $q = 8P$ is on the unit circle.

$$q = B_0 + B_1(c - r) + B_2(c - r)^2 + \ldots \tag{15.9}$$

where $B_k = 8P^{(k)}/k!$ What we need is a formula for c on the component edge in terms of a convenient point such as one on the unit circle. To get this formula we perform a series reversion to obtain a series for c in terms of $q = 8P$. I used the Lagrangian algorithm from Knuth.[11]

$$c = Q_0 + Q_1(q - B_0) + Q_2(q - B_0)^2 + \ldots \tag{15.10}$$

Now we have a Taylor series for points on the boundary of the period 3 components. The argument is $W = \exp(i\,\theta)$, which corresponds to points around a unit circle on the complex plane. Note that

$$c = \sum_{k=0}^{m} Q_k W^k \tag{15.11}$$

which is also a Fourier series in :

$$c = \sum_{k=0}^{m} Q_k \, exp(ik\theta) \qquad (15.12)$$

I think of each of these terms as circles or epicycles with the angle wrapping around k times. So the area of each is $\pi k Q_k^2$. Too easy!

$$A = \pi \sum_{k=0}^{m} k \, |Q_k|^2 \qquad (15.13)$$

Using 65 terms and 110-digit arithmetic gives the area of the small cardioid at -1.75 as

$A_{cardioid}$ = 0.00040641464713872120431940145891262007393230862 390575839940535386555080299823389660544023907073911373 21 73076544

and the area of each circle like bud as

$A_{circular\ bud}$ = 0.02806800174027133036562467727552262522903185 29199110560103175874123421869308503249919440129036378452 88293067213

The final result, the small cardioid plus the two buds, has area

A = 0.05654241812768138193556875600995787053199601446 3727 8704200405286902351768599345465893282648780148043087 59201466

ACCURATE EVALUATION OF COMPONENT AREAS OF ANY PERIOD

The procedure for finding the component area for period n follows that of the previous section except now we need to create the Taylor series in a more general manner. The quantity P (see equation 15.6) is the product of n iterates of $z_{k+1} = z_k^2 + c$ in the limit for large k. That is, after k iterations, when the iterations have settled (to sufficient precision) into a repeating cycle, we calculate

$$P = \prod_{j=1}^{n} z_{k+j} \qquad (15.13)$$

Derivatives of P and z can be built iteratively using the chain rule of multiplication. Here again, := will denote updating a variable during the iteration. Recall our basic iteration is $z^{(0)} := z^{(0)}z^{(0)} + c$. The superscript (0) means no derivative taken. $z^{(j)}$ means jth derivative, $0 \le j \le m$. The first step is an iteration until the repeating cycle has converged. This occurs immediately when a component root, r, is used for c. Before we can evaluate P and its derivatives, at least $m + n$ iterations are performed (until the derivatives converge) to calculate the first m derivatives of z. Note the derivative values are evaluated in order of high derivative to low.

$$z^{(j)} = 0, \text{ for } j = 0..m.$$

```
loop at least m + n times                                        (15.14)
z^(m)  := . . . ,
```

$$
\begin{aligned}
z^{(4)} &:= 2(z^{(0)}z^{(4)} + 4z^{(1)}z^{(3)} + 3z^{(2)}z^{(2)}), \\
z^{(3)} &:= 2(z^{(0)}z^{(3)} + 3z^{(1)}z^{(2)}), \\
z^{(2)} &:= 2(z^{(0)}z^{(2)} + z^{(1)}z^{(1)}), \\
z^{(1)} &:= 2z^{(0)}z^{(1)} + 1, \\
z^{(0)} &:= z^{(0)}z^{(0)} + c.
\end{aligned}
$$

```
end loop.
```

Now we initialize P and its derivatives and evaluate them using another n iterations.

$$P^{(0)} = 1.0, \quad P^{(j)} = 0.0, \quad j = 1..m.$$

```
loop n times                                                     (15.15)
z^(m)  := . . . ,
```

$$
\begin{aligned}
z^{(4)} &:= 2(z^{(0)}z^{(4)} + 4z^{(1)}z^{(3)} + 3z^{(2)}z^{(2)}), \\
z^{(3)} &:= 2(z^{(0)}z^{(3)} + 3z^{(1)}z^{(2)}), \\
z^{(2)} &:= 2(z^{(0)}z^{(2)} + z^{(1)}z^{(1)}), \\
z^{(1)} &:= 2z^{(0)}z^{(1)} + 1, \\
z^{(0)} &:= z^{(0)}z^{(0)} + c.
\end{aligned}
$$

```
P^(m)  := . . . ,
```

$$
\begin{aligned}
P^{(4)} &:= z^{(4)}P^{(0)} + 4z^{(3)}P^{(1)} + 6z^{(2)}P^{(2)} + 4z^{(1)}P^{(3)} + z^{(0)}P^{(4)}, \\
P^{(3)} &:= z^{(3)}P^{(0)} + 3z^{(2)}P^{(1)} + 3z^{(1)}P^{(2)} + z^{(0)}P^{(3)}, \\
P^{(2)} &:= z^{(2)}P^{(0)} + 2z^{(1)}P^{(1)} + z^{(0)}P^{(2)}, \\
P^{(1)} &:= z^{(1)}P^{(0)} + z^{(0)}P^{(1)}, \\
P^{(0)} &:= z^{(0)}P^{(0)}.
\end{aligned}
$$

```
end loop.
```

The equivalent of equation 15.9 is

$$q = B_0 + B_1(c - r) + B_2(c - r)^2 + \dots \tag{15.16}$$

where $B_k = 2^n P^{(k)}/k!$. After reversion we apply equations 15.10, 15.11, and 15.12 to obtain the area. Ten terms in equation 15.16 will usually get better than 15 digits accuracy. The appendix lists a program for calculating the first four terms of the Taylor's series and the component area.

John Stephenson kindly sent me his list of roots through period 11.[12] I computed the component areas using his list and ten terms in equation 15.12. Since there are so many components (a few thousand), I shall list the areas by period. (See table 15.1.) The results from GIPPP#3 are included for comparison. The first two areas are equal to $3\pi/8$ and $\pi/16$. The third was evaluated to many digits in "A Formula and Area for Period 3 Components." The remaining values were calculated with 80-bit floating point arithmetic (19 digits).

THE SEARCH FOR CENTERS
OF COMPONENTS OF HIGHER PERIOD

In this section I describe my search for components with period 12 and greater. To do this I shall use what is called symbolic dynamics.[13] Symbolic dynamics is usually applied to the logistic equation $x_{n+1} = r x_n(1-x_n)$, which only involves real arithmetic. This equation is closely related to the Mandelbrot set.

Table 15.1
Area by Period

period	total area (10 term)	GIPPP#3
1	1.17809724509617246	1.17809724509617246
2	0.19634954084936208	0.19634380805667473
3	0.05654241812768138	0.05642440719669048
4	0.02327513769630081	0.02320561843249499
5	0.01311506747889241	0.01306774668879608
6	0.00893479053124322	0.00889820871340634
7	0.00505792907074769	0.00503256427070825
8	0.00446801353188167	0.00444245093772218
9	0.00293643139498573	0.00291837082547177
10	0.00266092631691073	0.00264204766762939
11	0.00138279912887914	0.00137094334276880

Component centers are fixed points of the basic iteration

$$z := z^2 + c, \tag{15.17}$$

with initial $z = 0$. The component centers have values of c such that after n iterations z is again zero. We can expand this iteration algebraically without difficulty when the period is small. We get a polynomial whose roots are the component centers. Standard root-finding methods fail with periods of 8 or more, since the polynomials have 128 or more terms with very closely spaced roots. Small periods have low-degree polynomials. Period 3 results in a quartic $c(c^3 + 2c^2 + c + 1) = 0$, which has zero for one root and three other roots that we are interested in. The zero root is actually the period 1 root. It shows up here since a period of 1 also repeats with period 3. The pure period 3 polynomial is therefore $c^3 + 2c^2 + c + 1 = 0$.

Another method for finding component centers is simply to solve equation 15.17 in reverse using complex arithmetic.[14] For initial values, choose $z = 0$; c can be (almost) arbitrary. Using complex square root, we write equation 15.17 as what is called a preimage.

$$z := \pm \sqrt{(z - c)} \tag{15.18}$$

where the sign is chosen in a sequence peculiar to each component. At the end of each iteration sequence, use a value of c equal to the resulting z. Repeat until c converges to a root (or you run out of patience)! There are $2^{(n-1)}$ roots to equation 15.17 for period n, so there are $2^{(n-1)}$ possible sign sequences. A nice way to handle this numerically is to let the binary representation of the numbers from 1 to $2^{(n-1)}$ indicate the sign sequence. Starting from the right, choose + when the bit is 1; otherwise choose -. This method works fairly well except it does not always converge to a root. So we must limit the repetitions and also check that the final c indeed satisfies n iterations of equation 15.17. You will find that this procedure finds all the real roots without difficulty. But there are large gaps in the complex plane where roots seem to hide. Also, the algorithm may find more than one root, depending on the initial c. So I use several initial c values to find as many roots as possible. This requires running through $2^{(n-1)}$ sign sequences for each initial c.

We start to see difficulties with equation 15.18 even with $n = 8$. Experts say we are not supposed to expect the inverse iteration to work for complex roots. But this, I feel, should not stop us from trying, especially if we can make it work for most of the roots. To find even more roots from equation 15.18, I have examined the evaluation methods for the complex square root. If you use the "library" function on your favorite computer, your square root will be using a branch cut along the real line. There is really no preferred cut in this application. In fact, I have obtained additional roots using a more general branch cut definition. This cut goes through zero but at an angle (45 or 90 degrees) to the real line. This adds

Table 15.2
Area by Period

period	lower bound area (10 term)	GIPPP#3
12	0.0021014356785	0.00208578681061190
13	0.0008494394063	0.00084259103981510
14	0.0011312354040	0.00112252615290401
15	0.0009404763463	0.00093614172426903
16	0.0008909027712	0.00088226089739286

yet another dimension to the search. I have found all but five roots of period 12 and most roots through period 16. Table 15.2 extends table 15.1 to period 16, but is a lower bound on the area since a few small components are missing. Adding the areas in table 15.1 and table 15.2, we get the total area in components with periods 16 or less is at least 1.4987338.

AREAS OF CLUSTERS OF COMPONENTS

Considering all the difficulty involved with finding component centers by polynomial root finding, inverse iteration, or even graphical methods, it would seem obvious to evaluate areas of all the attached buds once a midget cardioid is found. This can be done by recursively evaluating for each component the centers of neighboring components. They can be located by evaluating equation 15.11 with $W = \exp(s + i\theta)$ where $s = (P/D)^2$ and $\theta = N/D$. Here P is the component period, D is the period of the attached component, and $N < D/P$ is an index number of the attached component. N and D must be relatively prime integers.

Each midget Mandelbrot, as some call the small island copies of the original, contains one cardioid component with many attached circular buds. I call these "clusters" and the area the "cluster area." See figure 15.2 for an example of a cluster. I have evaluated components up to period 100 for clusters whose cardioids have periods from 1 to 13. Table 15.3 shows the cumulative area of these clusters up to 13. This results in an improved lower bound to the Mandelbrot set area, 1.50585063, which is only 0.05 percent below Robert Manufo's area estimate.

RAPID COMPONENT PLOTTING

It is easy, given the component center, to plot the component boundary using equation 15.11. A precomputed table of several hundred points around a unit circle can be used for W. Attached components can be plotted also, using the

Table 15.3
Cumulative Cluster Area by Period

cardioid period	cluster area	cumulative cluster area
1	1.50463869	1.50463869
3	0.00050842	1.50514711
4	0.00020723	1.50535435
5	0.00015455	1.50550889
6	0.00008510	1.50559399
7	0.00006742	1.50566140
8	0.00005226	1.50571367
9	0.00004336	1.50575702
10	0.00002927	1.50578606
11	0.00002405	1.50581011
12	0.00002909	1.50583919
13	0.00001144	1.50585063

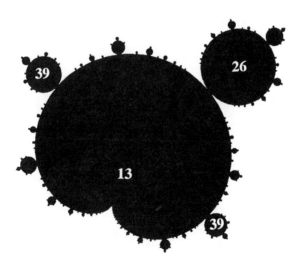

Figure 15.2   I named this COCKEYE since it is what my wife calls the most cockeyed midget she has seen. The picture is of a period 13 cardioid with "extreme" distortion; its components seem to be falling off to one side.

Table 15.4.
Distorted Cardioid Series

period = 13

	real part	imaginary part
c	0.314559489984262	0.029273969079785
Q_1	-9.642550543685020e-05	-1.329325256969317e-04
Q_2	6.727058433526614e-05	5.759791616535904e-05
Q_3	-1.205436806032880e-05	9.019929997995239e-06
Q_4	-1.048438553804530e-06	-2.353435019848357e-06

recursive evaluation of neighboring components. Examples of this are shown in figures 15.1 and 15.2. Figure 15.1 shows the cardioids and components through period 12. Also, we can use equation 15.16 as a rapid test for a point interior to a component. If q is in the unit circle, then c is in the component.

Another use of equation 15.11 is to explore quantitatively the distorted components. Figure 15.2 shows a period 13 cardioid (which is very distorted) and attached components up to period 208. The first four terms in equation 15.11 show quantitatively how distorted it is (table 15.4). The ratio Q_2/Q_1 for a perfect cardioid is $-\frac{1}{2}$ with all higher-order terms zero. This distorted cluster has Q_3 about 13 percent as large as Q_1.

SUMMARY

What began as a very interesting collaboration of more than four dozen enthusiastic volunteers around the world, whose computers are connect by Internet eventually resulted in a closer estimate of the very elusive area for the beautiful mathematical object, the Mandelbrot set. The pursuit has led to interesting side topics, such as special tests for a point being located in a component and rapid methods for plotting the Mandelbrot set. Alas, although several values suggested for the area, such as $\frac{3}{2}$ and $\pi / 2$, were ruled out, the precise nature and understanding of what we numerically estimate to be near 1.5066 must await deeper mathematical investigation. In the meantime, with my poems, I may finally get revenge on English teachers around the world who made mathematicians suffer through their poetry. In the future, I predict that more fractal research will rely on the parallel collaborative efforts of researchers and computers around the world.

APPENDIX

Accurate evaluation of the Taylor series for any component has been demonstrated using equations 15.13 through 15.16 where P and z were built iteratively. Another method that reduces this process to sums and products of the basic iteration formula and its derivatives was derived recently by Dante Giarrusso (private communication). His formula are algebraic expressions for the first four terms of the Taylor series, which required extensive algebraic manipulations to derive. In addition to the iterates z_k and its derivatives, he defines two more quantities and their derivatives,

$$\phi_n = \prod_{k=1}^{n-1} z_k$$

and

$$\Sigma = \phi_p \sum_{k=1}^{p-1} 2^k \, \phi_k/z_k$$

A C++ program (for the complex data type) that evaluates Giarrusso's formula and then the area of a component follows. To evaluate the area of a component, you need the component period and center. Here are the required values for some low period components. You can try values from this list:

p	$\mathrm{Re}(c)$	$\mathrm{Im}(c)$
1	0.0	0.0
2	-1.0	0.0
3	-1.75487766624669276000	0.0
3	-0.12256116687665362000	0.74486176661974423660
4	-1.94079980652948475220	0.0
4	-0.15652016683375506180	1.03224710892283180170
4	-1.31070264133683288360	0.0
4	0.28227139076691387970	0.53006061757852529950

Those are the easy ones. Here is a more difficult example, a period 152 cardioid. The command line and sample output are:

```
AREA 152 -0.101090652637403306326 8548 0.9562903787
08031184 7836702
```

Sample output:

```
period 152 cardioid
```

```
center -0.101090652637403300 0.956290378708031130
B[0] 7.664550938368819290e-11 -7.316280914437811590e-11
B[1]    -3.224342712491727540e+12 1.082589687028605840e+12
B[2]     4.637678973674128110e+24 -2.896961224435386240e+24
B[3]    -1.196494742220190850e+37 1.385153375250792490e+37
B[4]     3.088093083281463660e+49 -6.994010025140471340e+49
Q[1]    -2.787201786570228390e-13 -9.358173677130098170e-14
Q[2]     1.272639007915418050e-13 5.583303854146374010e-14
Q[3]     8.785954225591191240e-15 -5.394496555190365770e-15
Q[4]    -4.552609039525559740e-16 -5.823971083369849560e-16
Q2/Q1   -4.707867028009373180e-01 -4.225026440948358860e-02
2 term area:   3.929169094371931070e-25
3 term area:   3.939187028720043360e-25
4 term area:   3.939255697511832000e-25
```

```c
/*

    AREA:  Evaluates the area of Mandelbrot set components by Taylor

    series using the method in this chapter with Dante Giarrusso's

    coefficient formula.

    Program by Jay R. Hill, 1994.


    Usage: Area [p] [Re(c)] [Im(c)]

    where p = component period

       Re(c) = real part of component center

       Im(c) = imaginary part of component center
*/

#include <stdio.h>

#include <math.h>

#include <complex.h>

#include <stdlib.h>

#include <conio.h>

extern unsigned _stklen = 40000U; // stack variable for Borland C++

const int PMAX=400; // maximum period and dimension of arrays
```

```
void FirstFour(        // series expansions for Mandelbrot set component

    long p,            // input  = component period

    complex c0,        // input  = component center

    complex B[5],      // output = Taylor series coefficients

    complex Q[5])      // output = Taylor series coefficients

{

// c = sum(k=0..4) Q[k]*W^k.

// q = sum(k=0..4) B[k]*(c-c0).

  complex Z[PMAX+1],dZ[PMAX+1],ddZ[PMAX+1],dddZ[PMAX+1],ddddZ;

  complex PHI, dPHI, ddPHI, dddPHI, S, dS, SIGMA, dSIGMA;

  complex derZ, der2Z, der3Z, der4Z, derp, der2p, der3p;

  complex T1, T2, Q2;

  long j,k; long double n2;

  Z[0] = 0; dZ[0] = 0; ddZ[0] = 0; dddZ[0] = 0; ddddZ = 0;

  S = 0; dS = 0;  SIGMA = 0; dSIGMA = 0;

  PHI = 1; dPHI = 0; ddPHI = 0; dddPHI = 0;

  for(n2 = 1, k = 0; k<p; ++k, n2 *= 2){

    Z[k+1] = Z[k]*Z[k]+c0;  // basic iteration equation

    dZ[k+1] = 2*Z[k]*dZ[k]+1;

    ddZ[k+1] = 2*(Z[k]*ddZ[k]+dZ[k]*dZ[k]);

    dddZ[k+1] = 2*(Z[k]*dddZ[k]+3*dZ[k]*ddZ[k]);

    ddddZ = 2*(Z[k]*ddddZ+4*dZ[k]*dddZ[k]+3*ddZ[k]*ddZ[k]);

    if(k>0){

      S = S+n2*PHI/Z[k];

      dS = dS+n2*(dPHI/Z[k]-PHI*dZ[k]/(Z[k]*Z[k]));

      PHI *= Z[k];

    }

    dPHI = 0;

    ddPHI = 0;

    T1 = 0; T2 = 0;

    for(j = 1; j <= k; ++j){
```

```
    dPHI += dZ[j]/Z[j];

    ddPHI += ddZ[j]/Z[j]-dZ[j]*dZ[j]/(Z[j]*Z[j]);

    T1 += ddZ[j]/Z[j]-dZ[j]*dZ[j]/(Z[j]*Z[j]);

    T2 += dddZ[j]/Z[j]-3*ddZ[j]*dZ[j]/(Z[j]*Z[j])

       +2*dZ[j]*dZ[j]*dZ[j]/(Z[j]*Z[j]*Z[j]);

  }

    dPHI *= PHI;

    ddPHI *= PHI;

    ddPHI += (dPHI*dPHI)/PHI;

    dddPHI = 2*dPHI*ddPHI/PHI-dPHI*dPHI*dPHI/(PHI*PHI)+dPHI*T1+PHI*T2;

    if(k>0){

      SIGMA = PHI*S;

      dSIGMA = dPHI*S+PHI*dS;

    }

  }

  derp = dPHI;

  derZ = dZ[p];

  der2Z = ddZ[p]+n2*PHI*derZ*derZ;

  der3Z = dddZ[p] + 3*n2*(derp*derZ*derZ + PHI*derZ*der2Z);

  der4Z = ddddZ + n2*(6*ddPHI*derZ*derZ+12*derp*derZ*der2Z

     +3*PHI*der2Z*der2Z+4*PHI*derZ*der3Z+3*SIGMA*derZ*derZ*derZ*derZ);

  der2p = ddPHI+SIGMA*derZ*derZ;

  der3p = dddPHI+3*(dSIGMA*derZ*derZ+SIGMA*derZ*der2Z);

// Taylor series coefficients

  B[0] = Z[p];        // will be zero when c0 is a period p center

  B[1] = n2*PHI*derZ;

  B[2] = n2*(PHI*ddZ[p]+n2*PHI*derZ*PHI*derZ+2*derp*derZ)/2;

  B[3] = n2*(PHI*der3Z+3*(der2Z*derp+ddPHI*derZ+SIGMA*derZ*derZ*derZ))/6;

  B[4] = n2*(PHI*der4Z + 4*der3Z*derp + 6*der2Z*der2p + 4*derZ*der3p)/24;

  Q[0] = c0;
```

```
// Series reversion coefficients

  Q[1] = 1/B[1];

  Q2   = Q[1]*Q[1];

  Q[2] = -B[2]*Q2/B[1];

  Q[3] = (2*B[2]*B[2] - B[1]*B[3])*Q2*Q2/B[1];

  Q[4] = (5*B[1]*B[2]*B[3] - B[1]*B[1]*B[4]

     - 5*B[2]*B[2]*B[2])*Q2*Q2*Q2/B[1];

}


void main(int argc, char **argv){

  complex c, BA, D0[5], D1[5];

  long i,p;   double a,aA,aB,aC,aD,x,y;

  if(argc == 1) printf("Usage: Area [p] [Re(c)] [Im(c)]\n");

  if(argc > 1) p = atol(argv[1]); // read in p from command line

  else p = 3;

  if(p > PMAX) printf("p > PMAX\n"); else{

    if(argc > 2) x = atof(argv[2]); // read in Re(c) from command line
// Some roots for periods 1 through 4 are built in defaults

    else if(p == 1) x = 0;  // supply real root if not in command line

    else if(p == 2) x = -1;

    else if(p == 3) x = -1.7548776662466927600495088963585852;

    else if(p == 4) x = -1.9407998065294847522320909796552;

    else x = 0;

    if (argc > 3) y = atof(argv[3]); // read in Im(c) from command line

    else y = 0.0;

    // if p negative, supply a complex root as an example

    if(p <= -4){   p = 4;

      x = -0.1565201668337550617989866507878;

      y =  1.0322471089228318016716015324301;

    }
```

```
    if(p < 1){   p = 3;
      x = -0.1225611668766536199752455518207074;
      y =  0.7448617666197442365931704286043944039;
    }
    c = complex(x,y);
// Evaluate the series: period p, center c.
    FirstFour(p, c, D0, D1);
    BA = D1[2]/D1[1]; // This ratio is near -.5 for cardioids
    if(real(BA)<-.25) printf("period %ld cardioid\n",p);
    else printf("period %ld circle\n",p);
    printf("center %22.18lf %21.18lf\n",real(c),imag(c));
    for(i=0; i<5; ++i)
      printf("B[%ld]   %22.18le %21.18le\n",i,real(D0[i]),imag(D0[i]));
    for(i=0; i<5; ++i)
      printf("Q[%ld]   %22.18le %21.18le\n",i,real(D1[i]),imag(D1[i]));
    printf("Q2/Q1   %22.18le %21.18le\n",real(BA),imag(BA));
    aA = abs(D1[1]); aB = abs(D1[2]); aC = abs(D1[3]); aD = abs(D1[4]);
    a = M_PI*(aA*aA + 2*aB*aB);
    printf("2 term area:   %22.18le\n",a);
    a += M_PI*(3*aC*aC);
    printf("3 term area:   %22.18le\n",a);
    a += M_PI*(4*aD*aD);
    printf("4 term area:   %22.18le\n",a);
  }
}
```

Notes

I thank Keith Briggs, Bill Broadley, Yuval Fisher, Michael Frame, Dante Giarruso, Robert Manufo and John Stephenson for helpful and interesting discussions, all carried out using e-mail. Also I thank Kaufman Friedrich for his extended arithmetic package and Peter Montgomery for optimizing my series revision routine. Also, I thank Science Applications International Corporation (SAIC) for a connection to Internet and thank the dozens of Internet volunteers who helped in the GIPPP project. Finally, I would like to thank my family for letting me hog our 486/33 computer.

1. D. Rabenhorst, "What Is the Area of the Mandelbrot Set?" in *Journal of Chaos and Graphics* (ed. C. Pickover) 2 (1987): 26.

2. J. H. Ewing and G. Schober, "The Area of the Mandelbrot Set," *Numerical Mathematics* 61 (1992): 59-72.

3. Y. Fisher, "Exploring the Mandelbrot Set," in *The Science of Fractal Images,* H.-O. Peitgen and D. Saupe, ed. (New York: Springer-Verlag, 1991), 287-296.

4. D. Giarrusso and Y. Fisher, "A Parameterization of the Period 3 Hyperbolic Components of the Mandelbrot Set," *Proceeding of the American Mathematical Society* 123 (12) (December, 1995): 3731-3737.

5. Y. Fisher and J. Hill, "Bounding the Area of the Mandelbrot Set," MS.

6. J. Stephenson, "Formula for Cycles in the Mandelbrot Set," *Physica A* 177 (1991): 416-420. J. Stephenson and D. T. Ridgway, "Formula for Cycles in the Mandelbrot Set II," *Physica A* 190 (1992): 104-116.

7. Giarrusso and Fisher, "Parameterization of the Period 3 Hyperbolic Components."

8. E. Netto, *Theory of Substitutions* (Bronx, NY: Chelsea Publishing Company, 1892), 209.

9. J. Stephenson, "Formula for Cycles in the Mandelbrot Set III," *Physica A* 190 (1992): 117-129.

10. D. Giarrusso, private communication, 1994.

11. D. Knuth, *The Art of Computer Programming,* vol. 2 (Reading, MA: Addison-Wesley, 1981), 508-509.

12. J. Stephenson, private communication, 1993.

13. H.-O. Peitgen, H. Jurgens, D. Saupe, *Chaos and Fractals: New Frontiers of Science* (New York: Springer-Verlag, 1992), chapter 10.

14. Netto, *Theory of Substitutions,* 209.

16

Self-Similarity in Quasi-Symmetrical Structures

Arthur L. Loeb

―――――

In this brief informal chapter I show readers how to generate frieze designs from fractal-like quasi-symmetrical one-dimensional sequences of characters. Readers are urged to explore the intricate symmetries. I hope to see future designers make use of such patterns as ornaments in architecture and design.

―――――

The structure of apparently complex patterns often is clarified when the rule generating that pattern is revealed. Although seasoned mathematicians will find this observation obvious, exercises such as the one in this chapter are providing a fertile area for my students, artists as much as scientists, for experimentation. For example, figure 16.1 shows a frieze design comprising a horseshoelike motif in two orientations, which we shall refer to as *cup* and *cap*. There appears to be no regularity in the orientation pattern, but in point of fact, it is actually quite orderly, as we shall see.

Table 16.1 shows a series of strings of characters x and y; the strings are labeled S_0, S_1, S_2, ..., S_i, ... If x is identified with *cup* and y with *cap*, then the frieze pattern in figure 16.1 will be seen to correspond to S_7 in table 16.1.

Each successive string in table 16.1 was generated from the previous one by replacing x by $x\,y$, and replacing y by x. The process can be repeated indefinitely, ultimately achieving a string of infinitely many characters, which has no translational symmetry, and is called quasi-symmetrical.[1] Although other algorithms generate different quasi-symmetrical strings,[2] when I discuss such strings here, I limit myself to those strings generated by the transformation $x \rightarrow xy$, $y \rightarrow x$.

I also call each of the strings in table 16.1 quasi-symmetrical. In addition, I call the exact sequence of characters in each string the structure of that string. The number of characters, N_i, in a string S_i is called the length of that string. The strings in table 16.1 have the following regularities:

1. The length of each string, N_i, equals a Fibonacci number (1, 1, 2, 3, 5, 8, 13, ...)
2. In each string $N_i(x)$, the number of characters x, as well as $N_i(y)$, the number of characters y, equals a Fibonacci number.

These two observations are proven as follows. Since every symbol x in any quasi-symmetrical string S_i results from either x or y in the previous string S_{i-1},

$$N_i(x) = N_{i-1} \qquad (16.1)$$

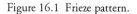

Figure 16.1 Frieze pattern.

Table 16.1

Generation of Quasi-Symmetric String

$$y = S_0$$
$$x = S_1$$
$$xy = S_2$$
$$xyx = S_3$$
$$xyxxy = S_4$$
$$xyxxyxyx = S_5$$
$$xyxxyxyxxyxxy = S_6$$
$$xyxxyxyxxyxxyxyxxyxyx = S_7$$
$$xyxxyxyxxyxxyxyxxyxyxxyxxyxyxxyxxy = S_8$$

Every symbol y results from an x in the previous string:

$$N_i(y) = N_{i-1}(x) \qquad (16.2)$$

from equation 16.1:

$$N_{i-1}(x) = N_{i-2} \qquad (16.3)$$

therefore:

$$N_i = N_i(x) + N_i(y) = N_{i-1} + N_{i-2}. \qquad (16.4)$$

Since $N_0 = 1$ and $N_1 = 1$, and equation 16.4 is the recursion relation generating Fibonacci numbers, the length N_i of any quasi-symmetrical string S_i is a Fibonacci number. This being the case, equations 16.1 and 16.2 tell us that the numbers of symbols x and y, $N_i(x)$ and $N_i(y)$, also equal Fibonacci numbers.

Note in table 16.1 that each string equals the previous string followed by the string prior to that previous string:

$$S_{i+1} = S_i \, S_{i-1}. \qquad (16.5)$$

If we were to define

$$X_i = S_i, \text{ and } Y_i = S_{i-1} \qquad (16.6x \text{ and } 16.6y)$$

then

$$X_{i+1} = X_i Y_i \qquad (16.7x)$$

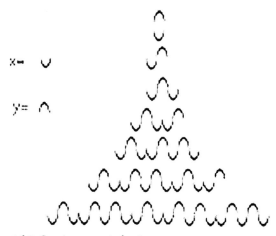

Figure 16.2 Quasi-symmetrical strings.

and

$$Y_{i+1} = X_i \qquad (16.7y)$$

These two equations are just the transformations of x and y used to generate table 16.1. That means that you can group any pair of successive quasi-symmetrical strings S_i and S_{i-1} together as if they had been the original characters x and y, and generate a new table 16.1, each *character* now representing a *string* of the old table. Conversely, each character x or y in the original table might be a composite string from an earlier construction.

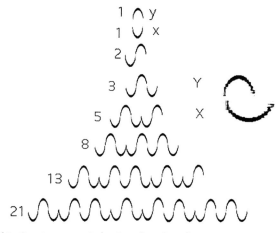

Figure 16.3 Quasi-symmetrical strings (continued).

Figure 16.2 is a pictorial representation of table 16.1, each *x* being represented by a cup, each *y* by a cap. (Note that the frieze design of figure 16.1 corresponds to string S_7). In figure 16.3 these strings are shown again. The structure of any quasi-symmetrical string S_i is determined solely by its length N_i, which is indicated to the left of each string in figure 16.3. To the right of the strings of lengths 5 and 3 are drawn, respectively, an elongated cup *X* and an elongated cap *Y*. The string at the bottom, of length 21, may be uniquely decomposed in terms of the strings of lengths 3 and 5 as follows:

$$xyxxyxyxxyxxyxyxxyxyx = (xyxxy)\ (xyx)\ (xyxxy)\ (xyxxy)\ (xyx) = XYXXY$$
$$(16.8)$$

This decomposition is shown graphically in figure 16.4. The original string of length 21 has become a string of length 5 in terms of *X* and *Y*, and as such, of course, it has the same structure as any other string of that same length. In particular, *X* itself has length 5, and it therefore has the same structure in terms of *x* and *y* as does the string of equation 16.8 in terms of *X* and *Y*. In turn, in figure

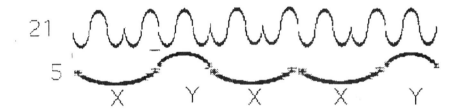

Figure 16.4 Quasi-symmetrical string and superstring.

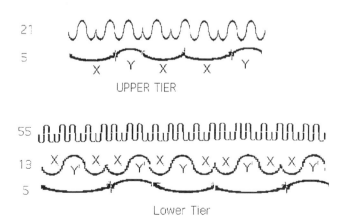

Figure 16.5 Superstring and superduper string.

16.5 the string of length 21 and its "superstructure" of length 5 have been split into a "superduper" X and Y (upper tier), out of which a new quasi-symmetrical string is constructed (lower tier), of length 5 in "superduper" terms, of length 13 in terms of X and Y, and of length 55 in terms of x and y.

The structure, superstructure, and superduper structure of the patterns in figure 16.5 show a striking analogy to fractals. Strings having the same structure are similar, and a string having the same structure as one of its component strings is *self-similar*. These examples may be generalized, but modifying the notation somewhat to do so is convenient. Since the structure of a quasi-symmetrical string was shown to depend solely on its length, such a string is entirely defined by its length and its characters. Therefore define:

$$S_i \equiv S(N_i , x, y). \tag{16.9}$$

Any string may be decomposed into its prior pair of strings:

$$S(N_{i+1}; x, y) = S(N_i; x, y)\, S(N_{i-1}; x, y),$$
$$\text{where } N_{i-1} + N_i = N_{i+1}. \tag{16.10}$$

Any string $S(N_{i+k} ; x, y)$ may be constituted from any prior pair of strings:

$$S(N_{i+k}; x, y) = S\{N_{k+1}; S(N_i; x, y), S(N_{i-1}; x, y)\}. \tag{16.11}$$

(Equation 16.10 actually is a special case of equation 16.11.) The latter may be interpreted as follows for figure 16.4:

$$N_{i+k} = 21, N_i = 5.$$

Self-similarity will occur whenever $k = i$-1 or $k = i$-2. Since 21 and 5 are respectively the seventh and fourth Fibonacci numbers, $i+k = 7$, $i = 4$, hence $k = 3$, so that equation 16.11 for this example becomes:

$S(21; x, y) = S\{5; S(5; x, y), S(3; x, y)\} =$
$S(5; x, y)\, S(3; x, y)\, S(5; x, y)\, S(5; x, y)\, S(3; x, y) =$
xyxxy xyx xyxxy xyxxy xyx.

(The spaces between groups of characters are for the convenience of identification only; they have no formal significance.)

Note that the condition for self-similarity is satisfied because $3 = k = i$-1 $= 4$-1.

In other words, the quasi-symmetrical string of 21 characters x and y may be decomposed into a quasi-symmetrical string of length 5, of which the component characters are themselves quasi-symmetrical strings of respective lengths 5 and 3, the structures of all of these strings being determined entirely by their length.

For the lower tier in figure 16.5:

$S(55; x, y) =$
$S[5; S\{3; S(5; x, y), S(3; x, y)\}, S\{2; S(5; x, y), S(3; x, y)\}]$
$= S\{3; S(5; x, y), S(3; x, y)\} S\{2; S(5; x, y), S(3; x, y)\}$
$S\{3; S(5; x, y) S(3; x, y)\} S\{3; S(5; x, y) S(3; x, y)\}$
$S\{2; S(5; x, y), S(3; x, y)\} =$
$S(5; x, y) S(3; x, y) S(5; x, y) S(5; x, y) S(3; x, y)$
$S(5; x, y) S(3; x, y) S(5; x, y) S(5; x, y) S(3; x, y) S(5; x, y)$
$S(5; x, y) S(3; x, y) =$
$XYXXYXYXYXXYXXY = xyxxy\ xyx\ xyxxy\ xyxxy\ xyx\ xyxxy\ xyx\ xyxxy\ xyxxy\ xyx\ xyxxy$
$xyxxy\ xyx$

Again, self-similarity occurs because several of the strings have the structure characteristic of length 5, namely *xyxxy*.

In general, *any* quasi-symmetrical string generated by the algorithm $x \rightarrow xy$, $y \rightarrow x$ may be uniquely decomposed into a string of which every component is one or the other of any successive pair of shorter quasi-symmetrical strings. (By *successive* strings I mean two strings whose lengths are equal to any two successive Fibonacci numbers.) This uniqueness is demonstrated as follows. The recursion relation generating Fibonacci numbers:

$$a_{n+1} = a_n + a_{n-1}$$

may, by repeated applications, be generalized to

$$a_{n+2} = a_{n+1} + a_n = (a_n + a_{n-1}) + a_n = 2a_n + a_{n-1} \text{ etc....}$$
$$a_{n+k} = a_k a_n + a_{k-1} a_{n-1} \tag{16.12}$$

This means that any Fibonacci number a_{n+k} may be expressed uniquely as a linear combination of two smaller successive Fibonacci numbers a_n and a_{n-1}. Since the structure of a quasi-symmetrical string $S(N_i)$ is determined solely by its length N_i, each string $S(a_{n+k})$ may be uniquely decomposed into any two successive shorter strings $S(a_n)$ and $S(a_{n-1})$: The coefficients a_k and a_{k-1}, themselves Fibonacci numbers, indicate how frequently each component string occurs. Therefore the length of quasi-symmetrical string $S(a_{n+k})$ is known in terms of quasi-symmetrical strings $S(a_n)$ and $S(a_{n-1})$; since its structure is determined solely

Figure 16.6 Puzzle string.

by its length, the positions of these component quasi-symmetrical strings are completely determined.

So much for the apparent lack of order in the frieze design of figure 16.1! My students are making use of these quasi-symmetrical patterns in architectural ornament as well as in music. String patterns are easy to generate with the use of computers, and games and puzzles may be constructed using cups and caps. Participants would try to guess the generating rules within a given period of time. The reader is invited to decode figure 16.6, generated from a single motif.

Notes

1. M. Senechal and J. Taylor, "Quasicrystals: The View from Les Houches," *Mathematical Intelligencer* 12 (1990): 54-64.
2. A. L. Loeb, *Concepts and Images,* chap. 21, (Boston: Birkhäuser Basel, 1992).

17

Fat Fractals in Lyapunov Space

Mario Markus and Javier Tamames

Fractals are usually defined by a noninteger dimension. The so-called fat fractals have recently become important and, in contrast, have an integer dimension. However, the coarse-grained measure of fat fractals varies exponentially with the scale length, allowing quantification via a "fatness exponent." In this chapter we show that regions of order within chaos on a plane defined by control parameters are fat fractals. We propose a method to determine the fatness exponent precisely and uniquely. Fat fractals, in this case, provide new practical and philosophical aspects regarding causality.

FAT FRACTALS

The best-known fractals are the so-called thin fractals, which are defined by having zero measure. The measure μ_o (length, surface, or volume) is defined in a D-dimensional space by considering a grid of D-dimensional cubes of length ε on a side. Letting $N(\varepsilon)$ be the number of cubes containing at least one point of the fractal, μ_o is the limit of $\mu(\varepsilon) = N(\varepsilon)\, \varepsilon^D$ as $\varepsilon \to 0$. Recently, fractals of non-zero measure (called "fat" fractals) have been discussed with increasing interest.

Note that $N(\varepsilon) \sim \varepsilon^{-F}$, where F is the fractal dimension. Thus, $\mu(\varepsilon) \sim \varepsilon^{D-F}$. Since $F \le D$, μ_o can only be nonzero if $F = D$, which is an integer. In an attempt to characterize the fractality of the sets investigated in this work, we checked the scaling $N(\varepsilon) \sim \varepsilon^{-F}$ and, in all cases, we found $F = 2(= D)$, indicating that these sets are fat.

The difference between thin and fat fractals can be easily illustrated by *Cantor sets* for $D = 1$. (See figure 17.1.) In order to produce a thin Cantor set, you delete the central third of an interval, then delete the central third of each remaining subinterval, and so on. The resulting set has zero measure (length) and fractal dimension log2/log3. To "fatten" this fractal, you delete first the central third, then one-ninth of the remaining intervals, then one-twenty-seventh, and so on. In this set, the holes decrease in size sufficiently fast so that the resulting limit set has nonzero measure μ_o and thus fractal dimension one. Such fractals also can be called dusts with positive volume.

Because their fractal dimension is an integer, fat fractals are not properly characterized by this dimension. Nonetheless, like all fractals, their apparent size depends on the scale of resolution. To quantify this dependence in a *D*-dimensional space, you consider that fat fractals follow the scaling law

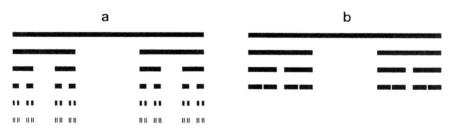

Figure 17.1 Example for the construction (depicted downwards) of one-dimensional fractals. (a): Thin fractal: in each step one-third of each remaining subset is removed. (b): Fat fractal: In each step m, $1/3^m$ of each remaining subset is removed.

$$\mu\ (\varepsilon) = \mu_o + C\varepsilon\gamma \qquad (17.1)$$

The fatness exponent γ is used to characterize the fat fractal.[1] You should keep in mind, however, that different ways to define fat fractals have been proposed. Readers interested in these alternative definitions and the relationships between them may consult Grebogi, Ott, and Yorke and Eykholt and Umberger.[2]

We now list fat fractals reported so far in the literature:

1. Parameter intervals of one-dimensional quadratic maps leading to chaos.[3]
2. Chaotic solutions of two-dimensional area-preserving maps, which behave as Poincaré sections of orbits resulting from two coupled conservative oscillators.[4]
3. The human body minus all blood vessels.[5]
4. Quantum wave functions in an exchange-coupled three-spin system.[6]
5. The set between Arnold tongues for the sine map.[7]
6. Chaotic attractors for two-dimensional locally conservative hyperbolic maps, such as the "butcher's" map.[8]
7. Basins of attraction of orbits coexisting in phase space.[9]

LYAPUNOV SPACE

One of us (MM) has worked for many years on Lyapunov exponents for one-dimensional maps that are driven periodically in different ways.[10] Dewdney called the investigated space of control parameters "Lyapunov space" in his *Scientific American* column.[11] Only now have we discovered that the periodic subsets of this space are fat fractals scaling as equation 17.1. We will now review the ideas and equations relevant to Lyapunov spaces.

One of the equations we have studied is the logistic equation

$$x_{n+1} = r\, x_n\, (1 - x_n), \qquad (17.2)$$

which is known to have universal properties found in many physical, biological and chemical systems. (See the references in Rössler and associates and in Markus.[12]) This universality requires the maximum of the map to be parabolic. Thus it is interesting to determine the effect of a small discontinuity at the maximum of equation 17.2 as in the following map presented with its physical applications in Markus.[13]

$$x_{n+1} = r\,x_n(1 - x_n) \qquad \text{if } x_n > 0.5$$

$$x_{n+1} = r\,x_n\,(1 - x_n) + \tfrac{1}{4} \times (\alpha - 1)(r - 2) \qquad \text{if } x_n \le 0.5 \qquad (17.3)$$

In this chapter, we study for the first time the effect of periodic mapping by introducing the maps

$$x_{n+1} = b\,\sin^{2k}(x_n + r), \qquad\qquad (17.4)$$

where b and k are constants throughout the iteration ($b > 0$, $k = 1$ or 2 in this work).

We periodically modulate the parameter r in all maps. Thus, we replace r by r_n. In natural systems, this corresponds to environmental conditions (for example, light or temperature) changing in time. The parameter r_n varies according to some predefined periodic sequence, such as BABA . . . or BBABA BBABA . . . . For quantification of results, the *Lyapunov exponent* is estimated by the equation

$$n\lambda = \frac{1}{N} \sum_{n=1}^{N} \log \left| \frac{dx_{n+1}}{dx_n} \right| \qquad\qquad (17.5)$$

The derivative dx_{n+1}/dx_n can be calculated analytically in equations 17.2 through 17.4. For example, the map given by equation 17.4 yields

$$\lambda = \frac{1}{N} \sum_{n=1}^{N} \log\ 2bk \, | \sin^{2k-1}(x_n + r) \cos(x_n + r) |$$

Equation 17.5 is valid in the limit $N \to \infty$. We set $N = 4 \times 10^3$ after letting 600 iterations pass to allow transients to die away. Chaos is indicated by $\lambda > 0$, and periodicity by $\lambda < 0$. λ is represented on the A-B plane by using black and white. Gray levels or colors are convenient for the display of the whole range of complexity in the parameter plane. Unless stated otherwise, $x_0 = 0.5$ (starting value of the iteration), A is shown on the ordinate and B on the abscissa. For visualization improvement, some graphs were rotated counterclockwise by 45 degrees, so that the line A = B appears vertical. (In this chapter, we simply use the word "rotated.")

Figure 17.2 Periodicity ("structures" in the picture foreground) and chaos (picture background) on the A-B-plane for the logistic equation 17.2, r_n-sequence: BABA. . . .

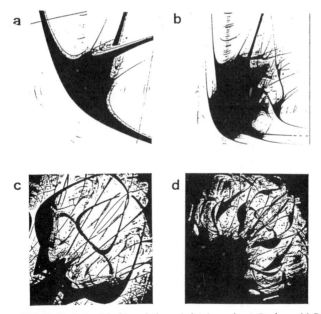

Figure 17.3 Periodicity (black) and chaos (white) on the A-B-plane. (a) Period-3-island of figure 17.2. (b) Same as (a), but sequence BBABA BBABA. . . . (c) Equation 17.4, $k = 1$, $B^3A^3B^3A^3 \ldots$, $b = 2.05$ (d) Equation 17.4, $k = 1$, sequence $B^6A^6B^6A^6 \ldots$, $b = 1.95$.

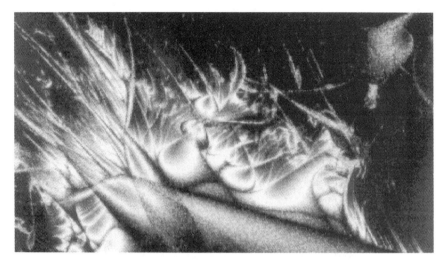

Figure 17.4 Periodicity ("structures" in the picture foreground) and chaos (picture background) on the A-B-plane for the discontinuous equation 17.3, sequence BBABABA BBABABA ..., α = 0.957.

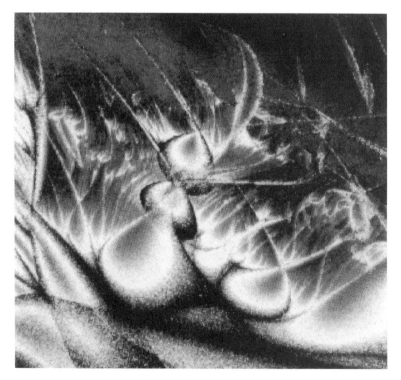

Figure 17.5 Fragment of figure 17.4.

INVESTIGATED GRAPHS

Here we list and number the analyzed graphs.

 I Equation 17.2, sequence: BABA . . . ,$2 \leq A, B \leq 4$ (figure 17.2).

 II Equation 17.2, sequence: BABA . . . , $2 \leq B \leq 3, 3 \leq A \leq 4$ (upper left part of figure 17.2).

 III Equation 17.2, sequence: BABA . . . , $3.808 \leq A, B \leq 3.862$ (figure 17.3a; period-3-island in upper right part of figure 17.2).

 IV Equation 17.2, sequence: BBABABA BBABABA . . . , $3.8215 \leq A, B \leq 3.8553$ (figure 4 in Markus[14]).

 V Equation 17.2, sequence: BBABA BBABA . . . , $3.8225 \leq A \leq 3.8711, 3.8218 \leq B \leq 3.8607$.(Figure 17.3b.)

 VI Equation 17.2, sequence: BBABA BBABABA . . . , $3.212 \leq A \leq 4, 2.759 \leq B \leq 3.744$. (Figure 6 in Markus[15].)

 VII Equation 17.2, sequence: $B^6A^6 \; B^6A^6$ . . . , $3.1313 \leq A \leq 4, 2.516 \leq B \leq 3.647$.(Figure 5 in Markus[16].)

 VIII Equation 17.2, $x_o = 0.8315$, sequence: $B^7A^2B^9(BA)^9A^7B^2A^7$ $B^7A^2B^9(BA)^9A^7B^2A^7$ . . . , $1.1518 \leq B \leq 1.2002, 3.769 \leq A \leq 3.8175$. (Figure 11 in Markus[17].)

 IX Equation 17.3, $\alpha = 0.907$, $x_o = 0.499$, sequence: BABA . . . , $3.8109 \leq A, B \leq 3.8207$ (figure 17 in Markus[18]).

 X Equation 17.3, $\alpha = 0.8616$, $x_o = 0.499$, sequence: BABA . . . , $3.7902 \leq A, B \leq 3.8276$ (rotated: figure 24 in Markus[19]).

 XI Equation 17.3, $\alpha = 0.908$, $x_o = 0.499$, sequence: BABA . . . , $3.7980 \leq A, B \leq 3.828$(figure 22a in Markus[20]).

 XII Equation 17.3, $\alpha = 0.9935$, $x_o = 0.7$, sequence: BABA... , $3.8142 \leq B \leq 3.8720, 3.8136 \leq A \leq 3.8714$ (rotated: figure 23 in Markus[21]).

 XIII Equation 17.3, $\alpha = 0.907$, $x_o = 0.499$, sequence: BBABABA BBABABA . . . , $3.8097 \leq A \leq 3.8175, 3.8120 \leq B \leq 3.8198$ (figure 17.4).

 XIV Equation 17.3, $\alpha = 0.907$, $x_o = 0.499$, sequence: BBABABA BBABABA . . . , $3.8144 \leq B \leq 3.8177, 3.8121 \leq A \leq 3.8155$ (figure 17.5).

 XV Equation 17.4, $b = 2.5$, $k = 1$, sequence: BABA . . . , $0 \leq A, B \leq 10$ (figure 17.6).

 XVI Equation 17.4, $b = 2.8$, $k = 1$, sequence: BABA . . . , $0 \leq B \leq 10, 0 \leq A \leq 5.6$ (figure 17.8).

 XVII Equation 17.4, $b = 2.05$, $k = 1$, sequence: $B^3A^3B^3A^3$ . . . , $0 \leq A, B \leq 1.115$ (figure 17.3c; rotated: figure 17.9).

 XVIII Equation 17.4, $b = 1.95$, $k = 1$, sequence: $B^6A^6B^6A^6$ . . . , $0 \leq A, B \leq 1.128$ (figure 17.3d; rotated: figure 17.10).

 XIX Equation 17.4, $b = 2$, $k = 2$, sequence: BABA... , $1 \leq A, B \leq 7$ (figure 17.7).

Note that graphs I through VIII are based on the logistic equation 17.2, graphs IX through XIV on the discontinuous equation 17.3, and graphs XV through XIX on the periodic equation 17.4. Graphs I through III are different windows of the same pictures; this is also true for graphs XIII and XIV.

METHODS OF FRACTAL ANALYSIS

In all cases, we investigated the subsets of the parameter plane for which λ 0 (periodicity). These subsets are shown in black in figure 17.3. In this figure, as well as in figures 17.2 and 17.4 through 17.10 (with gray levels) these sets correspond

Figure 17.6   Periodicity ("structures" in the picture foreground) and chaos (picture background) on the A-B-plane for the periodic map 17.4 with $k = 1$, $b = 2.5$, sequence BABA. . . .

to the "structures" in the picture foreground. (The "background" of the pictures corresponds to $\lambda > 0$, that is, to chaos).

In previous investigations of fat fractals, the parameters μ_0, C, and γ have been determined by two methods: (a) determination of μ_0 by extrapolating μ for $\varepsilon \to 0$, and then linear optimization of log $(\mu - \mu_0)$ vs. log ε and (b) nonlinear optimization.

We found that method (a) yields highly uncertain estimates of μ_0 and thus of C and γ. We found that method (b) is also not convenient, as often more than one local minimum of the sum of squares is found; the problem of finding a global minimum is, in fact, common to all nonlinear optimization problems. Here we solve the problem by carrying out a (quasi-)continuum of linear optimizations, as follows.

For a fixed (arbitrary) γ we consider μ as the dependent and ε^γ as the independent variable, fitting the points $(\mu, \varepsilon^\gamma)$ by linear optimization. Let us call SSQ the minimum sum of squares obtained from this optimization. We then change γ in very small steps (quasi-continuously) and determine the minimum

Figure 17.7 Same as figure 17.6, but $k = 2$, $b = 2$.

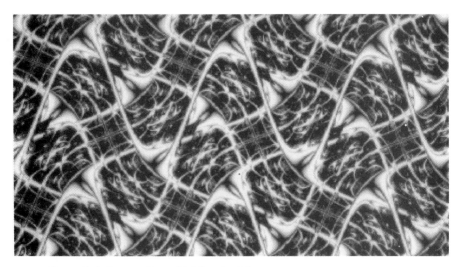

Figure 17.8 Same as figure 17.6, but b = 2.8.

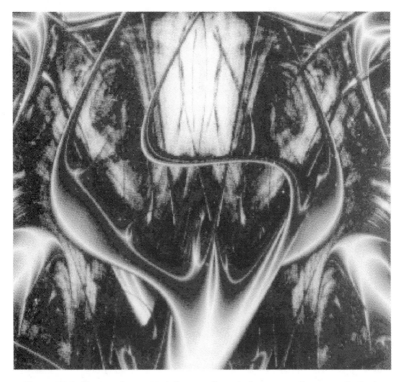

Figure 17.9 Same as figure 17.6, but smaller window, rotated counterclockwise by 45 degrees, b = 2.05 and sequence $B^3A^3B^3$. . . .

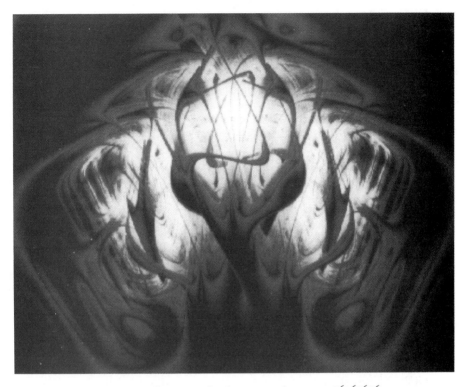

Figure 17.10 Same as figure 17.9, but b = 1.95 and sequence $B^6A^6B^6A^6$ . . . .

of SSQ (γ). This is exemplified in figure 17.11, which corresponds to graph III (figure 17.3a). The linear dependence of $\log(\mu(\varepsilon) - \mu_o)$ vs. $\log \varepsilon$ corresponding to the optimal γ of figure 17.11 is given in figure 17.12. For all cases considered, we obtain a clear single minimum of SSQ, such as that in figure 17.11, and could thus determine γ uniquely.

If not stated otherwise, the images were calculated with a resolution of 2160 x 2160 points. The number 2160 has the advantage of being divisible by a large number of integers, all of which were used as ε's: 1, 2, 3, 4, 5, 6, 8, 9, 10, 12, 15, 16, 20, 24, 27, 30, 36, 40, 45, 48, 54, and 60. (Because of the divisibility by these numbers, no residuals were left when covering the graphs with these ε's.)

The error of γ was determined (with a confidence of 95 percent) by considering it composed of two parts: a deviation propagating from linear regression, and a deviation stemming from the curvature (see figure 17.11) around the minimum of SSQ vs. γ. The latter was determined by using the F-test, which yields a quotient of SSQ's (ordinate) corresponding to an error interval of γ (abscissa); according to the F-test, γ-values within this interval have not significantly different variances, that is, SSQs.

Table 17.1

Graph	Equation	Figure	γ
I	17.2	17.2	0.837 ± 0.13
II	17.2	Part of 17.2	0.875 ± 0.086
III	17.2	17.3a	0.853 ± 0.128
IV	17.2	4 of Markus (1990)*	0.711 ± 0.06
V	17.2	17.3b	0.799 ± 0.051
VI	17.2	6 of Markus (1990)*	0.671 ± 0.071
VII	17.2	5 of Markus (1990)*	0.544 ± 0.055
VIII	17.2	11 of Markus (1990)*	0.451 ± 0.063
IX	17.3	17 of Markus (1990)*	0.612 ± 0.061
X	17.3	24 of Markus (1990)*	0.483 ± 0.037
XI	17.3	22a of Markus (1990)*	0.885 ± 0.148
XII	17.3	23 of Markus (1990)*	0.859 ± 0.161
XIII	17.3	17.4	0.401 ± 0.068
XIV	17.3	17.5	0.345 ± 0.085
XV	17.4	17.6	0.823 ± 0.064
XVI	17.4	17.8	0.738 ± 0.059
XVII	17.4	17.3c	0.572 ± 0.051
XVIII	17.4	17.3d	0.300 ± 0.047
XIX	17.4	17.7	0.416 ± 0.069

*M. Markus, "Chaos in Maps with Continuous and Discontinuous Maxima," *Computers in Physics* (September/October 1990): 481-494.

In our analyses we eliminated all isolated single points—points for which the sign of λ is opposite to that in all neighbors . For such points, the sign of λ was set equal to the surrounding sign. Large errors are obtained if this is not done, because points that appear isolated correspond to complex fractal structures that are not resolved.

RESULTS AND DISCUSSION

The obtained fatness exponents γ are given in table 17.1 for the investigated graphs. We do not obtain equal γ's, even for the same equation. Thus we cannot conjecture, as was done for a one-dimensional system in Farmer,[22] that the value of γ is universal. We do, however, obtain the same values of γ (within the error) for different parts of the same picture, as long as these parts contain borders of the

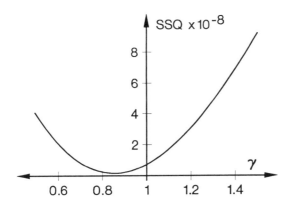

Figure 17.11 Minimum of sum of squares (SSQ) showing dependence on γ after a linear fit of μ *vs.* $\varepsilon\gamma$ (computed for graph III; see text).The minimum of the SSQ on this figure determines the optimal, as it is given in table 17.1

fractal set (we are referring here to the triad I, II, and III as well as to the pair XIII and XIV). This invariance should, however, be checked by further investigation. Note that no scaling law is obtained—that is, $\mu = \mu_o$ for all ε (see equation 17.1) if the graph contains only λ-values below zero (no borders of the set).

In order to find out if our results are dependent on the choice of the smallest ε, we repeated calculations (for two graphs) using a higher resolution, namely 4320 $\times$ 4320, instead of 2160 $\times$ 2160, thus halving the absolute length of the ε's. In the first case (graph III), we obtained $\gamma = 0.826 \pm 0.094$. In the second case (graph II), we obtained $\gamma = 0.858 \pm 0.079$. A comparison of these results at high resolution with those in table 17.1 shows no significant difference.

CONCLUSION

We have shown in this chapter our equations exhibit a scale-independent, sensitive dependence on control parameters. In fact, the self-similarity of the borders and islands in the graphs implies that an arbitrarily small change in parameters (A and B) may cause a completely different behavior. "Different" here means either periodic within a broad range of stability, as quantified by negative λ's (points of the fractal set), or within a broad range of unpredictability, as quantified by positive λ's (points outside the set). At any level of resolution, you may stumble on a periodic island in the "sea of chaos," each island containing all λ-values between $-\infty$ (so-called superstability: immediate recovery after a per-

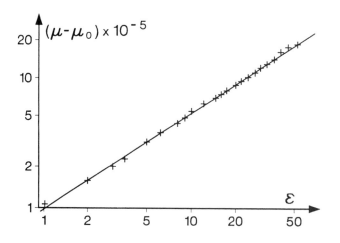

Figure 17.12 Log-log plot of $\mu - \mu_0$ vs. ε (computed for the γ at which the SSQ is minimum in figure 17.11).

turbation) and 0 ("critical slowing down": infinitely long recovery after a perturbation).

This conclusion reminds us of the well-known butterfly effect within a single strange attractor. Our results have one thing in common with the butterfly effect: Arbitrarily small changes may lead to changes as large as the whole system. Taking this as a generalized definition of the butterfly effect, we may set up a "butterfly effects list" resulting from what we have observed in this chapter and from previous works:

1. Sensitive dependence on initial conditions within a strange attractor ("classical" butterfly effect). The Lyapunov exponent describes the exponential growth of perturbations. As the perturbations grow larger, they increase more slowly than exponentially, until they overshadow the whole attractor.

2. Sensitive dependence on initial conditions due to having fractal basins of attraction in phase space. Different initial conditions may lead to different attractors, that is, to completely different periodic, quasi-periodic, or chaotic dynamics. For the type of equations investigated here fractal basins were found[23] (for example, for equation 17.2, sequence: BBABABA BBABABA . . . , see figures 7-9 in Markus[24]).

3. Sensitive dependence on the algebraic form of the map, as shown here and in Markus.[25] Equations 17.2 and 17.3 exemplify this point where,

when values (close to 1) are only slightly different, there are very different results. In fact, the period-3-island obtained by the logistic equation 17.2 and displayed in figure 17.3a (graph III) drastically changes by introducing a discontinuity in the parabola (equation 17.3). To see this, one may just compare figure 17.3a ($\alpha = 1$) with graph XII, where $\alpha = 0.9935$. Somewhat larger perturbations yield graph XI ($\alpha = 0.908$), graph IX ($\alpha = 0.907$), and graph X ($\alpha = 0.8616$). (Note that not only the shape of the periodic island changes, but also its location.) These changes are due to the breakdown of the Feigenbaum-scenario in the absence of a parabolic maximum.

4. Sensitive, scale-independent dependence on parameters, as shown in this chapter for fat fractals in the plane defined by the control parameters A and B. Arbitrarily small deviations of A or B may cause the system to reach any λ-value between $-\infty$ (superstability) and positive values (chaos).

OUTLOOK

Our results suggest that analyses of fat fractals in parameter space and fractal basins of attraction in phase space, as well as the study of discontinuous functions, will widen the concept of unpredictability. In the past, unpredictability had been mainly related to the deterministic-chaotic butterfly effect within one strange attractor. However, fractals in parameter space, fractal basins, and the dramatic features of discontinuous equations all show us that strong causality (that is, similar causes yield similar effects) can break down for more reasons than just by the widely known fluctuation enhancement due to deterministic chaos. Due to this fact, we will face even more restrictions in our ability to foresee dynamic developments in all sciences. On the other hand, novel techniques for controlling chaos should not be limited to the consideration of just a single attractor.

Notes

This work was supported by the Deutsche Forschungsgemeinschaft (Grant Nr. MA 629/4-1).

1. J. D. Farmer and D. K. Umberger, "Response to Grebogi, Ott and York, 'Comment on "Sensitive Dependence,"'" *Physical Review Letters* 56 (3) (1986): 267.

2. C. Grebogi, E. Ott, and J. A. Yorke, "Comment on 'Sensitive Dependence on Parameters in Nonlinear Dynamics' and on 'Fat Fractals on the Energy Surface,'" *Physical Review Letters* 56 (3) (1986): 266. R. Eykholt and D. K. Umberger,

"Characterization of Fat Fractals in Nonlinear Dynamical Systems," *Physical Review Letters* 57 (19) (1986): 2333-2336. R. Eykholt and D. K. Umberger, "Extension of the Fat Fractal Exponent β to Arbitrary Sets in D Dimensions," *Physics Letters A* 163 (1992): 409-414.

3. J. D. Farmer, "Sensitive Dependence on Parameters in Nonlinear Dynamics," *Physical Review of Letters* 55 (4) (1985): 351-354

4. D. K. Umberger and J. D. Farmer, "Fat Fractals on the Energy Surface," *Physical Review Letters* 55 (7) (1985): 661-664.

5. C. Grebogi, S. W. McDonald, E. Ott, and J. A. Yorke, "Exterior Dimension of Fat Fractals," *Physics Letters A* 110 (1) (1985): 1-4.

6. K. Nakamura, Y. Okazaki, and A. R. Bishop, "Fat Fractals in Quantum Chaos," *Physics Letters A* 117 (9) (1986): 459-464.

7. R. E. Ecke, J. D. Farmer, and D. K. Umberger, "Scaling of the Arnold Tongues," *Nonlinearity* 2 (1989): 175-196.

8. J. Theiler, G. Mayer-Kress, and J. B. Kadtke, "Chaotic Attractors of a Locally Conservative Hyperbolic Map with Overlap," *Physica D* 48 (1991): 425-444.

9. See references in Grebogi, Ott, and Yorke, "Comments on 'Sensitive Dependence.'"

10. M. Markus and B. Hess, "Lyapunov Exponents of the Logistic Map with Periodic Forcing," *Computers & Graphics* 13 (1989): 553-558. J. Rössler, M. Kiwi, B. Hess, and M. Markus, "Modulated Nonlinear Processes and a Novel Mechanism to Induce Chaos," *Physical Review A* 39 (11): 5954-5960. M. Markus, "Chaos in Maps with Continuous and Discontinuous Maxima," *Computers in Physics* (September/October 1990): 481-494.

11. A. K. Dewdney, "Leaping into Lyapunov Space," *Scientific American* (September 1991): 178-180.

12. See references in Rössler et al., "Modulated Nonlinear Processes" and Markus, "Chaos in Maps."

13. Markus, "Chaos in Maps."

14. Ibid.

15. Ibid.

16. Ibid.

17. Ibid.

18. Ibid.

19. Ibid.

20. Ibid.

21. Ibid.

22. Farmer, "Sensitive Dependence."

23. M. Markus and B. Hess, "Properties of Modulated One-Dimensional Maps," in *A Chaotic Hierarchy,* ed. G. Baier and M. Klein (Singapore: World Scientific, 1991), 267-283.

24. Markus, "Chaos in Maps."

25. Ibid.

About the Contributors

William P. Beaumont received his Ph.D. in computer science in 1976. Until 1990 he taught computer science at the University of Adelaide and in the Philippines, with special interests in operating systems and concurrent programming. He is now working in educational computing at the University of South Australia. His interests include computer graphics, fractal geometry, and the use of visualization techniques to enhance education. He spends much of his time trying to convince the world of education that computers can be used for things other than word processing. E-mail: Beaumont@UniSA.edu.au.

Ronald R. Brown has been a software engineer since 1979. His interest in the Knight's tour problem goes back to 1972, and he began creating art from chess moves in 1988. He can be reached at: Ronald Brown, 569 Lake Warren Road, Upper Black Eddy, PA 18972. E-mail: rrbrown@epix.net

Kevin Dooley is an associate professor of mechanical engineering at the University of Minnesota (125 Mechanical Engineering, Minneapolis, MN 55455). He has research and teaching interests in quality engineering, quality management, organizational behavior, and complex systems.
E-mail: kdooley@maroon.tc.umn.edu,
Web: http://servme.me.umn.edu:70/1/home/kdooley.

Glenda Eoyang is president of Chaos Limited (50 East Golden Lake Road, Circle Pines, MN 55014), a chaos-based organizational consulting firm. She is also founder of Excel Instruction, specializing in customized training. E-mail: eoyang@delphi.com.

David Fowler is an assistant professor of instructional technology in the Teachers College at the University of Nebraska-Lincoln. His research interests include the origins of fractal intuition in children and educational applications of the symbolic mathematics program Mathematica. E-mail: dfowler@unlinfo.unl.edu,
Web: http://www.unl.edu/tcweb/Faculty/dfHomePage.html.

Michael Frame is a professor of mathematics at Union College in Schenectady, New York. He has taught courses in fractals and chaos as an introduction to science for humanities students and as a branch of mathematics to science students at Union and at Yale. With physicist David Peak, he wrote *Chaos Under Control: The Art and Science of Complexity* (W. H. Freeman). His research interests include renormalization, symbolic dynamics, and applications of fractals to problems in topology. E-mail: `FrameM@gar.union.edu`.

Danielle Gaines, president, AVES, Inc. is both a designer and a writer. AVES includes collections of fractal activewear incorporating computer-generated fractal art. She is also a founding director of THE FRACTAL TRIBE, a consortium of professionals with expertise in art, design, science, computer technology, and global communication. THE FRACTAL TRIBE promotes the understanding, appreciation, and practical use of fractal design, beauty, and technology. E-mail: `danielle@netscape.com`.

Timothy D. Greer is currently employed by IBM, testing software for the VM operating system. Playing with fractals is strictly a sideline, although some of his work has been published as invention disclosures published in the *IBM Technical Disclosure Bulletin.* Tim has M.S. degrees in math from Georgia Institute of Technology and in industrial engineering from University of Illinois, and has found attempts to relate these two backgrounds using fractals to be his most fruitful area of research. E-mail: `tim_greer@vnet.ibm.com`.

Mario Hilgemeier has a doctorate in nuclear astrophysics. In 1991 he codesigned and developed megatelVISOR, a new geographical information system. One of his primary current interests is digital cartography. Contact: Mario Hilgemeier, megatel Informations und Kommunikationssyteme GmbH, Wiener Strasse 3, D-28359 Bremen, Germany. E-mail: `mhi@vent.winnet.de` or `mhi@megatel.de`.

Jay R. Hill has had a long interest in fractals and parallel processing. He is founder of the Grand Internet Parallel Processing Project, a large collaboration of researchers around the world who compute ever-closer estimates of the area of the Mandelbrot set. He received his Ph.D. from the University of California at San Diego, working in space physics, studying planetary magnetospheres, and dusty plasmas. He now works at Science Applications International Corporation, specializing in orbital mechanics and satellite communications. He can be reached at: Jay R. Hill, Science Applications International Corporation, 10770 Wateridge Circle, San Diego, CA 92121-2785. E-mail: `JAY.R.HILL@cpmx.saic.com`.

Gabriel Landini is a research fellow in oral pathology at the School of Dentistry, The University of Birmingham, U.K. He received his dental degree from the Republic University (Montevideo, Uruguay) in 1983 and his Ph.D. in oral pathology from Kagoshima University (Kagoshima, Japan) in 1991. His research interests include image analysis in cancer research, analysis of biological pattern formation, and computer modelling. E-mail: G.Landini at bham.ac.uk.

Arthur L. Loeb is senior lecturer on visual and environmental studies and a member of the faculty in the Graduate School of Education at Harvard University, where he teaches design science and visual mathematics. He is a visual artist and musician, and a Life Fellow of the Royal Society of Arts and the American Institute of Chemistry. He received his B.S.Ch. from the University of Pennsylvania and his A.M. (physics) and Ph.D. (chemical physics) from Harvard University.

Jonathan Mackenzie has a Ph.D. in chaos theory and sound modeling from King's College, University of London. His research interests are the application of science and technology to music composition and the theory of nonlinear dynamical systems, chaos, and complexity. He is currently working as a research fellow at the University of Westminster on balanced system models for audio applications. He can be contacted at: Dr. Jonathan Mackenzie, School of Electronic and Manufacturing Systems Engineering, University of Westminster, London W1M 8JS, UK, E-mail: mackenj1@wmin.ac.uk.

Mario Markus is author of numerous articles on the subjects of chaos and fractals. He works on computer graphics, chemical turbulence and cellular automata. He can be reached at: Max-Planck-Institut fur molekulare Physiologie, Postfach 10 26 64, 44026 Dortmund, Germany. E-mail: mario.markus@mpi-dortmund.mpg.de.

Clifford A. Pickover is a research staff member at the IBM Watson Research Center and received his Ph.D. from Yale University's Department of Molecular Biophysics and Biochemistry. He is author of numerous books, including: *Keys to Infinity* (Wiley); *Black Holes, A Traveler's Guide* (Wiley); *Chaos in Wonderland: Visual Adventures in a Fractal World* (St. Martin's); *Mazes for the Mind* (St. Martin's); *Computers and the Imagination* (St. Martin's); *Computers, Pattern, Chaos and Beauty* (St. Martin's); and *The Pattern Book* (World Scientific). He is associate editor for *Computers and Graphics, Computers in Physics,* and *Theta, a Mathematics Journal,* and brain-boggler columnist for *Discover* magazine.
E-mail: cliff@watson.ibm.com. Web: http://sprott.physics.wisc.edu/pickover/home.htm. He can be reached at: Dr. Cliff Pickover, P.O. Box 549, Millwood, NY 10546-0549.

Manfred Schroeder is author of *Fractals, Chaos, Power Laws* (Freeman) and *Number Theory in Science and Communication* (Springer). He served as a distinguished member of the research staff of AT&T Bell Laboratories for 33 years. A holder of 45 U.S. patents for inventions in various fields, he won the Gold Medals of the Acoustical Society of America and Audio Engineering Society and the Lord Rayleigh Medal in 1991, 1972, and 1987, respectively. He can be reached at: Drittes Physikalische Institut, Universität Göttingen, Bürgerstr 42-44, D-37073 Göttingen Germany. E-mail: MRS@PHYSIK3.GWDG.DE.

J. C. Sprott is professor of physics at the University of Wisconsin (1150 University Avenue, Madison, WI 53706). His interests are in experimental plasma physics and nonlinear dynamics. He is author of the book *Strange Attractors: Creating Patterns in Chaos* (M&T Books). E-mail: sprott@juno.physics.wisc.edu, Web: http://sprott.physics.wisc.edu/.

Javier Tamames works in biophysical chemistry. He can be reached at: Departamento de Bioquimica, Laboratorio de Biofisica, Facultad de Quimicas, Universidad Complutense, 28040 Madrid, Spain. E-mail: tamames@gredos.cnb.uam.es.

Bruce J. West is a professor of physics and the founding director of the Center for Nonlinear Science at the University of North Texas (Denton, Texas 76203). Before joining the faculty at UNT in 1989, he was the associate director of the Center for Studies of Nonlinear Dynamics, a division of the LaJolla Institute, which in the late 1970s was quite alone in systematically applying nonlinear dynamical techniques both inside and outside the physical sciences. He recently received the Decker Scholar Award and the UNT President's Award for his research efforts and is editor-in-chief of the book series *Studies of Nonlinear Phenomena in the Life Sciences* for World Scientific. E-mail: bwest@soliton.phys.unt.edu.

Douglas Winsand is supervisor of the Advanced Development group in Advanced Systems Engineering at GDE Systems Inc. in San Diego, California. He designs systems that extract map features and terrain data from digital stereo aerial imagery. He received an M.S. in applied mathematics and a B.A. in physics from the University of Colorado, and a B.A. in philosophy from Pennsylvania State University. E-mail: winsand@gdesystems.com.

INDEX